A Ha...

Mechanical Engineering

Revised & Updated

Contains well illustrated formulae
& key theory concepts

— *For* —

IES, GATE, PSUs

& OTHER COMPETITIVE EXAMS

MADE EASY
Publications

MADE EASY
Publications

MADE EASY Publications

Corporate Office: 44-A/4, Kalu Sarai (Near Hauz Khas Metro Station), New Delhi-110016
E-mail: infomep@madeeasy.in
Contact: 011-45124612, 0-9958995830, 8860378007
Visit us at: www.madeeasypublications.org

A Handbook on Mechanical Engineering

First Edition: 2012
Reprint: 2013
Reprint: 2014
Second Edition : 2015

ISBN 978-93-5147-125-7

9 789351 471257

Typeset at: MADE EASY Publications, New Delhi-110016

Director's Message

B. Singh (Ex. IES)

In the Current Era of globalization and international competition in the field of Science and Technology, its challenging task to ensure Indian participation and contribution through skilled technical professionals. Constant efforts and desire are required to excel and achieve top positions.

I firmly believe that every candidate has the ability to succeed but competitive and quality guidance are required to attain sky high goals. At MADE EASY, we help you to discover your hidden talent and success quotient to make you reach your dreams. In my opinion IAS, IES, GATE and PSU's exams are the tools to bring out true potential of serving the Nation. The actual application of knowledge and talent lies in the successful accomplishment of assigned roles and responsibilities in the working arena. We at MADE EASY ensure that you are trained to become a winner in your life and achieve job satisfaction in your chosen field.

Right since its inception, MADE EASY alumnae have been sharing their winning stories of success and expressed their gratitude towards the quality guidance provided by MADE EASY. Our students have not only secured All India First Ranks in IES, GATE and PSUs but also secured top positions in their career. I invite you to become a part of MADE EASY to explore and achieve ultimate aims of your life. I promise to provide you quality guidance with competitive environment which is far advanced and beyond the reach of other institutes. I ensure you to give a comprehensive guidance, support and inspiration that you need to reach the pinnacle of your career.

I have true desire to serve the Society and Nation by contributing to the field of education. Needless to say, the endeavor to nurture and even further enhance the quality of education will be our constant feature.

After a very long experience of teaching in Mechanical Engineering, MADE EASY team has realized that there is a need of good Handbook which can provide the crux of Mechanical Engineering in a concise form for the students to brush up the formulae and important concepts required for IES, GATE, PSUs and other competitive examinations. This handbook contains all the formulae and important aspects of Mechanical Engineering. It provides much needed revision aid and study guidance before the examinations.

B. Singh (Ex. IES)
CMD, MADE EASY Group

A Handbook on

Mechanical Engineering

C O N T E N T S

○○○○

A Handbook on
Mechanical Engineering

1

Fluid Mechanics &
Fluid Machines

CONTENTS

◀ ▢ ▶

Introduction 1

1. Ideal Fluid and Real Fluid

- **Ideal fluid**
  A fluid is said to be ideal if it is assumed to be both incompressible and non-viscous. Its bulk modulus is infinite.
- **Real fluid**
  Real fluid have viscosity, finite compressibility and surface tension.

Remember: ....
- Ideal fluid has no surface tension.
- Ideal fluid are imaginary and do not exist in nature.

2. Specific Weight, Specific Volume, Specific Gravity

- **Specific weight (ω) or weight density**

$$= \frac{\text{Weight}}{\text{Volume}}$$

$$= \frac{mg}{V} = \rho g$$

Here, ρ = Density
$\quad\quad\quad$ g = Acceleration due to gravity
Specific weight of water = 9810 N/m^3

- **Specific Volume**

$$= \frac{1}{\text{Density}}$$

- **Specific gravity (S) or Relative density**

$$\text{Specific gravity} = \frac{\text{Density of fluid}}{\text{Density of standard fluid}}$$

$$= \frac{\text{Specific weight of fluid}}{\text{Specific weight of standard fluid}}$$

Remember

- Specific gravity for water is 1.0 at 4°C and for mercury it is 13.6
- Specific gravity varies with temperature therefore it should be determined at specified temperature (4°C or 27°C).

3. Newton's Law of Viscosity

$$\tau = \mu \cdot \frac{du}{dy} = \mu \frac{d\theta}{dt}$$

τ = shear stress

μ = coefficient of viscosity or absolute viscosity or dynamic viscosity

Here, $\dfrac{du}{dy}$ = Velocity gradient

$\dfrac{d\theta}{dt}$ = Rate of angular deformation or Rate of shear strain

- For newtonian fluid, coefficient of viscosity remain constant.

4. Viscosity/Kinematic Viscosity

Due to viscosity a fluid offer resistance to flow

(i) Dynamic Viscosity (μ):
- Its SI unit is pascal-second or **N-sec/m²**
- Its CGS unit is Poise = Dyne-sec/cm²
- 1 poise = **0.1** N-s/m²

(ii) Kinematic Viscosity

$$\nu = \frac{\text{Dynamic viscosity } (\mu)}{\text{Mass density } (\rho)}$$

- Its SI unit is m²/s.
- Its CGS unit is cm²/s or stoke.
- 1 stoke = 10^{-4} m²/s

Remember

- Viscosity of **liquids** decreases with temperature whereas viscosity of **gases** increases with increase in temperature.
- Liquids with increasing order of viscosity are gasoline, water, crude oil, castor oil.
- Viscosity of **water** at 20°C is 1 centipoise.
- Viscosity of liquids is due to **cohesion** *and molecular momentum transfer.*

5. Type of Fluid

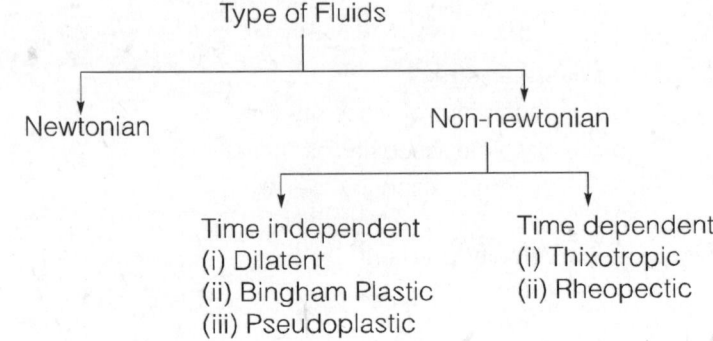

- **Non-Newtonian Fluids**

  These do not follow Newton's law of viscosity. The relation between shear stress and velocity gradient is

  $$\tau = A\left(\frac{du}{dy}\right)^n + B$$

  where A and B are constants depending upon type of fluid and condition of flow.

  (i) For Dilatant Fluids: $n > 1$ & $B = 0$,
  Ex. Butter, Quick sand.

  (ii) For Bingham Plastic Fluids: $n = 1$ & $B \neq 0$
  Ex. Sewage sludge, Drilling mud, tooth paste and gel.
  These fluids always have certain minimum shear stress before they yield.

  (iii) For Pseudoplastic Fluids: $n < 1$ & $B = 0$
  Ex. Paper pulp, Rubber solution, Lipsticks, Paints, Blood, Polymetric solutions etc.

  (iv) For Thixotropic Fluids: $n < 1$ & $B \neq 0$
  Viscosity increases with time.
  Ex. Printers ink and Enamels.

  (v) For Rheopectic Fluids: $n > 1$ & $B \neq 0$
  Viscosity decreases with time.
  Ex. Gypsum solution in water & Bentonite solution.

-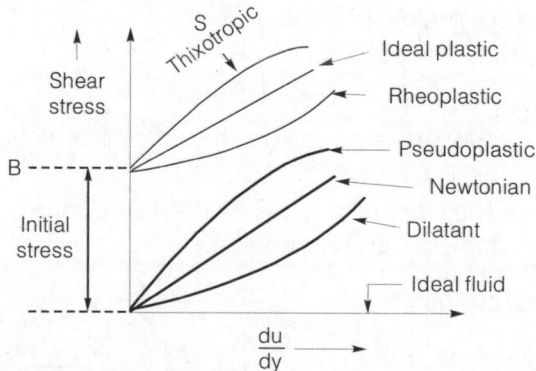

6. Compressibility (β), Isothermal Bulk Modulus (k_T) and Adiabatic Bulk Modulus

- Compressibility (β)

  It is inverse of bulk modulus of elasticity.

$$\beta = \frac{1}{k} = \frac{-dv}{vdp}$$

$$\beta = \frac{d\rho}{\rho \cdot dP}$$

Here, k = bulk modulus of elasticity

ρ = Density

v = Volume

- Isothermal bulk modulus (k_T)

$$k_T = P_{final} = \rho RT$$

- Adiabatic bulk modulus

$$K_a = \gamma \cdot P_{final}$$

Here, $\gamma = \dfrac{C_p}{C_v}$

C_p = Specific heat at constant pressure

C_v = Specific heat at constant volume

7. Surface Tension/Pressure Inside Drop, Bubble and Jet

Surface tension occur at the interface of liquid and a gas *or* at the interface of two liquid. Surface tension is inversely proportional to temperature and it also acts when fluid is at rest.

- Pressure inside drop (Solid like sphere)

$$P = \frac{4\sigma}{d}$$

- Pressure inside bubble

$$P = \frac{8\sigma}{d}$$

- The pressure inside the droplet of soap bubble will be higher than P_{atm}.
- The higher the pressure inside the soap bubble the smaller the size of soap bubble.

Remember

- Pressure inside jet

$$P = \frac{2\sigma}{d}$$

Here   d = Diameter of drop

P = Gauge pressure

- It is a *surface* phenomenon
- It is force per unit length (N/m)
- For *water-air* interface at 20°C its value is 0.0736 N/m and Air-mercury Interface σ = 0.480 N/m
- At critical point, liquid-vapuor state are same thus surface tension = 0
- It is due to *cohesion* only

Remember

8. Capillary Action

- Height of water in capillary tube

$$h = \frac{4\sigma\cos\theta}{\rho g d}$$

Where,   h = rise in capillary

σ = surface tension of water & glass

d = dia of tube

θ = angle of contact between the liquid and the material.

θ = 0° for water and glass (clean)

θ = 128° for mercury and glass (clean)

- When a liquid surface supports another liquid of density "ρ_b", then rise in capillary is given as

$$h = \frac{4\sigma\cos\theta}{(\rho - \rho_b)gd}$$

- Capillary action is due to adhesion and cohesion, *both*.
- For capillary action diameter of tube should be *less* than 3 cm.

■■■■

Manometry

2

1. Pascal's Law

- The intensity of fluid at any point in a stationary fluid is same in all directions.

$$p_x = p_y = p_z$$

- Pressure varies **only with depth** in stationary fluids, whereas if fluids is in motion pressure may vary in horizontal direction also.
- Fluid pressure is measured in Force/Area and it is expressed in Pascal (N/m^2) or Bar.

  1 Bar = 10^5 N/m^2

  1 MPa = 10 Bar
- Barometer shows **atmospheric** pressure.
- 1 kgf = 9.81 Newton.
- Pressure is a scaler quantity.

2. Absolute Pressure

Pressure measured with reference to absolute zero. Absolute pressure cannot be negative

Asbolute pressure = gauge pressure + local atmospheric pressure

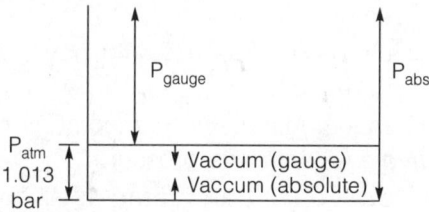

- $P_{gauge} = \rho g h$

  Here, ρ = Density of fluid

  g = Acceleration due to gravity

  h = Height
- **Gauge** pressure can be positive, negative or zero.
- Atmospheric pressure varies with **altitude**, **temperature** and **local** conditions.
- At **mean sea level** atmospheric pressure is 1.01×10^5 Pascal or 1 Bar or 10.3 mts. of height of water or 76 cm height of **mercury**.

3. Hydrostatic Law

- For downward 'h'

$$\frac{dP}{dh} = w$$

Atmoshperic

δ h ↓downward

- For upward 'h'

$$\frac{dP}{dh} = -w$$

- Hydrostatic pressure distn. flows linear variation of depth below the free surface.

Remember

4. Conversion of one Fluid Column to Another Fluid Column

$$\rho_1 h_1 = \rho_2 h_2$$

$$s_1 h_1 = s_2 h_2$$

Here, ρ = Density of fluid

      s = Relative density

Remember

- Piezometer is suitable for *small* and *positive* pressure measurement.
- The manometric liquid should have *high density* and *vapour pressure*.
- Simple manometer/U-tube manometer can measure both *positive* and *negative* pressure.
- Aneroid/Mecury barometer used to measure *local* atmospheric pressure on *absolute* scale.
- Density of mercury = 13.6×10^3 kg/m^3
  Density of air = 1.24 kg/m^3

■■■■

Hydrostatic Force

3

1. Hydrostatic Force on Submerged Surface

Case	Force	Center of pressure (h)
Horizontal Position	$\omega A\bar{x}$	$h = \bar{x}$
Vertical Position	$\omega A\bar{x}$	$h = \bar{x} + \dfrac{I_G}{A\bar{x}}$
Inclined Position	$\omega A\bar{x}$	$h = \bar{x} + \dfrac{I_G}{A\bar{x}} \sin^2\theta$

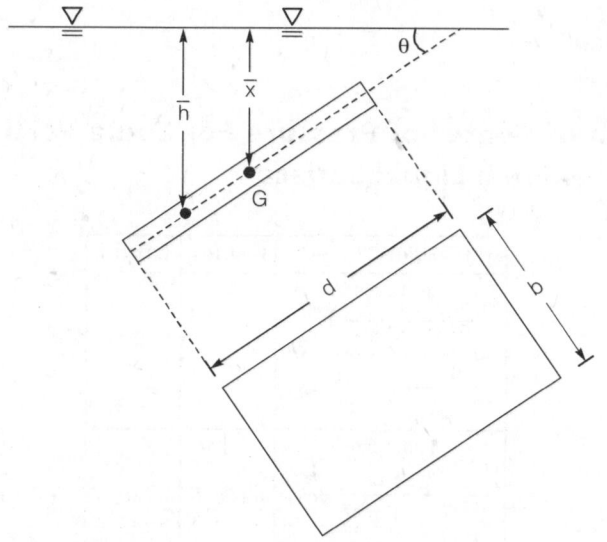

$$I_G = \frac{bd^3}{12} \qquad \text{(For rectangular plate)}$$

$$I_G = \frac{\pi}{64}(\text{diameter})^4 \qquad \text{(For circular plate)}$$

Here, A = Area of surface touching fluid

I_G = Area moment of inertia about centroidal axis and parallel to free axis.

$\bar{x}$ = Vertical distance of C.O.G. of body from free surface.

ω = Specific weight

θ = Angle at which the surface is inclined with horizontal

2. Pressure Force on Curved Surface

- ### Horizontal Force (F_H)

    Horizontal component of the resultant hydrostatic force 'F_x' of curved surface may be computed by projecting the surface upon a vertical plane and multiplying the projected area by the pressure at its own centre of area.

- ### Vertical Force (F_V)

    Vertical component of force 'F_y' is equal to the weight of the liquid block lying above the curved surface upto free surface.

- ### Resultant Force (F)

$$F = \sqrt{(F_H)^2 + (F_V)^2}$$

    Angle of line of action of resultant force with the horizontal is given

    by $\tan\theta = \dfrac{F_y}{F_x}$

3. Depth of Center of Pressure For Some Vertical Plane Surfaces From Liquid Surface

SURFACE	C.G. ($\bar{x}$)	C.P. ($\bar{h}$)
Rectangle	$\dfrac{h}{2}$	$\dfrac{2h}{3}$
Trapezium	$\dfrac{a+2b}{a+b} \cdot \dfrac{h}{3}$	$\dfrac{a+3b}{a+2b} \cdot \dfrac{h}{2}$
Triangle (a)	$\dfrac{2h}{3}$	$\dfrac{3h}{4}$
(b)	$\dfrac{h}{3}$	$\dfrac{h}{2}$

Circle D	$\dfrac{D}{2}$	$\dfrac{5D}{8}$
Semi Circle ⟵ D ⟶	$\dfrac{2D}{3\pi}$	$\dfrac{3\pi D}{32}$
Parabola (a) ⟵ b ⟶ , h	$\dfrac{3h}{5}$	$\dfrac{5h}{7}$
(b) ⟵ b ⟶ , h	$\dfrac{2h}{5}$	$\dfrac{4h}{7}$

Remember

- In case of vertical surface, when depth of immertion ($\bar{x}$) is very large then centre of pressure = centre of gravity or $\boxed{h = \bar{x}}$.

■■■■

Buoyancy and Floatation 4

1. Archimedes Principle

When a body is submerged either fully of partially then it is acted upon by a force of buoyancy vertically up which is equal to weight of liquid displaced by the body.

Remember: ....

- This force of buoyancy always acts through the centroid of liquid displaced.
- Centre of Buoyancy is that point through which buoyant force act.

2. Principal of Flotation

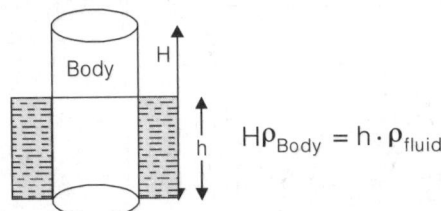

$$H\rho_{Body} = h \cdot \rho_{fluid}$$

Here, H = Height of body

h = Height of body that is submerge in fluid

3. Condition for Equilibrium for Floating/Submerged Body

- For stable equilibrium
  - In case of floating body, metacenter should be above centre of gravity.

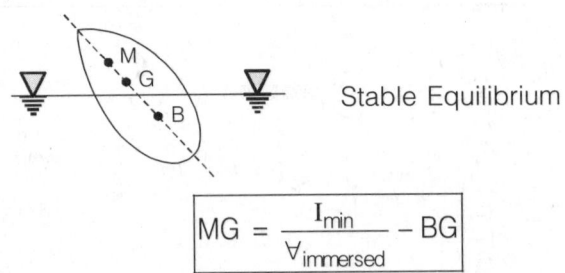

Stable Equilibrium

$$MG = \frac{I_{min}}{\forall_{immersed}} - BG$$

  - In case of submerged body, center of buoyancy should be above centre of gravity.
  - Distance between metacenter and centre of buoyancy (B.M.)

$$= \frac{I_{min}}{V_{immersed}}$$

Here,

I_{min} = Moment of inertia of top view of floating body about longitudinal axis

V = Volume of body immersed in liquid

Remember

- Metacentric height for rolling condition will be less then metacentric height for pitching condition.
- For Neutral equilivrium $\boxed{M = G}$.

4. Time Period of Oscillation

If a floating body oscillates then its time period of transverse oscillation is given by

$$T = 2\pi \sqrt{\frac{K_G^2}{g.GM}}$$

Here, K_G = Least radius of gyration

GM = Meta-centric height

5. For cone the center of gravity lies at $\frac{3}{4}H$ from the pointed end.

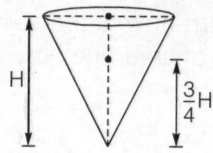

■■■■

Fluid Kinematics

- Lagrangian concept is for single fluid particle
- Eularian concept is for particular section or point

1. Steady and Unsteady Flow

If the fluid and flow characteristics (such as density, velocity, pressure etc.) *at a point* do not change with time, the flow is said to be steady flow.

$$\boxed{\frac{dv}{dt} = 0, \frac{dP}{dt} = 0, \frac{d\rho}{dt} = 0 \quad \begin{array}{l} v = \text{velocity} \\ \rho = \text{density} \end{array}} \quad \textbf{For steady flow.}$$

- It is applicable for *all* properties.
- If the fluid and flow variables at a point may **change** with time, the flow will be **unsteady**.

2. Uniform and Non-Uniform Flow

If the velocity vector at all points in the flow is same *at any instant of time*, the flow is **uniform flow**. If the velocity vector **varies** from point to point at any instant of time, the flow will be **non-uniform**.

$$\frac{dv}{ds} = 0$$

- It is applicable *only* for velocity.

3. Laminar and Turbulent Flow

In laminar flow, the particles moves in layers sliding smoothly over the adjacent layers while in turbulent flow particles have the random and erratic movement, intermixing in the adjacent layers.

4. Streakline

When a dye is injected in a liquid or smoke in a gas so as to trace the subsequent motion of fluid particles passing a fixed point, the path followed by the dye or smoke is called the streakline.

5. Pathline

A pathline is a curve traced by a single fluid particle during its motion.

6. Streamline

A streamline is an imaginary line drawn in a flow field such that a tangent drawn at any point on this line represents the direction of velocity vector at that point.

Remember

- There is no *velocity* component normal to stream lines.
- In steady flow Streakline, Pathline and Streamline are Identical.

7. Equation of stream line

Tangent to stream line gives velocity

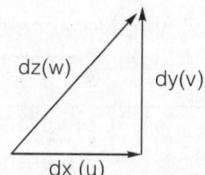

dz(w)   dy(v)

dx (u)

$$\frac{dx}{u} = \frac{dy}{v} = \frac{dz}{w}$$

Here,   u, v, w = Components of velocity in x, y, z direction

8. Continuity Equation (Conservation of Mass)

$$\rho_1 A_1 V_1 = \rho_2 A_2 V_2$$

Here,
ρ = Density
A = Area
V = Velocity

- For incompressible fluid density will be constant thus continuity equation will be

$$A_1 V_1 = A_2 V_2$$

9. General Continuity Equation

$$\frac{\partial \rho}{\partial t} + \frac{\partial}{\partial x}(\rho u) + \frac{\partial}{\partial y}(\rho v) + \frac{\partial}{\partial z}(\rho w) = 0$$

Special Case:

- If flow is steady then $\left(\frac{\partial \rho}{\partial t} = 0\right)$

Thus continuity equation will be

$$\frac{\partial}{\partial x}(\rho u) + \frac{\partial}{\partial y}(\rho v) + \frac{\partial}{\partial z}(\rho w) = 0$$

- For steady, incompressible density will be constant thus

$$\frac{\partial u}{\partial x} + \frac{\partial v}{\partial y} + \frac{\partial w}{\partial z} = 0$$

10. Total Acceleration of fluid

$$a_x = \underbrace{u\frac{\partial u}{\partial x} + v\frac{\partial u}{\partial y} + w\frac{\partial u}{\partial z}}_{\text{Convective acceleration}} + \underbrace{\frac{\partial u}{\partial t}}_{\substack{\text{Temporal or}\\\text{local acceleration}}}$$

$$a_y = u\frac{\partial v}{\partial x} + v\frac{\partial v}{\partial y} + w\frac{\partial v}{\partial z} + \frac{\partial v}{\partial t}$$

- Total Acceleration = Covective acceleration with respect to space + local acceleration with respect to time

Type of flow	Convective Acceleration	Temporal Acceleration
Steady & uniform	0	0
Steady & non-uniform	Exists	0
Unsteady & uniform	0	Exists
Unsteady & non-uniform	Exists	Exists

11. Rotational Component/Vorticity/Circulation

- Rotational component (ω)

$$\omega = \frac{1}{2}\begin{vmatrix} i & j & k \\ \dfrac{\partial}{\partial x} & \dfrac{\partial}{\partial y} & \dfrac{\partial}{\partial z} \\ u & v & w \end{vmatrix}$$

$$\omega_x = \frac{1}{2}\left(\frac{\partial w}{\partial y} - \frac{\partial v}{\partial z}\right)$$

For irrotational flow

$$\omega_x = \omega_y = \omega_z = 0$$

- Vorticity (ξ)

$$= 2 \times \text{rotational component}$$

- Circulation (Γ)

It is line integral of tangential component of velocity around a closed curve.

$$\Gamma = \text{Vorticity} \times \text{Area}$$

Remember

- In irrotational flow, the vorticity is zero, at all points in the flow region while for rotational flow, vorticity is non-zero.
- Flow outside the boundary layer has irrotational characteristic while that within the boundary layer is rotational characteristic.

12. Velocity Potential Function (ϕ)

$$u = \frac{-\partial\phi}{\partial x}, \quad v = \frac{-\partial\phi}{\partial y}, \quad w = \frac{-\partial\phi}{\partial z}$$

Remember: ....

- Flow always occurs in the direction of decreasing potential.
- Velocity potential function exists only for irrotational flow.
- Laplace equation is given by

$$\frac{\partial^2\phi}{\partial x^2} + \frac{\partial^2\phi}{\partial y^2} = 0, \quad \boxed{\nabla^2\phi = 0}$$

- If velocity potential function satisfies Laplace equation then it also satisfies continuity equation and hence the **flow** is **possible**.

13. Streamline Function (ψ)

$$v = \frac{\partial\psi}{\partial x}, \quad u = -\frac{\partial\Psi}{\partial y}$$

- If Ψ exists, then it satisfies continuity equation and flow can be rotational or irrotational.
- If stream function satisfies Laplace equation then it is case of

  Irrotational flow. $\boxed{\nabla^2\Psi = 0}$

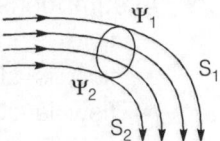

- Discharge per *unit* length $= \psi_1 - \psi_2$

  $\psi_1 > \psi_2$

14. Cauchy-Reimann equation

$$\frac{\partial\phi}{\partial x} = \frac{\partial\psi}{\partial y} = u$$

$$-\frac{\partial\phi}{\partial y} = \frac{\partial\psi}{\partial x} = v$$

- Equipotential lines and constant stream function lines are *orthogonal*.

■■■■

Fluid Dynamics & Flow Measurements

6

1. Different Kind of Force Acting on Fluid Particle

- Pressure force
- Gravity force
- Viscous force

Remember: ....

- If all the three force are taken into account then equation obtained is known as Naiver stokes equation.

2. Euler's Equation

It represents momentum equation in a 2-D, *inviscid* steady flow.

$$\frac{dP}{\rho} + g.dz + v.dv = 0$$

- No viscous effects are considered.

3. Bernoulli's Equation

- Assumptions in Bernoulli's equation:
    - fluid is ideal
    - flow is steady
    - flow is continuous
    - fluid is incompressible
    - flow is non-viscous
    - flow is irrotational
    - applicable along a stream line

This equation is obtained by integrating Euler's equation.

$$\frac{P}{\omega} + z + \frac{V^2}{2g} = C$$

Here, $\frac{P}{w} + z$ = Pressure head + gravitational head

= Piezometric head

- Bernoulli's equation is applicable across the streamlines, if flow is **Irrotational**.

Remember

4. Venturimeter

- General proportion of venturimeter

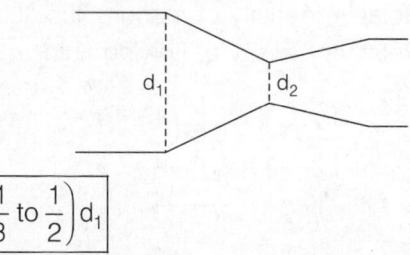

$$d_2 = \left(\frac{1}{3} \text{ to } \frac{1}{2} \right) d_1$$

Angle of convergence = 15 – 20°
Angle of divergence = 6 – 7° and it should be not greater than 7° to avoid *flow separation*.

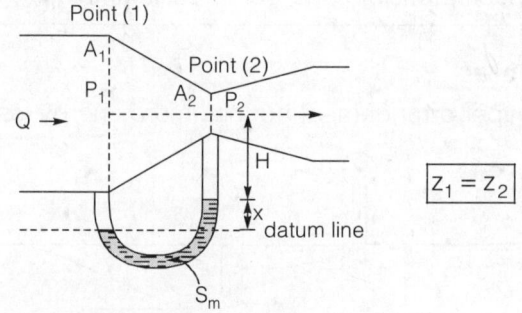

$$Z_1 = Z_2$$

- It is used for measuring discharge

$$Q_{ideal} = \frac{A_1 A_2 \sqrt{2gh}}{\sqrt{A_1^2 - A_2^2}}$$

- h = Piezometric head difference

$$h = \left(\frac{P_1}{\gamma} + Z_1 \right) - \left(\frac{P_2}{\gamma} + Z_2 \right)$$

$$h = \frac{P_1 - P_2}{\omega} \qquad (\because \text{gravitational head difference} = 0)$$

$$= \frac{V_2^2 - V_1^2}{2g}$$

$$h = x \left(\frac{S_m}{S} - 1 \right)$$

Here,

x = manometric deflection

s_m = Relative density of manometric fluid

s = Relative density of flowing fluid

• $Q_{actual} = C_{dv} \cdot Q_{ideal}$

For venturimeter C_{dv} = 0.94 – 0.98

Here, C_{dv} = coefficient of discharge

$$C_{dv} = \sqrt{\frac{h - h_l}{h}}$$

h_l = Head loss in convergent divergent section

• Venturimeter is used for measuring rates of flow in both incompressible and compressible fluids.

5. Orifice Meter

• It is cheaper arrangement but has more energy loss.

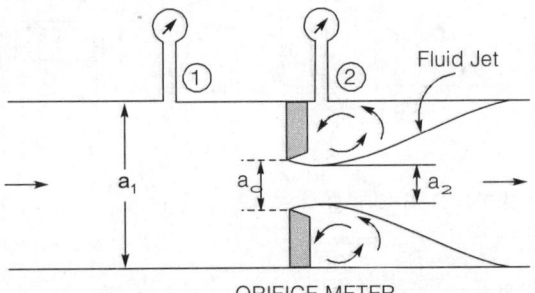

ORIFICE METER

$$Q = \frac{C_d \, a_1 a_o}{\sqrt{a_1^2 - a_o^2}} \sqrt{2gh}$$

$$a_1 = \frac{\pi}{4} D^2$$

$$a_o = \frac{\pi}{4} d_o^2$$

d_o = diameter of orifice

Remember

- It is used to measure **discharge**
- $C_d = 0.64 - 0.76$
- For orifice, $C_c = \dfrac{C_d}{C_v}$

Here, C_C = Coefficient of contraction
C_d = Coefficient of discharge
C_v = Coefficient of velocity'

- A **flow nozzle** is essentially an orifice meter in which jet contraction is eliminated by smooth entrance boundary and thus result in much **smaller losses** than orifice meter.

6. Pitot Tube

- It is based on principle of conversion of kinetic head into pressure head. The point at which velocity reduces zero is called **stagnation point**.

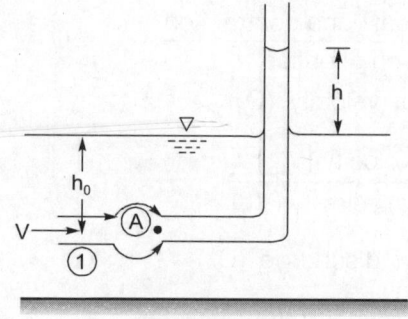

Simple Pitot Tube

$$V_{th} = \sqrt{2gh}$$

$$= \sqrt{2g\left(\frac{P_s - P_o}{\rho g}\right)}$$

$$V_{ac} = C_V V_{th}$$
C_V = Coefficient of velocity (0.98)

$P_s/\rho g$ = stagnation head &

$P_o/\rho g$ = static head.

- Velocity head is indicated by the difference in liquid level between the Pitot tube and the piezometer. The **Pitot tube** measures the total head and therefore known as **total head tube**.

- Application of Pitot Tube in Pipes

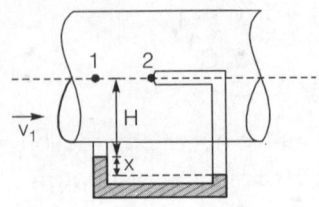

$$v_1 = \sqrt{2gh}$$

$$h = x\left(\frac{s_m}{s} - 1\right)$$

Here, s_m = Relative density of manometric fluid
         s = Relative density of flowing fluid

8. Hydraulic Coefficients

- Contraction coefficient (C_c)

$$= \frac{\text{Area of jet at vena contraction}}{\text{Area of orifice}}$$

- Coefficient of velocity (C_v)

$$= \frac{\text{Actual velocity } (V_{ac})}{\text{Theoretical velocity } (V_{th})}$$

- Coefficient of discharge (C_d)

$$= \frac{\text{Actual discharge } (Q_{ac})}{\text{Theoretical discharge } (Q_{th})}$$

9. Devices and Their Uses

Device	Measurement
Venturimeter	rate of flow (discharge)
Flow nozzle	rate of flow
Orifice meter	rate of flow
Bend meter	rate of flow
Rotameter	rate of flow
Pitot tube	velocity
Hot wire anemometer	air & gas velocity
Current meter	velocity in open channels

■ ■ ■ ■

7

Viscous Flow of Incompressible Fluid

In viscous flow fluid particles move along straight parallel paths in *layers*. It occurs at low velocity and *viscous force* predominates inertial force.

1. Reynold's Number (R$_e$)

$$R_e = \frac{\text{Inertia force}}{\text{Viscous force}} = \frac{\rho v d}{\mu}$$

Here, ρ = Density
d = characteristics length
V = average velocity

2. Nature of Flow According to Reynold's Number For Pipe and Open Channel Flow

The limiting values of Reynold's Number corresponding to which flow is Laminar is given by:

Flow Condition	Pipe flow	Open channel flow
Laminar Flow	Re ≤ 2000	Re ≤ 500
Transitional Flow	2000 < Re < 4000	500 < Re < 1000
Turbulent Flow	Re > 4000	Re > 1000

1. Entrance Length (L$_e$)

* Entrance length (L$_e$)

  The length of pipe from its entrance upto the point where flow attains fully developed velocity profile and which remains unaltered beyond that known as entrance length.

* The entrance length required to establish fully developed *laminar flow* is given by

$$\frac{L_e}{D} = 0.07 \, R_e$$

* The entrance length for fully developed *turbulent flow* is given by

$$\frac{L_e}{D} = 50$$

4. Laminar Flow Through Circular Pipe (Hagen-Poiseulle Flow)

- Shear stress (τ) distribution

$$\tau = \left(-\frac{\partial P}{\partial X}\right)\cdot\frac{r}{2}$$

The negative sign on $\dfrac{\partial p}{\partial x}$ indicates decrease in pressure in the direction of flow. The pressure must decrease because pressure force is the only means available to compensate for resistance to the flow, the potential and kinetic energy remain constant.

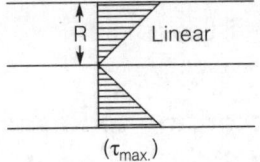

$(\tau_{max.})$

Remember: ....

- Above equation is valid for steady and uniform flow.
- The maximum value of stress occur at $r = R$.
- In laminar flow shear stress is entirely due to **viscous** action.
- **Velocity distribution**

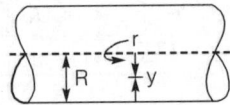

r is the distance measured from axis

$$U = \frac{1}{4\mu}\left(-\frac{\partial P}{\partial x}\right)(R^2 - r^2)$$

At center, $r = 0$ thus

$$U_{max} = \frac{1}{4\mu}\left(-\frac{\partial P}{\partial x}\right)\cdot R^2$$

Here, U_{max} = Maximum velocity

$$U = U_{max}\left(1 - \frac{r^2}{R^2}\right)$$

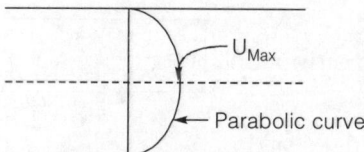

- **Discharge (Q) through a pipe**

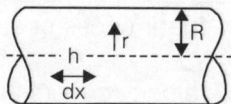

  Q ≠ AV, because 'V' is changing

- For *viscous* flow

$$Q = \frac{\pi}{128\mu} \cdot \left(\frac{-\partial P}{\partial x} \right) \cdot D^4$$

Mean velocity $(V_{avg.}) = \dfrac{U_{max}}{2}$

- The maximum velocity occur at axis.
- The point where local velocity is equal to mean velocity is given by

$$r = \frac{R}{\sqrt{2}} = 0.707\ R$$

- **Pressure drop in pipe**

$$P_1 - P_2 = \frac{32\mu V_{avg.}\ L}{D^2}$$

$$\boxed{hf = \frac{32\mu VL}{\gamma D^2}}$$

5. Laminar Flow Between Two Fixed Parallel Plate

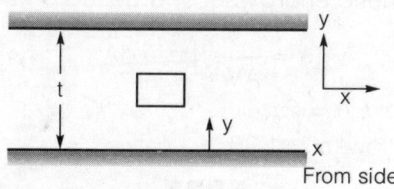

From side

- Shear stress distribution $\boxed{\tau = -\dfrac{dP}{dx}\left(\dfrac{t}{2} - y\right)}$

- Velocity $u = \dfrac{1}{2\mu}\left(-\dfrac{\partial P}{\partial x}\right)(ty - y^2)$

Remember

- The maximum shear stress occurs at $y = 0$.

- The maximum velocity occurs at $y = \dfrac{t}{2}$ and it is given by,

$$u_m = \dfrac{t^2}{8\mu}\left(-\dfrac{\partial p}{\partial x}\right)$$

- $V_{avg.} = \dfrac{2}{3}u_{max} = 0.66\, u_{max}$

- Discharge per unit width

$$Q = \dfrac{1}{12\mu}\left(-\dfrac{\partial P}{\partial x}\right)t^3$$

- Pressure drop in given length

$$\boxed{P_1 - P_2 = \dfrac{12\mu VL}{t^2}} \quad \boxed{h_f = \dfrac{12\mu VL}{\gamma\, t^2}}$$

6. Momentum correction factor (β)

It is defined as the ratio of momentum/sec based on *actual* velocity to the momentum/sec based on *average* velocity.

$$\beta = \dfrac{1}{AV^2}\int u^2 \cdot dA$$

V = Average velocity
u = Local velocity at distance r

- For laminar flow, $\beta = 1.33$
- For turbulent flow, $\beta = 1.2$

9. Kinetic energy correction factor (α)

It is defined as the ratio of kinetic energy/second based on *actual* velocity to the kinetic energy/second based on *average* velocity.

$$\alpha = \dfrac{1}{AV^3}\int u^3 \cdot dA$$

- For laminar flow, $\alpha = 2$
- For turbulent flow, $\alpha = 1.33$

■■■■

Flow Through Pipes

8

1. Friction Loss/Darcy Weisbach Equation
- Darcy Weisbach equation

$$h_f = \frac{fLV^2}{2gD}$$

  L = Length of pipe

  D = Dia of pipe

  V = Mean velocity of flow

  f = Friction factor (0.02 to 0.04 for metals)

  h_f = Head loss due to friction

- **For laminar flow**

$$f = \frac{64}{R_e}$$

- Friction factor is inversely proportional to diameter of pipe.
- **Turbulent flow**

$$f = \frac{0.316}{Re^{1/4}}$$

- Friction factor $\propto \dfrac{1}{(\text{diameter of pipe})^{1/4}}$

- Friction loss in pipe whose end is closed and flow takes place through sides at regular interval

$$h_f = \frac{1}{3}\frac{fLV^2}{2gD}$$

2. Chezy's Formula

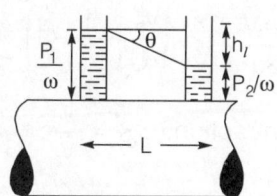

$$V = C\sqrt{mi}$$

C = Chezys const.

m = Hydraulic mean depth

$$= \frac{Area}{Wetted\,Perimeter}$$

$$i = Hydraulic\ slope\ = \frac{h_L}{L} = \tan\theta$$

- Relation between Chezys constant and friction factor.

$$f = \frac{8g}{C^2}$$

3. Different Type of Minor Losses in Pipe

- Losses due to sudden expansion

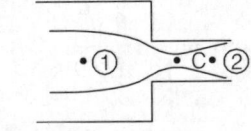

A_1, V_1 A_2, V_2

Eddy region

$$h_{le} = \frac{(V_1 - V_2)^2}{2g} = \frac{V_1^2}{2g}\left(1 - \frac{A_1}{A_2}\right)^2$$

Remember

- For sudden expansion, the optimum ratio of diameters of the pipe so that the pressure loss is minimum.

$$\frac{d_1}{d_2} = \frac{1}{\sqrt{2}}$$

- Losses due to sudden contraction.

$$h_{lc} = \frac{(V_c - V_2)^2}{2g} = \frac{V_2^2}{2g}\left(\frac{1}{C_c} - 1\right)^2$$

Coefficiency of contraction $(C_c) = \dfrac{V_2}{V_c}$

If C_c is not given

$$h_{lc} = \frac{0.5 V_2^2}{2g}$$

- Losses at exit of pipe

$$h_l = \frac{V^2}{2g}$$

- Losses at entrance to pipe

$$h_l = \frac{0.5 V^2}{2g}$$

Here, V = **Mean** velocity of flow in pipe

- Loss due to gradual expansion

$$h_L = K_L \frac{(V_1 - V_2)^2}{2g} = K_L \frac{V_1^2}{2g}\left(1 - \frac{A_1}{A_2}\right)^2$$

K_L depends upon angle of expansion.

- Losses due to bends

$$h_l = \frac{KV^2}{2g}$$

K ≃ 1.2 for 90° bend
K ≃ 0.4 for 45° bend

4. Hydraulic Gradient Line (HGL) and Total Energy Line (TEL)

HGL: It joins Piezometric head $\left(\dfrac{P}{\omega} + z\right)$ at various points.

TEL: It joins total energy head at various points

$$\left(\frac{P}{w} + z + \frac{V^2}{2g}\right)$$

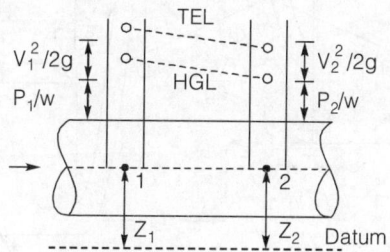

Remember: ....
- HGL is always parallel and lower than TEL.

5. Power Transmission Through Pipe (P)

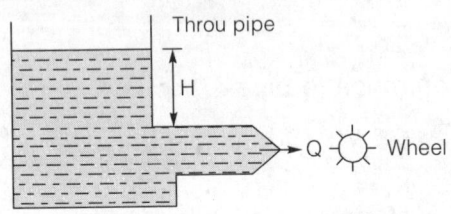

- $P_{ideal} = \rho g Q H$
- $P_{actual} = \rho g Q (H - h_l)$

Here,

h_l = Head loss

- Efficiency$(\eta) = \dfrac{P_{actual}}{P_{ideal}} = \dfrac{H - h_l}{H}$

- Power delivered by a given pipeline is **maximum** when the flow is such that **one third** of static head is consumed in pipe friction. Thus efficiency is limited to only 66.66% (For η_{max}, h_l = H/3).

6. Siphon Action of Pipe

A pipe which raises above its hydraulic grade line has negative pressure and is known as a siphon.

- Formation of air lock at highest portion of pipe.

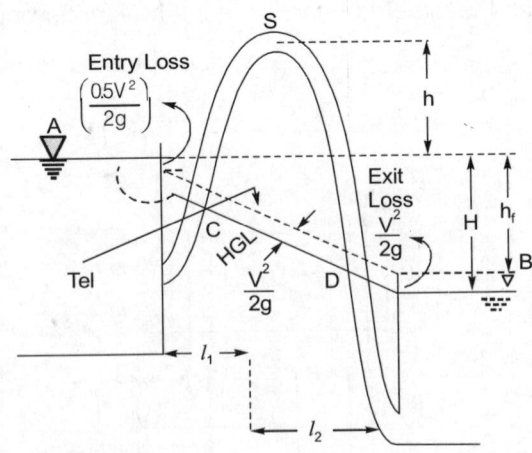

- Summit pressure (p_s) should not be less than vapour pressure of liquid to avoid air lock formation at summit. (For water it is 2.5 m at 20°C)

- Pressure at C and D = atmospheric pressure

- Pressure below C and D = Positive

- Pressure above C and D = Negative (Suction pressure).

- It total head loss between A and B is H then,
  H = entry loss + friction loss + exit loss

$$H = 0.5\frac{V^2}{2g} + \frac{fLV^2}{2gD} + \frac{V^2}{2g} \qquad ...(i)$$

- If h is height of summit above reservoir A, then by applying Bernoulli's equation between A and S,

$$\frac{p_A}{\omega} = \left(\frac{p_s}{\omega} + h\right) + \frac{V^2}{2g} + 0.5\frac{V^2}{2g} + \frac{fl_1V^2}{2gD} \qquad ...(ii)$$

from (i) and (ii) p_s and h can be calculated.

7. Water Hammer in Pipe

- There are three cases of water hammer
  (i) Gradual closure of valve
  (ii) Sudden closure of valve and pipe is rigid
  (iii) Sudden closure of valve and pipe is elastic

- The valve closure is gradual if $t > \dfrac{2L}{C}$

- The valve closure is sudden if $t < \dfrac{2L}{C}$

where, L = Length of pipe
        t = time in sec.
        C = velocity of pressure wave

- **Gradual Closure**

  In this case the pressure head is given by

$$\frac{p}{\omega} = \frac{LV}{gt} \qquad ...(i)$$

where,   L = Length of pipe
            V = Velocity of flow
            t = time in second to close the valve.

- **Sudden Closure With Rigid Pipe**

  The pressure head in this case is given by

  $$\frac{p}{\omega} = \frac{VC}{g} \text{(Allievi formula)} \qquad \qquad ...(ii)$$

  where, $C = \sqrt{\frac{K}{\rho}}$ = Velocity of pressure wave

  K = Bulk modulus of water

  ρ = density of water.

- **Sudden Closure With Elastic Pipe**

  The pressure head is given by:

  $$\frac{p}{\omega} = \frac{V}{\sqrt{\rho g^2 \left(\frac{1}{K} + \frac{D}{tE} \right)}} \qquad \qquad ...(iii)$$

  where,

  D = Diameter of pipe

  t = Thickness of pipe

  E = Modulus of elasticity of pipe material

  (In deriving equation (iii), the Poisson's ratio of pipe material is assumed as 0.25.)

■ ■ ■ ■

Vortex Motion

9

A *whirling* mass of fluid is called vortex flow.

1. Free Vortex Flow

In this flow fluid mass rotates due to conservation of angular momentum.

The velocity profile is inversely proportional with the radius.

$$v\,r = constant$$

The point at the centre of rotation is called **singular point**, where velocity approaches to infinite.

Example of free vortex motion are whirling mass of liquid in wash basin, whirlpool in rivers etc.

Remember: ....

- No *external* torque or energy is required.
- In free vortex flow *Bernoulli's equation* can be applied.

2. Forced Vortex Flow

When a fluid is rotated about a vertical axis at constant speed, such every particle of its has the same angular velocity, motion is known as the forced vortex.

$$V = r\,\omega$$

$$h = \frac{\omega^2 r^2}{2g}$$

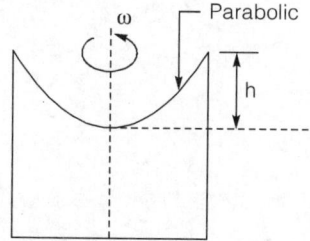

where h is height of paraboloid, and r is radius of cylinder.

$$\text{Volume of paraboloid} = \frac{1}{2}\pi r^2 h$$

$$= \tfrac{1}{2}\text{ of volume of circumscribing cylinder}$$

Remember: ....

- The surface profile of forced vortex flow is *parabolic*.
- Forced vortex requires constant supply of *external* energy/torque. Example of forced vortex flow is rotating cylinder & flow inside centrifugal pump.

3. Variation of Pressure

- Valid for Free as Well as Forced Vortex

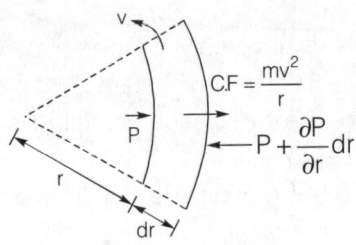

$$dp = \frac{\rho v^2}{r} \cdot dr - \rho g\, dz \qquad\qquad \text{z in upward direction}$$

■ ■ ■ ■

Boundary Layer Theory

10

Boundary layer is a region in the immediate vicinity of the boundary surface in which the velocity of flowing *fluid increases gradually* from zero at the boundary surface to the velocity of the main stream.

1. Development of Boundary Layer Region

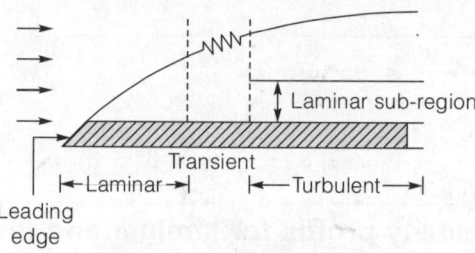

2. Various Terms Associated With Boundary Layer Theory

- **Boundary layer thickness (δ)**
  It is defined as the distance from the boundary surface in which the velocity reaches the 99% of the velocity of the main stream.

  $$y = \delta$$
  for $\quad v = 0.99 \, V_0$

- **Displacement Thickness (δ^*)**
  It is the distance measured normal to the boundary, by which the solid boundary should be displaced in order to compensate for the reduction in mass flow rate due to boundary layer growth.

  $$\delta^* = \int_0^\delta \left(1 - \frac{v}{V_0}\right) dy$$

  The quantity $(V_0 - v)$ known as the velocity defect.

- **Momentum Thickness (θ)**

  $$\theta = \int_0^\delta \frac{v}{V_0}\left(1 - \frac{v}{V_0}\right) dy$$

- **Energy Thickness (δ_e)**

  $$\delta_e = \int_0^\delta \frac{v}{V_0}\left(1 - \frac{v^2}{V_0^2}\right) dy$$

- If $\dfrac{V}{V_0} = \left(\dfrac{y}{\delta}\right)^{1/n}$,

- $H = \dfrac{\delta^*}{\theta} = \dfrac{n+2}{n}$

- The ratio of displacement thickness to momentum thickness is called the **shape factor** (H). Its value should be always greater than 1.

Remember

- For $\dfrac{V}{V_0} = \dfrac{y}{\delta}$, here n = 1.

- $\delta^* = \dfrac{\delta}{2}$, $\theta = \dfrac{\delta}{6}$, $\delta_e = \dfrac{\delta}{4}$ thus $\delta^* > \delta_e > \theta$

---

3. General velocity profile for laminar and turbulent flow

- For laminar flow $\dfrac{u}{u_\infty} = \dfrac{3}{2}\left(\dfrac{y}{\delta}\right) - \dfrac{1}{2}\left(\dfrac{y}{\delta}\right)^3$

- For turbulent flow $\dfrac{u}{u_\infty} = \left(\dfrac{y}{\delta}\right)^{1/2}$

4. Reynold number for different types of flow over flat plate

$$R_e = \dfrac{\rho V x}{\mu}$$

Here   x = Distance from where solid surface starts

$R_e < 5 \times 10^5$ – laminar flow

$R_e > 5 \times 10^5$ – turbulent flow

5. Blassius Experiment Results/When Velocity Profile is Not Given

Laminar	Turbulent
1. $\dfrac{\delta}{x} = \dfrac{5}{\sqrt{R_{ex}}}$	1. $\dfrac{\delta}{x} = \dfrac{0.576}{(R_e x)^{1/5}}$
2. $C_{fx} = \dfrac{0.664}{\sqrt{R_{ex}}}$	2. $C_{fx} = \dfrac{0.059}{(R_{ex})^{1/5}}$
3. $C_d = \dfrac{1.328}{\sqrt{R_{el}}}$	3. $C_d = \dfrac{0.074}{(R_{el})^{1/5}}$

For laminar flow

$$\delta \propto \sqrt{x}$$

$$\tau_0 \propto \frac{1}{\sqrt{x}}$$

Here, δ = Boundary layer thickness
τ_0 = Shear stress at solid surface
x = Distance from where solid surface starts

6. Hardness

Local skin friction coefficient

$$C_{fx} = \frac{\tau_0}{\frac{1}{2}\rho u_\infty^2}$$

Average drag coefficient

$$C_d = \frac{F_d}{\frac{1}{2}\rho A U_\infty^2}$$

7. Condition For Boundary Layer Separation

It is caused by **adverse pressure** gradient $\left(\dfrac{dp}{dx} > 0\right)$

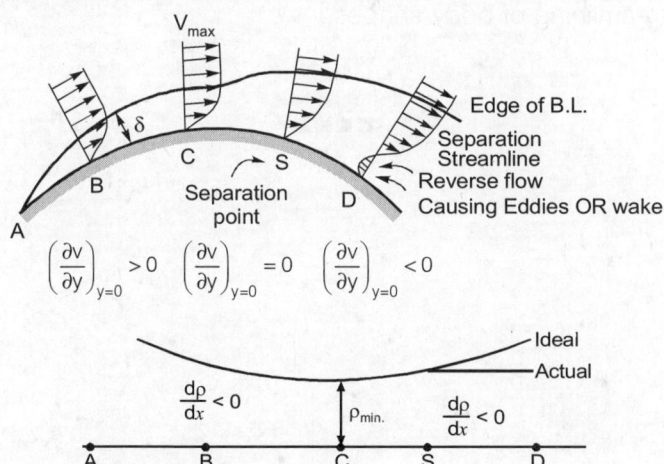

$$\left(\frac{\partial v}{\partial y}\right)_{y=0} > 0 \quad \left(\frac{\partial v}{\partial y}\right)_{y=0} = 0 \quad \left(\frac{\partial v}{\partial y}\right)_{y=0} < 0$$

- **Location of Separation point**
  The separation point S is determined from the condition

  $$\left(\frac{\partial v}{\partial y}\right)_{y=0} = 0.$$

For a given velocity it can be determined whether the B.L. has separated or on the verge of separation or will not separate from the following conditions:

- If $\left(\dfrac{\partial v}{\partial y}\right)_{y=0} < 0$: Flow has separated.

- If $\left(\dfrac{\partial v}{\partial y}\right)_{y=0} = 0$: Flow is on verge of separation.

- If $\left(\dfrac{\partial v}{\partial y}\right)_{y=0} > 0$: Flow is attached with the surface

8. Methods of Preventing Separation

- Rotating the boundary in the direction of flow.
- Suction of the slow moving fluid by a suction slot.
- Supplying additional energy from a blower.
- Providing a bypass in the slotted wing.
- Providing guide blades in a bend.
- Injecting fluid into boundary layer.
- Streamlining of body shapes.

■■■■

Turbulent Flow

11

1. Shear Stress in Turbulent Flow

$$\tau = \mu \frac{d\overline{v}}{dy} + \eta \frac{d\overline{v}}{dy}$$

where,

μ = dynamic coefficient of viscosity (*fluid* characteristic)

$$\eta = \rho l^2 \left(\frac{du}{dy} \right) \text{ eddy viscosity coefficient (\textit{flow} characteristic)}$$

where, l = ky called **mixing length**.

- Eddy viscosity comes in picture due to the turbulence effect.

2. Hydro Dynamically Smooth And Rough Pipes

- If the *average height of irregularities* (k) is greater than the thickness of *laminar* sublayer (δ'), then the boundary is called hydrodynamically smooth.
- If the average height of irregularities (k) is less than the thickness of laminar sublayer (δ'), then the boundary is called hydrodynamically rough.
- On the basis of NIKURADSE'S EXPERIMENT the boundary is classified as:

  Hydrodynamically smooth: $\frac{k}{\delta'} < 0.25$

  Boundary in transition: $0.25 < \frac{k}{\delta'} < 6.0$

  Hydrodynamically Rough: $\frac{k}{\delta'} > 6.0$

- $\frac{R}{k}$ is known as specific roughness. where 'k' is average height of roughness and 'R' is radius of the pipe.

3. Velocity Distribution For Turbulent Flow In Pipes

- Prandtl's universal velocity distribution equation

$$v = v_{max} + 2.5V^* \log_e \left(\frac{y}{R} \right)$$

where

$$V^* = \sqrt{\frac{\tau_0}{\rho}} = \text{shear or friction velocity.}$$

y = distance from pipe wall

ρ = Density of fluid

Remember:

- The above equation is valid for both smooth and rough pipe boundaries.
- Karman - Prandtl Velocity distribution equation

(i) Hydro Dynamically Smooth pipe

$$\frac{v}{V^*} = 5.75 \log_{10}\left(\frac{V^*y}{\nu}\right) + 5.5$$

(ii) Hydro Dynamically Rough pipe

$$\frac{v}{V^*} = 5.75 \log_{10}\left(\frac{y}{k}\right) + 8.5$$

where

V^* = shear velocity

y = distance from pipe wall

k = average height of roughness

ν = kinematic viscosity.

v = Average velocity

- Velocity distribution in terms of mean velocity

$$\frac{v - V^*}{V^*} = 5.75 \log_{10}\left(\frac{y}{R}\right) + 3.75$$

The above equation is for *both rough and smooth* pipes.

4. Friction Factor

- Friction factor 'f' for laminar flow:

$$f = \frac{64}{Re} \text{ where Re = Reynolds number}$$

- Friction factor (f) for turbulent flow in smooth pipes

- $$f = \frac{0.316}{(Re)^{1/4}} \quad (4 \times 10^3 < Re < 10^5) \text{ Blasius equation.}$$

- $$\frac{1}{\sqrt{f}} = 2.0 \log_{10}\left(Re\sqrt{f}\right) - 0.8$$

  $$\left(5 \times 10^4 < Re < 4 \times 10^7\right)$$

- **Friction factor (f) for turbulent flow in Rough pipes**

  $$\frac{1}{\sqrt{f}} = 2.0 \log_{10}\left(\frac{R}{k}\right) + 1.74$$

  This equation shows that for rough pipes friction factor depends

  only on $\dfrac{R}{K}$ (Relative smoothness) and not on Reynold's number

  (Re)

- **Friction factor for commercial pipes**

  $$\frac{1}{\sqrt{f}} - 2.0 \log_{10}\left(\frac{R}{k}\right) = 1.74 - 2.0 \log_{10}\left(1 + 18.7\right)\frac{R/k}{Re\sqrt{f}}$$

  In this equation, K = equivalent sand grain roughness.

Remember

- There is no specific relationship between f and Re for transition flow in pipes.
- In turbulent flow for **smooth pipe** friction factor depends on **Reynold number.**
- In general, turbulent flow for **rough pipe** friction factor depends on relative roughness $\left(\dfrac{R}{k}\right)$ and Reynold number (Re) **both.** But in fully developed turbulent flow for **rough pipe** friction factor depends on relative roughness $\left(\dfrac{R}{k}\right)$ **only.**

■■■■

Dimensional Analysis

12

- Velocity potential $= [L^2\ T^{-1}]$
  Stream function $= [L^2\ T^{-1}]$
  Acceleration $= [LT^{-2}]$
  Vorticity $= [T^{-1}]$

- Total no. of variables influencing the problem is equal to the no. of independent variables plus one, one being the no. of dependent variable.

- **Buckingham π theorem** states that if all the n-variable are described by m fundamental dimensions, they may be grouped into (n - m) dimensionless π terms.

- Selection of 3 repeating variables from the geometry of flow, fluid properties and fluid motion.

- **Geometric similarity - similarity of shape**
  Kinematic similarity - similarity of motion
  Dynamic similarity - similarity of forces

Number	Equation	Significance
Reynolds No.	$\dfrac{F_i}{F_v} = \dfrac{\rho V L}{\mu}$	Flow in closed conduit pipe
Froude No.	$\sqrt{\dfrac{F_i}{F_g}} = \dfrac{V}{\sqrt{gL}}$	where a free surface is present, structure eg. weirs spillway, channels, etc. where *gravity* force is predominant.
Eulers No.	$\sqrt{\dfrac{F_i}{F_p}} = \dfrac{V}{\sqrt{\dfrac{p}{\rho}}}$	In cavitation studies.
Mach No.	$\sqrt{\dfrac{F_i}{F_e}} = \dfrac{V}{C}$	where fluid compressibility is important.
Weber No.	$\sqrt{\dfrac{F_i}{F_\sigma}} = \dfrac{v}{\sqrt{\sigma/\rho L}}$	In capillary studies.

Here, F_i = Inertia force
F_v = Viscous force
F_p = Pressure force
F_e = Elastic force
F_σ = Surface tension force

1. Reynolds Model Law

$$(Re)_m = (Re)_p$$

$$\frac{\rho_r V_r L_r}{\mu_r} = 1$$

- **Applications of Reynold's Model Law**
  - Flow through small sized pipes.
  - Low velocity motion around automobiles and aeroplane.
  - Submarines completely under water.
  - Flow through low speed turbo machines.
  - Flow through venturimeter, orifice meter.

2. Froude's Model Law

$$(F_r)_m = (F_r)_p$$

$$\frac{V_r}{\sqrt{L_r \, g_r}} = 1$$

- **Applications of Froude's Model Law**
  - Open channels
  - Notches and weirs
  - On bridge piers and offshore structure.
  - Spill ways and dams
  - Liquid jets from orifice
  - Ship partially submerged in rough and turbulent sea

■■■■

Impact of Jets

13

Impact of jet means the force exerted by a jet on a plate which may be stationary or moving. This force is obtained from *impulse-momentum* equation.

1. Force Excreted by Jet on a Stationary Plate

- ### Plate is vertical to the jet
  Jet Strikes normal to the flat stationary plate.

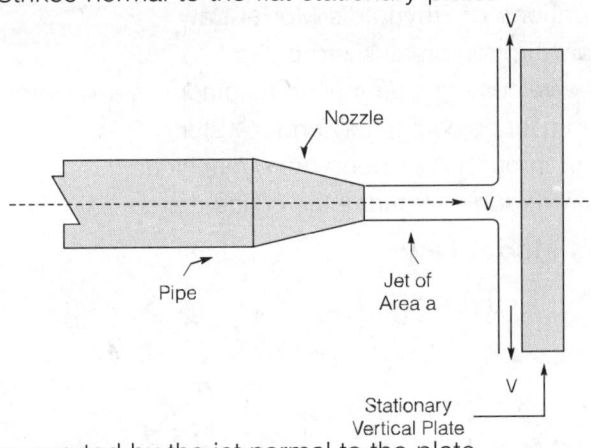

Force exerted by the jet normal to the plate

$$P_n = \rho\, a\, V^2$$
$$a = \text{area of jet}$$
$$V = \text{velocity of jet}$$

- ### Plate is inclined to the jet
  Jet strikes on an inclined stationary plate.

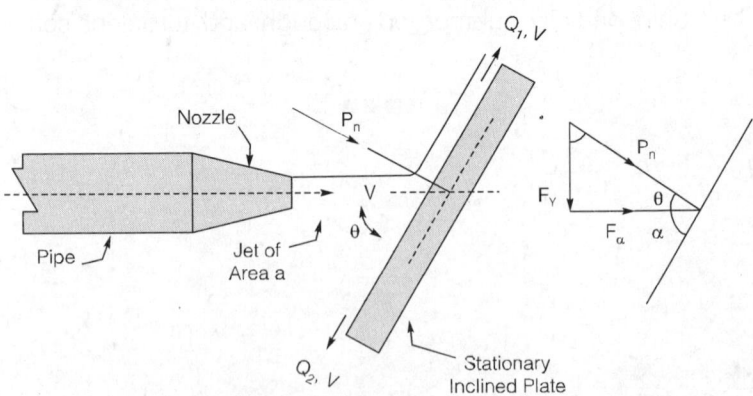

Force exerted by the jet normal to the plate

$$P_n = \rho \, a \, V^2 . \sin\theta$$

$$Q_1 = \frac{Q}{2}(1 + \cos\theta)$$

$$Q_2 = \frac{Q}{2}(1 - \cos\theta)$$

$$\frac{Q_1}{Q_2} = \frac{1 + \cos\theta}{1 - \cos\theta}$$

$$P_x = \rho a V^2 \sin^2\theta$$
$$P_y = \rho a \, V^2 \sin\theta \cos\theta$$

- **Plate is curved**

  Jet striking on a symmetrical stationary curved plate

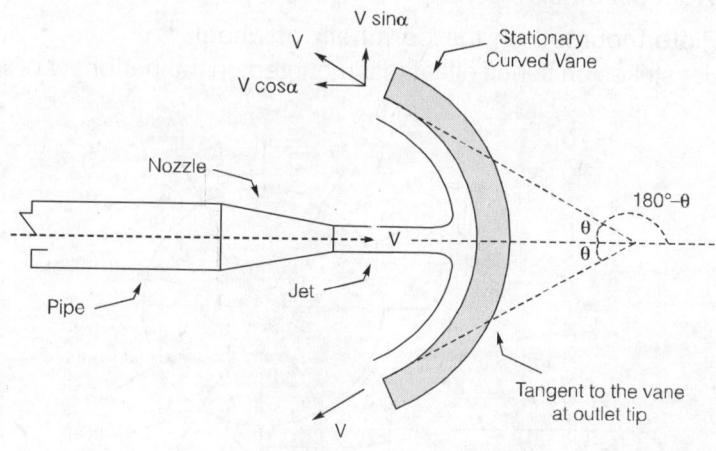

$$P_n = \rho a V^2 (1 + \cos\theta)$$

- Force exerted by a jet in its direction of flow on a curved vane is *always greater* than that exerted on a flat plate.
- Angle of deflection $= (180 - \theta°)$

2. Force Exerted by Jet On a Moving Plate

- **Plate is vertical to the jet**

  Force exerted by jet on moving flat plate normal to jet.

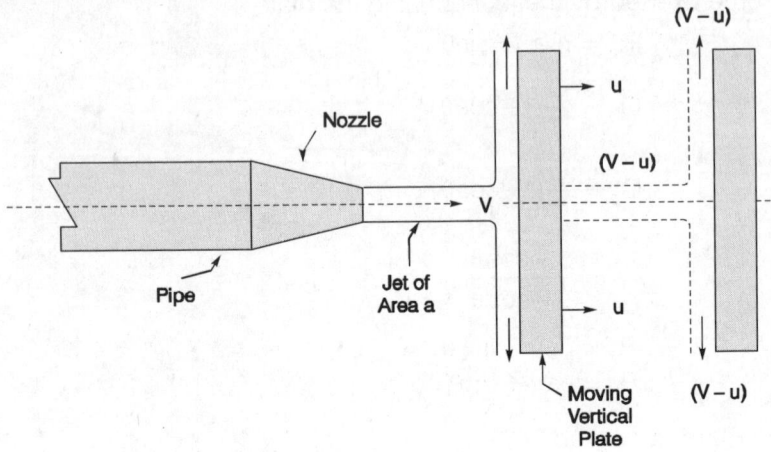

$$u = \text{plate velocity}$$
$$P_n = \rho a(V - u)^2$$

Work done per second $(W) = P_n \times u = \rho a[V - u]^2 \times u$

• **Plate mounted on the periphery of wheel**

Jet strikes on series of flat plat mounted on the periphery of wheel.

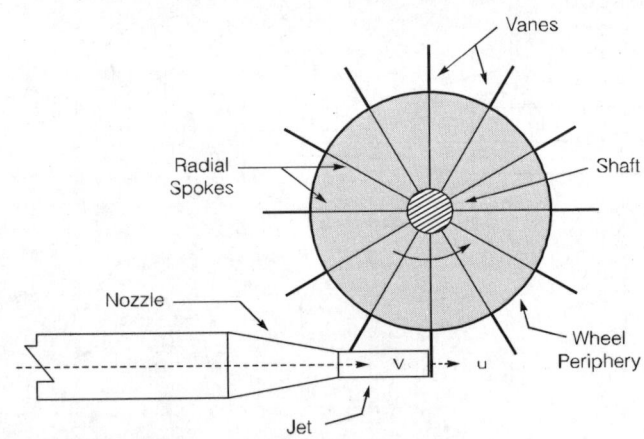

$$P_n = \rho a V(V - u)$$

Work done by the jet $= P_n \times u$

$$W = \rho a V(V - u)u$$

Efficiency of the work done of wheel $\eta = \dfrac{2u(V - u)}{V^2}$

- When peripheral velocity will be half of the velocity of jet i.e. $u = \dfrac{V}{2}$ then efficiency will be maximum i.e. $\eta_{max} = 50\%$

- **Curve plate when the plate is moving in the direction of jet**

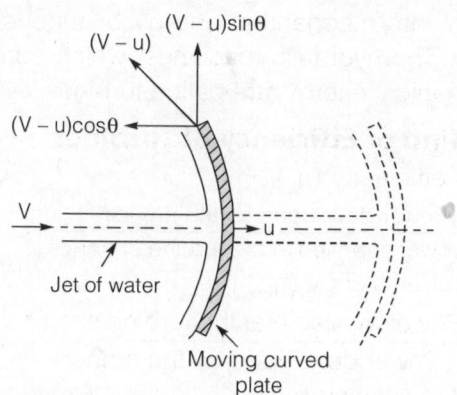

Jet of water

Moving curved
plate

Force excreted by the jet of water on the curved plate in the direction of the jet.

$$P = \rho a (V - u)^2 (1 + \cos\theta)$$

Work done by the jet on the plate per second

$$W = \rho a (V - u)^2 \times u [1 + \cos\theta]$$

3. Force Exerted By a Jet on a Hinged Plate

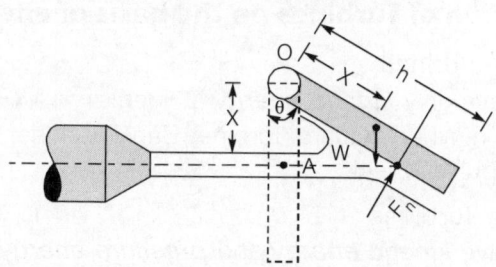

- Force due to jet of water, normal to the plate

$$P_n = \rho a V^2 \sin(90° - \theta)$$

- For equilibrium of plate

$$\sin\theta = \dfrac{\rho a V^2}{W}$$

■■■■

Hydraulic Turbine

14

Hydraulic machines are defined as those machines which convert either hydraulic energy into mechanical energy or mechanical energy into hydraulic energy. The hydraulic machines, which convert the hydraulic energy into mechanical energy, are called turbines.

1. Different Kind of Efficiency of Turbines

- Hydraulic efficiency (η_h)

$$\eta_h = \frac{\text{Power developed by the runner}}{\text{Net power supplied at the turbine entrance}}$$

- Mechanical efficiency (η_{mech})

$$\eta_{mech} = \frac{\text{Power available at the turbine shaft}}{\text{Power developed by the runner}}$$

- Volumetric efficiency (η_{vol})

$$\eta_{vol} = \frac{\text{Quantity of water actually striking the runner}}{\text{Quantity of water supplied to the turbine}}$$

- Overall efficiency (η_o)

$$\eta_o = \frac{\text{Shaft Power}}{\text{Hydraulic Power}}$$

$$\therefore \quad \eta_o = \eta_{hyd} \times \eta_{mech} \times \eta_{vol}$$

2. Classification of Turbines on the basis of energy at inlet

- Impulse Turbines

  These have only *kinetic energy* at inlet.
  Ex.: Pelton Wheel, Guard Turbine, Banki Turbine, Jonval Turbine, Turgo Impulse Wheel.

- Reaction Turbines

  These have *kinetic energy* and *pressure energy* both at inlet.
  Ex.: Francis Turbine, Propeller Turbine, Kaplan Turbine, Thomson Turbine, *Fourneyron* Turbine.

3. Classification of Turbines on the basis of type of flow within turbine

- Tangential Flow Turbine

  Ex.: Pelton Wheel, Turgo Impulse Wheel.

- **Radial Flow Turbines**
  Ex.: Inward-radial flow turbines are Francis turbine, Thomson Turbine, Gurard Turbine. Outward-radial flow turbine is Fourneyron turbine

- **Axial Flow Turbines**
  Ex.: Jonval Turbine, Propeller Turbine, Kaplan Turbine

- **Mixed Flow Turbine**
  Ex.: *Modern* Francis Turbine

4. **Classification of Turbines on the Basis of Value of Specific Speed**

Turbine	Specific speed (MKS Unit)
Pelton turbine	10-35
Francis turbine	60-300
Kaplan & propeller turbine	300-1000

5. **Classification of Turbines on the Basis of Value of head Speed**

Turbine	Head
Pelton turbine	Above 250 m
Francis turbine	60 m-250 m
Kaplan & propeller turbine	Below 60 m

6. **Analysis of Pelton Turbine**

- Work done per second $= \rho Q\left(V_{w_1} + V_{w_2}\right)u$

$$= \rho Q(V_1 - u)(1 + k\cos\phi)u$$

$$= \rho a V_1 (V_1 - u)(1 + k\cos\phi)u$$

V_1 = Velocity of jet coming out of nozzle

$u_1 = u_2 = u$ = tangential velocity of vane

K = Bucket friction coefficient

- **Efficiency of Pelton wheel**

  - Nozzle efficiency

$$\eta_{Nozzle} = \frac{K.E. \text{ of jet per sec}}{\text{Water power at the base of nozzle}}$$

- Hydraulic efficiency

$$\eta_{hyd} = \frac{\text{Runner Power}}{\text{Kinetic Energy per sec}}$$

$$\eta_{hyd} = \frac{2(V_1 - u)(1 + k\cos\phi)u}{V_1^2}$$

$$\eta_{hyd\,max} = \left(\frac{1 + k\cos\phi}{2}\right) \qquad \left[\text{when } u = \frac{V_1}{2}\right]$$

K = Friction factor for blades

- Mechanical efficiency

$$\eta_{mech} = \frac{\text{Shaft Power}}{\text{Runner power}}$$

- Overall efficiency

$$\eta_o = \eta_N \times \eta_h \times \eta_{mech}$$

- Velocity of jet, $V_1 = C_v \sqrt{2gH}$

C_v = coeff of velocity = 0.97 – 0.99

- Speed Ratio $K_u = \dfrac{u}{\sqrt{2gH}}$, where $u = \dfrac{\pi DN}{60}$

K_u = 0.43 – 0.47

- Jet ratio (m) = $\dfrac{\text{dia of wheel (D)}}{\text{jet dia (d)}}$

A jet ratio of 12 is normally adopted

- No. of Buckets (z) = $\dfrac{D}{2d} + 15 = \dfrac{m}{2} + 15 = 18$ to 25

This formula is called Tygun formula.

- It is *tangential flow impulse* turbine.
- It is *high head low* discharge turbine.
- The bucket of a pelton wheel is *double* semi-ellipsoidal in shape. The jet of water impinges at the centre of the bucket & deflects through **160-170°**.
- The advantage of having double cup-shaped buckets is that the axial thrust *neutralizes* each other.
- If P is power developed by the pelton wheel when working under head (H) & having one jet only, the power developed by the same wheel will be **(n.P)** if n jets are used under the *same head*.

7. Analysis of Francis Turbine

- Work done per sec $= \rho Q\left[V_{w_1}.u_1 - V_{w_2}.u_2\right]$

  For best efficiency flow should be radial at outlet, $V_{w_2} = 0$

  i.e. work done per sec $= \rho Q V_{w_1} u_1$

- Degree of reaction

  $$= \frac{\text{Pressure head drop in the runner}}{\text{Hydraulic work done on the runner/unit wt of water}} = \frac{\left(\dfrac{p_1}{\rho g} - \dfrac{p_2}{\rho g}\right)}{\left(\dfrac{V_{w_1}.u_1 - V_{w_2}.u_2}{g}\right)}$$

  V_{w_1} & V_{w_2} are whirl velocities at inlet and outlet

  u_1 & u_2 are peripheral velocities of blades at inlet and outlet.

- Flow ratio $= \psi = \dfrac{V_{f_1}}{\sqrt{2gH}} = 0.15 - 0.30$

- Speed ratio $= k_u = \dfrac{u_1}{\sqrt{2gH}} = 0.60\text{-}0.90$

- Discharge through the turbine
  $$Q = \pi D_1 B_1 V_{f_1}$$
  $$= \pi D_2 B_2 V_{f_2}$$
  B_1, B_2 are width at inlet & outlet respectively.
  D_1, D_2 are diameter at inlet & outlet respectively.
  V_{f_1}, V_{f_2} are velocities of flow at inlet & outlet respectively.

Remember

- It is *inward radial* flow reaction turbine.
- *Scroll casing* is used to evenly distribute the water along the periphery & maintaining the constant velocity for the water.
- The basic function of draft tube is to convert *kinetic head into the pressure head*.
- The angle of *straight divergent* type draft tube should *not be more than 8°* otherwise eddies will be formed & efficiency will be reduced.

8. Some Important Facts of Different Turbines

- **Kaplan Turbine**
  - Its *velocity triangle* and calculation of *efficiency* is similar to Francis turbine.
  - It is *axial flow* reaction turbine.
  - Runner of kaplan turbine has *four to six* blades.
  - Runner blades of propeller turbines are fixed but of kaplan turbine can be *turned* about axis.
  - *At part load*, high efficiency can be obtained in kaplan Turbine.

- **Diagonal Turbine**
  - This is *intermediate* between the *mixed flow* and *axial flow* turbine because the flow of water as it passes through the runner is at an angle of 45° to the axis and hence is known as diagonal turbine.
  - *Head range* is 30 m to 150 m.
  - Blades are *adjustable*.
  - It can be used as turbine as well as *pump*.

- **Tubular Turbine**
  - Adjustable or non-adjustable *runner blades*.
  - Head range **3 m to 15 m**.
  - Scroll casing is *not* provided.

9. Performance of Turbines/Unit Quantities

- Unit speed

$$N_u = \frac{N}{\sqrt{H}}$$

- Unit discharge

$$Q_u = \frac{Q}{\sqrt{H}}$$

- Unit power

$$P_u = \frac{P}{H^{3/2}}$$

10. Specific Speed/Shape Number

It is the speed in rpm of a turbine, geometrically similar to the actual turbine but of such a size that under corresponding conditions it will develop unit power when working under unit head.

$$N_s = \frac{N\sqrt{P}}{H^{5/4}}$$

Remember

- Specific speed is *not* a dimensionless parameter, its dimension is $[M^{1/2}L^{-1/4}T^{-5/2}]$.
- Dimensionless form of specific speed is called shape number (S).

$$S = \frac{N\sqrt{P/\rho}}{(gH)^{5/4}}$$

11. Model Laws of Turbines

Similarity of model (m) and prototype (p) turbines are based on the assumption that efficiency of model is equal to that of the prototype. But the efficiency of prototype is higher than the models efficiency. This is known as *scale effect*.

- $$\left(\frac{Q}{ND^3}\right)_m = \left(\frac{Q}{ND^3}\right)_p$$

  The parameter $\frac{Q}{ND^3} = C_Q$ is called flow coefficient.

- $$\left(\frac{Q}{D^2\sqrt{H}}\right)_m = \left(\frac{Q}{D^2\sqrt{H}}\right)_p$$

- $$\left(\frac{H}{N^2D^2}\right)_m = \left(\frac{H}{N^2D^2}\right)_p$$

  The parameter $\frac{H}{N^2D^2} = C_H$ is called head coefficient.

- $$\left(\frac{P}{D^5N^3}\right)_m = \left(\frac{P}{D^5N^3}\right)_p$$

  The parameter $\frac{P}{N^3D^5} = C_P$ is called power coefficient.

- $$\left(\frac{P}{H^{3/2}D^2}\right)_m = \left(\frac{P}{H^{3/2}D^2}\right)_p$$

■■■■

Hydraulic Pump

15

Centrifugal pump is *reverse of inward flow* reaction turbine. It works on principle of forced vortex motion. It has high discharging capacity and can be used for lifting highly viscous liquids.

1. Priming

It is an operation in which liquid is completely filled in the chamber of pump so that air or gas or vapour from the portion of pump is driven out & no air pocket is left.

2. Head Vs Discharge and Power Vs Discharge Relationship

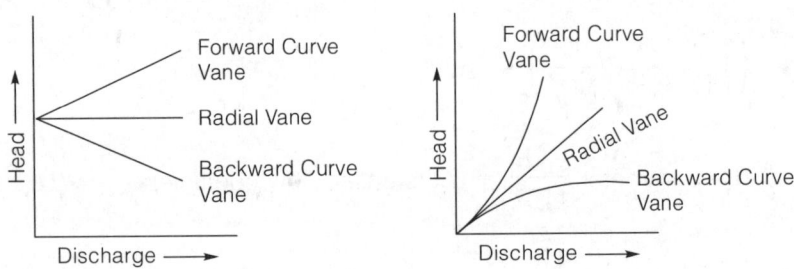

3. Different Types of Pump Based on Head

Types of Pump	Range of Head
Low head pump	upto 15 m head
Medium head pump	15 m to 40 m
High head pump	above 40 m

4. Type of Pump Based on Specific Speed

Pump	Specific Speed
Radial Flow	10 to 80
Mixed Flow	80 to 160
Axial Flow	160 to 450

5. Different Efficiency of Pump

- Manometric Efficiency (η_{man})

$$\eta_{man} = \frac{gH_m}{\left(V_{w_2}.u_2\right)} = \frac{H_m}{H_m + \text{Losses in pump}}$$

- Volumetric Efficiency (η_{vol})

$$\eta_{vol} = \frac{Q}{(Q+\Delta Q)}$$

- Mechanical Efficiency (η_{mech})

$$\eta_{mech} = \frac{\text{Impeller Power}}{\text{Shaft Power}}$$

- Overall Efficiency (η_o)

$$\eta_o = \eta_{man} \times \eta_{vol} \times \eta_{mech}$$

6. Different Parameters of Pump

- **Work done** per second $= \rho Q\left(V_{w_2}.u_2\right)$

  (If discharge enters without whirl & shock, $V_{w_1} = 0$)
- **Static head** (H_s) $= h_s + h_d =$ static suction lift + static delivery lift.
- **Manometric Head** (H_m)- it is the head against which pump has to do the work.

$$H_m = \frac{V_{w_2}.u_2}{g} - \text{Losses of head in the pump}$$

(or)

$$H_m = \left(\frac{p_d}{\rho g} + \frac{V_d^2}{2g} + Z_d\right) - \left(\frac{p_s}{\rho g} + \frac{V_s^2}{2g} + Z_s\right)$$

(or)

$$H_m = h_s + h_d + h_{fs} + h_{fd} + \frac{V_d^2}{2g}$$

where
p_d & p_s are pressures at delivery & suction points.
V_d & V_s are velocities at delivery & suction points.
Z_d & Z_s are position head of delivery & suction points.

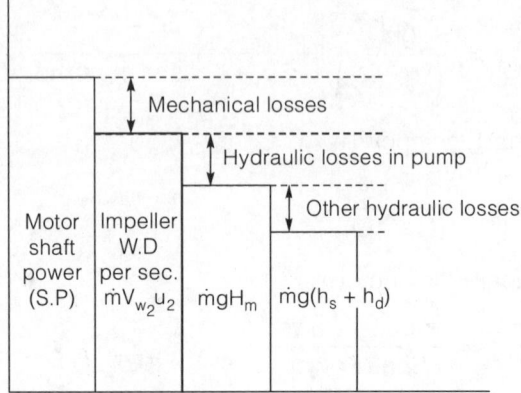

- *Speed ratio* $(k_u) = \dfrac{u_2}{\sqrt{2gH}}$ = 0.95 to 1.25 .

- Flow ratio $(\psi) = \dfrac{V_{f_2}}{\sqrt{2gH}}$ = 0.1 to 0.25

Remember: ....

- For optimum efficiency impeller should be designed such that whirl velocity *at inlet* is zero. It means discharge should enter in the pump radially $\left(V_{w_1} = 0\right)$.

- Pump will start pumping only when pressure head will be greater than or equal to *manometric head*.

7. Specific Speed of Pump

The specific speed of a centrifugal pump may be defined as the speed in revolution per minute of a geometrically similar pump of such a size that under corresponding conditions it would deliver 1 liter of liquid per second against of a head of 1 m.

- Specific speed $N_s = \dfrac{N\sqrt{Q}}{\left(H_m\right)^{3/4}}$

  N_s has dimension of $\left[L^{3/4}T^{-3/2}\right]$

- Dimenionless form specific speed for pump: $S_p = \dfrac{N\sqrt{Q}}{\left(gH_m\right)^{3/4}}$

- For multi stage pump $H_m = \dfrac{\text{Total head}}{\text{No. of stage}}$

8. Pump in Series/Parallel

- **Pumps in series**

  The pumps are connected in series in order to increase the head at *constant discharge*.

  $$Q_1 = Q_2 = Q_3 = \ldots$$
  $$H = H_1 + H_2 + H_3 + \ldots$$

- **Pumps in parallel**

  The pumps are connected in parallel in order to increase the discharge at *constant head*.

  $$Q = Q_1 + Q_2 + Q_3 + \ldots$$
  $$H = H_1 = H_2 = H_3 = \ldots$$

9. Model Laws in Pumps

- $$\left(\frac{N\sqrt{Q}}{H_m^{3/4}}\right)_m = \left(\frac{N\sqrt{Q}}{H_m^{3/4}}\right)_P$$

- $$\left(\frac{\sqrt{H_m}}{DN}\right)_m = \left(\frac{\sqrt{H_m}}{DN}\right)_P$$

- $$\left(\frac{Q}{D^3N}\right)_m = \left(\frac{Q}{D^3N}\right)_P$$

- $$\left(\frac{P}{D^5N^3}\right)_m = \left(\frac{P}{D^5N^3}\right)_P$$

  Here, D = Outer diameter of impeller

  P = Shaft power

  H_m = Manometric head

  N = Impeller speed in rpm

10. Cavitation in Pump

A cavitation parameter σ called as Thoma number

$$\sigma = \frac{\dfrac{P_{s(Absolute)}}{\gamma} + \dfrac{V_s^2}{2g} - \dfrac{P_{v(Absolute)}}{\gamma}}{H_m} = \frac{NPSH}{H_m}$$

$\sigma > \sigma_c$. To avoid cavitation.

The following factors contribute towards onset of cavitation.

- High runner speed
- High temperature
- Less available NPSH

11. Different Parameter/Operating Characteristic Curve of Reciprocating Pump

- Volume of water discharged per second,

$$Q = \frac{ALN}{60} m^3/sec$$

A = Area of cylinder (in m^2)
L = Length of cylinder (in m)
N = Crank speed (in rpm)

To increase discharge and to maintain it more uniform, double acting reciprocating pumps are used.

$$Q \approx \frac{2ALN}{60}, \text{ thus power also gets doubled.}$$

- Work done per second $= \gamma Q(h_s + h_d)$ $[\gamma = \rho g]$

- If the head against which water is to be lifted is
$H_s = (h_s + h_d)$
h_s = suction head
h_d = delivery head
- Slip in Percentage is given by

$$\% \text{ slip} = \frac{Q_{th} - Q_{act}}{Q_{th}} \times 100 = \left(1 - \frac{Q_{act}}{Q_{th}}\right) \times 100 = (1 - C_d) \times 100$$

where C_d = coefficient of discharge
Slip is negative when delivery pipe is *small* and suction pipe is *long* and Pump is running at very *high speed*.

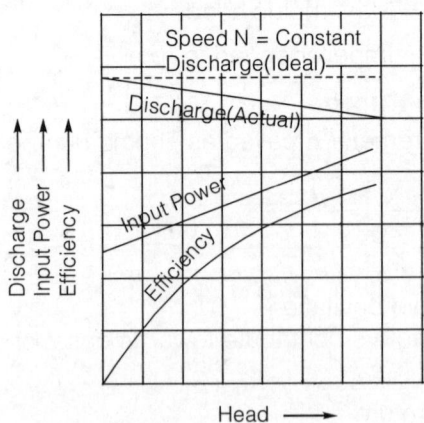

Remember: ....

- Reciprocating pumps are used to lift water against high head at low discharge.
- **Air Vessel** is used to obtain continuous supply of water at uniform rate, to save a considerable amount of work and to run the pump at a high speed without separation.
- Advantage of multicylinder pumps are that the pump even without air vessels deliver liquid more uniformly as compared to single cylinder pump.
- Percentage of work saved is 84.8% when single acting pump with air vessel is used while this saving is only 39.2% when air vessel is used in double acting pump.
- The length of suction pipe in a pump cannot exceed 10 m.
- Net positive suction head (NPSH) is the pressure head above vapour pressure of liquid at the impeller entry.

■■■■

A Handbook on
Mechanical Engineering

2

Heat and Mass Transfer

CONTENTS

Conduction

"Conduction" is the transfer of heat from one part of a substance to another part of the same substance, or from one substance to another in physical contact with it, without appreciable displacement of molecules forming the substance.

1. Conduction Mechanism in Solid, Liquid, Gas

- **Solid**

  In solids heat is conducted by two mechanisms.

  (a) Lattice vibrations

  (b) Transport of free electrons

- **Liquid and Gas**

  Heat is conducted by two mechanism

  (a) Collisions

  (b) diffusion.

- **Gases**

  In case of gases, the mechanism of heat conduction is simple. The kinetic energy of a molecule is a function of temperature. The kinetic energy of the molecules is due to there random *translational* motion as well as there *vibrational* and *rotational* motions. These molecules are in a continuous random motion exchanging energy and momentum. When a molecule from the high temperature region collides with a molecule from the low temperature region, it loses energy by collisions.

- **Liquids**

  In liquids, the mechanism of heat is nearer to that of gases. However, the molecules are more closely spaced and intermolecular forces come into play.

2. Fourier's Law of Heat Conduction

The rate of heat conduction through a medium depends on the geometry of the medium, its thickness and material of the medium as well as temperature across the medium.

- **Assumptions:**

  The following are the assumptions on which Fourier's law is based:

  (i) Conduction of heat takes place under steady state conditions.

  (ii) The heat flow is unidirectional.

(iii) The temperatures gradient is constant and the temperature profile is linear.

(iv) There is no internal heat generation.

(v) The bounding surfaces are isothermal in character.

(vi) The material is homogeneous and isotropic (i.e., the value of thermal conductivity is constant in all directions).

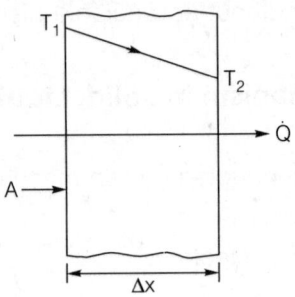

- **Law**

  Mathematically, it can be represented by the equation:

  $$Q = -KA \frac{dt}{dx}$$

  Where, $\dot{Q}$ = Heat flow through a body per unit time (in watts), W,

        A = Surface area of heat flow (perpendicular to the direction of flow), m²,

        dt = Temperature difference of the faces of block (homogeneous solid) of thickness 'dx' through which heat flows, °C or K, and

        dx = Thickness of body in the direction of flow, m.

        k = Constant of proportionality and is known as thermal conductivity of the body.

- It is applicable to all matter (may be solid, liquid or gas).

- Heat conduction in a medium is said to be steady when the temperature does not vary with time, and unsteady or transient when it does.

- The rate of heat transfer per unit surface area is called heat flux.

$$q = \frac{Q}{A} W/m^2$$

heat flux may vary with time as well as position on the heat transfer surface.

3. Thermal Conductivity of Material

The thermal conductivity of a material can be defined as the rate of heat transfer through a unit thickness of the material per unit area per unit temperature difference.

- **Thermal conductivity (k) of different material**

Diamond	—	2300 W/m-K (Highest)
Silver	—	405 W/m-k
Copper	—·	385 W/m-k
Aluminium	—	200 W/m-k
Steels	—	15-35 W/m-k
Mercury	—	8 W/m-k
Glass	—··	1.2 W/m-k
Air	—	0.024 W/m-k
Water	—	0.6 W/m-k
Ice	—	2.25 W/m-k
Asbestos	—	0.2 W/m-k
Freon-12	—	0.0083 W/m-k (Lowest)

Remember

- $$K_{(Air)} = .024 < K_{(Water)} = 0.6 < K_{(Ice)} = 2.25 \text{ unit } W/m-k$$

- Thermal conductivity of a gas increases with increasing temperature and decreasing molecular weight.
- Thermal conductivity is independent of pressure except in extreme cases as for e.g. vacuum.
- Thermal conductivity of liquid can decreases with temperature except water, glycerine.
- Thermal conductivity of metal decreases with increase in temperature. Exception Mercury, Aluminium and Uranium.
- In pure metals the electrons contribution to conduction heat transfer dominate. While in non conductors and semiconductors, the lattice vibration phenomenon contribution is dominant.
- For crystalline non metals solids such as diamond and beryllium oxide, the contribution to thermal conductivity by lattice vibrations can be quite large exceeding values of conductivity associated with good conductors such as aluminium.
- Thermal conductivity of non metallic liquids generally decrease with increasing temperature.
- Thermal conductivity of liquids is usually insensitive to pressure except near critical point.
- The thermal conductivity of a substance is highest in the solid phase and lowest in the gas phase.

4. Thermal Resistance

- By analogy, the heat flow equation (Fourier's equation) may be written as

$$\text{Heat flow rate}(Q) = \frac{\text{Temperature difference (dt)}}{\left(\dfrac{dx}{kA}\right)}$$

By comparing eq. we find that I is analogous to, Q, dV is analogous to dt and R is analogous to the quantity $\left(\dfrac{dx}{kA}\right)$. The quantity $\dfrac{dx}{kA}$ is called Thermal conduction resistance $(R_{th})_{cond.}$ i.e.,

$$(R_{th})_{cond.} = \frac{dx}{kA}$$

- The reciprocal of the thermal resistance is called thermal conductance.
- It may be noted that rules for combining electrical resistances in series and parallel apply equally well to thermal resistances.

Remember

5. Thermal Diffusivity

- Thermal Diffusivity

$$\alpha = \frac{\text{heat conducted}}{\text{heat capacity}} = \frac{k}{\rho c_p} \text{m}^2/\text{s}$$

- The larger the thermal diffusivity, the faster the propagation of heat into the medium.
- K represents how well a material conducts heat and heat capacity (ρC_P) represents how much energy a material stores per unit volume.

6. General Heat conduction Equation for Constant Thermal Conductivity

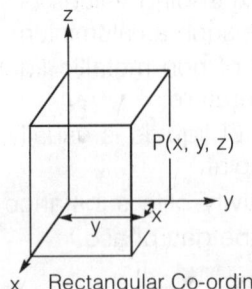

Rectangular Co-ordinate

$$\frac{\partial^2 t}{\partial x^2} + \frac{\partial^2 t}{\partial y^2} + \frac{\partial^2 t}{\partial z^2} + \frac{q_g}{k} = \frac{\rho c}{k} \cdot \frac{\partial t}{\partial \tau} = \frac{1}{\alpha} \cdot \frac{\partial t}{\partial \tau}$$

(Fourier-biot equation)

- **Special Cases**
  (i) For the case when no internal source of heat generation is present, eq. reduces to

$$\frac{\partial^2 t}{\partial x^2} + \frac{\partial^2 t}{\partial y^2} + \frac{\partial^2 t}{\partial z^2} = \frac{1}{\alpha} \cdot \frac{\partial t}{\partial \tau}$$

[Unsteady state $\left(\dfrac{\partial t}{\partial \tau} \neq 0\right)$ heat flow with no internal heat generation]

or, $\boxed{\nabla^2 t = \dfrac{1}{\alpha} \cdot \dfrac{\partial t}{\partial \tau}}$ (Diffusion equation)

(ii) Under the situations when temperature does not depend on time, the conduction then takes place in the **steady state** i.e., $\left(\dfrac{\partial t}{\partial \tau} = 0\right)$ and the eq. reduces to

$$\frac{\partial^2 t}{\partial x^2} + \frac{\partial^2 t}{\partial y^2} + \frac{\partial^2 t}{\partial z^2} + \frac{q_g}{k} = 0$$

or $\boxed{\nabla^2 t + \dfrac{q_g}{k} = 0}$ (Poisson's equation)

(iii) In the absence of internal heat generation, eq. reduces to

$$\frac{\partial^2 t}{\partial x^2} + \frac{\partial^2 t}{\partial y^2} + \frac{\partial^2 t}{\partial z^2} = 0$$

or $\boxed{\nabla^2 t = 0}$ (Laplace equation)

(iv) Steady state and one-dimensional heat transfer

$$\frac{\partial^2 t}{\partial x^2} + \frac{q_g}{k} = 0$$

(v) Steady state, one-dimensional, without internal heat generation

$$\frac{\partial^2 t}{\partial x^2} = 0$$

(vi) Unsteady state, one dimensional, without internal heat generation

$$\frac{\partial^2 t}{\partial x^2} = \frac{1}{\alpha} \cdot \frac{\partial t}{\partial \tau}$$

7. Conduction Equation in Cylindrical Coordinates

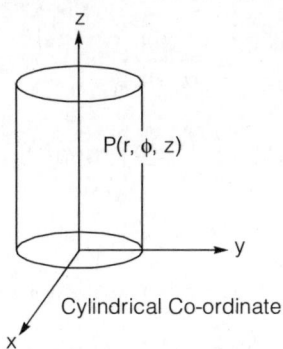

Cylindrical Co-ordinate

$$\frac{\partial^2 t}{\partial r^2} + \frac{1}{r}\left(\frac{\partial t}{\partial r}\right) + \frac{1}{r^2}\left(\frac{\partial^2 t}{\partial \phi^2}\right) + \frac{\partial^2 t}{\partial z^2} + \frac{\dot{q}}{k} = \frac{1}{\alpha}\left(\frac{\partial t}{\partial \tau}\right)$$

If temperature varies only in r-direction

$$\frac{1}{r}\frac{\partial}{\partial r}\left(r\frac{\partial t}{\partial r}\right) + \frac{\dot{q}}{k} = \frac{1}{\alpha}\left(\frac{\partial t}{\partial \tau}\right)$$

8. Conduction Equation in Spherical Coordinates

$$\frac{1}{r^2}\frac{\partial}{\partial r}\left(\frac{r^2 \partial t}{\partial r}\right) + \frac{1}{r^2 \sin\psi}\frac{\partial}{\partial \psi}\left(\sin\psi\frac{\partial t}{\partial \psi}\right) + \frac{1}{r^2 \sin^2\psi}\left(\frac{\partial^2 t}{\partial \phi^2}\right) + \frac{\dot{q}}{k}$$

$$= \frac{1}{\alpha}\left(\frac{\partial t}{\partial \tau}\right)$$

If Temperature varies only in r-direction

$$\frac{1}{r^2}\frac{\partial}{\partial r}\left(\frac{r^2 \partial t}{\partial r}\right) + \frac{\dot{q}}{k} = \frac{1}{\alpha}\left(\frac{\partial t}{\partial \tau}\right)$$

$$\frac{\partial^2 t}{\partial r^2} + \frac{2}{r}\left(\frac{\partial t}{\partial r}\right) + \frac{\dot{q}}{k} = \frac{1}{\alpha}\left(\frac{\partial t}{\partial \tau}\right)$$

9. Heat Conduction Through a Plane Wall

- **Assumption**
    - (i) The contact between the adjacent layers is perfect,
    - (ii) At the interface there is no fall of temperature, and
    - (iii) At the interface the temperature is continuous, although there is discontinuity in temperature gradient.
- **Uniform Thermal Conductivity**
    Consider a plane wall of homogeneous material through which heat is flowing only in x-direction.
    Let,   L = Thickness of the plane wall
             A = Cross-sectional area of the wall (perpendicular to x-direction)
             k = Thermal conductivity of the wall material
             t_1, t_2 = Temperatures maintained at the two face 1 and 2 or the wall, respectively.

$$\therefore \qquad Q = \frac{-kA(t_2 - t_1)}{L} = \frac{t_1 - t_2}{(R_{th})_{cond.}}$$

Here
$(R_{th})_{cond.}$ = Thermal resistance to heat conduction
Heat is conducted in the direction of decreasing temperature and the temperature gradient becomes negative when temperature decreases with increasing x. The negative sign ensure that heat transfer in the positive x direction is a positive quantity.

- **Variable thermal conductivity with temperature**

$$\dot{Q} = k_m A \left[\frac{t_1 - t_2}{L} \right]$$

$$k_m = k_0 (1 + \beta t)$$

Here, k_0 = Thermal conductivity at zero temperature
         k_m = Mean thermal conductivity
         β = Slope of temperature variation line

- Temperature variation in slab for different values of β.

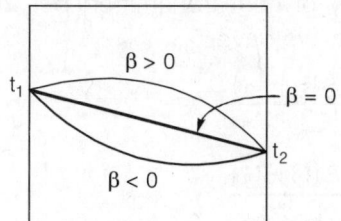

$$K_m = \left[1 + \beta\left(\frac{t_1 + t_2}{2}\right)\right] K_0$$

10. Heat Conduction Through a Composite Wall

Let,

L_A, L_B, L_C = Thickness of slabs A, B and C respectively

k_A, k_B, k_C = Thermal conductivities of the slabs A, B and C respectively

$t_1, t_4 (t_1 > t_4)$ = Temperatures at the wall surfaces 1 and 4 respectively and

t_2, t_3 = Temperatures at the interfaces 2 and 3 respectively

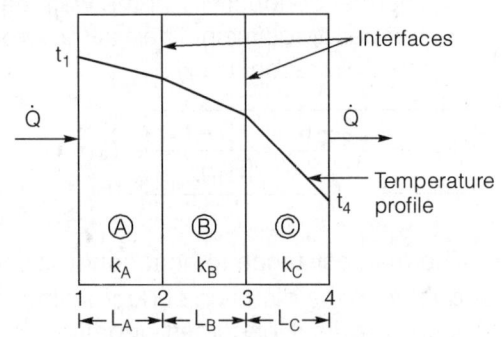

$$R_{th.A} = \frac{L_A}{k_A \cdot A}$$

$$R_{th.B} = \frac{L_B}{k_B \cdot A}$$

$$R_{th.C} = \frac{L_C}{k_C \cdot A}$$

Since the quantity of heat transmitted per unit time through each slab/layer is same, we have,

$$\dot{Q} = \frac{k_A \cdot A (t_1 - t_2)}{L_A}$$

$$= \frac{k_B \cdot A (t_2 - t_3)}{L_B}$$

$$= \frac{k_C \cdot A(t_3 - t_4)}{L_C}$$

$$\dot{Q} = \frac{(t_1 - t_4)}{\left[\dfrac{L_A}{k_A \cdot A} + \dfrac{L_B}{k_B \cdot A} + \dfrac{L_C}{k_C \cdot A}\right]}$$

$$= \frac{(t_1 - t_4)}{[R_{th-A} + R_{th-B} + R_{th-C}]}$$

- **Thermal Contact Resistance**

  Due to surface roughness and void spaces (usually filled with air) the contact surfaces touch only at discrete locations. Thus there is not a single plane of contact, which means that the area available for the flow of heat at the interface will be small compared to geometric face area. Due to this reduced area and presence of air voids, a large resistance to heat flow at the interface occurs. This resistance is known as **thermal contact resistance** and it causes temperature drop between two materials at the interface.

  For unit area of interface the contact resistance is

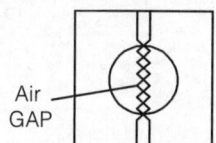
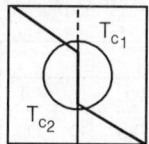

$$(R)_{contact} = \frac{\text{Temperature drop at interface}}{(\dot{Q}/A)} = \left[\frac{T_{c_1} - T_{c_2}}{Q/A}\right]$$

11. The overall Heat-Transfer Coefficient (U)

Let,   L = Thickness of the metal wall

   k = Thermal conductivity of the wall material

   t_1 = Temperature of the surface-1

   t_2 = Temperature of the surface-2

   t_{hf} = Temperature of the hot fluid

   t_{cf} = Temperature of the cold fluid

   h_{hf} = Heat transfer coefficient from hot fluid to metal surface

   h_{cf} = Heat transfer coefficient from metal surface to cold fluid

The suffix 'hf' and 'cf' stand for hot fluid and cold fluid respectively.

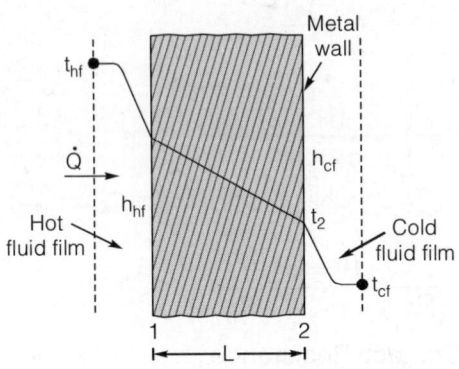

$$\dot{Q} = UA \, \Delta T = \frac{\Delta T}{R_{total}}$$

$$U = \frac{1}{A(R_{total})}$$

$$Q = \frac{1}{\dfrac{1}{h_{hf}} + \dfrac{L}{k} + \dfrac{1}{h_{cf}}}$$

12. Heat Conduction Through a Hollow Cylinder

Consider a hollow cylinder made of material having constant thermal conductivity and insulated at both ends.

Let, r_1, r_2 = Inner and outer radii

$\quad\quad t_1$, t_2 = Temperatures of inner and outer surfaces, and

$\quad\quad\quad k$ = Constant thermal conductivity within the given temperature range.

Temperature distribution

$$\frac{t - t_1}{t_2 - t_1} = \frac{\ln(r/r_1)}{\ln(r_2/r_1)}$$

$$\dot{Q} = \frac{t_1 - t_2}{\dfrac{\ln(r_2/r_1)}{2\pi k L}}$$

Remember

• Resistance offered by cylinder :

$$R_{cylinder} = \frac{\ln(r_2 / r_1)}{2\pi KL}$$

• The temperature distribution is *logarithmic*.
• Temperature at any point in the cylinder can be expressed as a function of *radius only*.

13. Logarithmic Mean Area (A_m) for the Hollow Cylinder

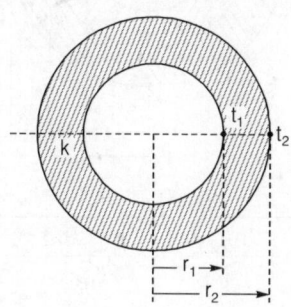

$$A_m = \frac{A_0 - A_i}{\ln\left(\frac{A_0}{A_i}\right)}$$

Here A_i = Area of inside surface of cylinder
        A_0 = Area of outside surfaces of cylinder

Logarithmic Mean Radius (r_m)

$$r_m = \frac{r_2 - r_1}{\ln\left(r_2/r_1\right)}$$

Here, r_1 = Inside radius
         r_2 = Outside radius

14. Heat Conduction Through a Composite Cylinder

Consider flow of heat through a composite cylinder as shown in figure. Let,

t_{hf} = The temperature of the hot fluid flowing inside the cylinder
t_{cf} = The temperature of the cold fluid (atmospheric air)
k_A = Thermal conductivity of the inside layer A
k_B = Thermal conductivity of the outside layer B

t_1, t_2, t_3 = Temperatures at the points 1, 2 and 3

L = Length of the composite cylinder, and

h_{hf}, h_{cf} = Inside and outside heat transfer coefficients.

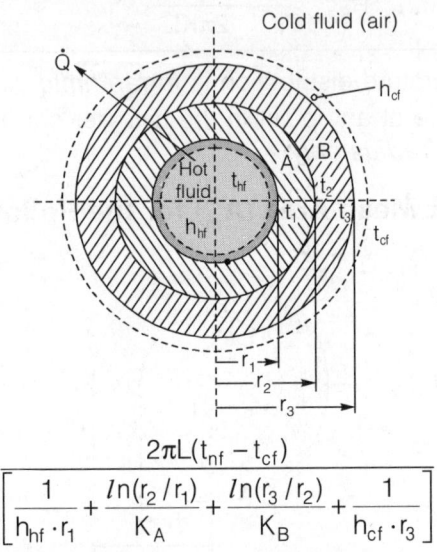

$$\dot{Q} = \frac{2\pi L(t_{nf} - t_{cf})}{\left[\dfrac{1}{h_{hf} \cdot r_1} + \dfrac{In(r_2/r_1)}{K_A} + \dfrac{In(r_3/r_2)}{K_B} + \dfrac{1}{h_{cf} \cdot r_3} \right]}$$

15. Heat Conduction Through Hollow Sphere

Consider a hollow sphere made of material having constant thermal conductivity.

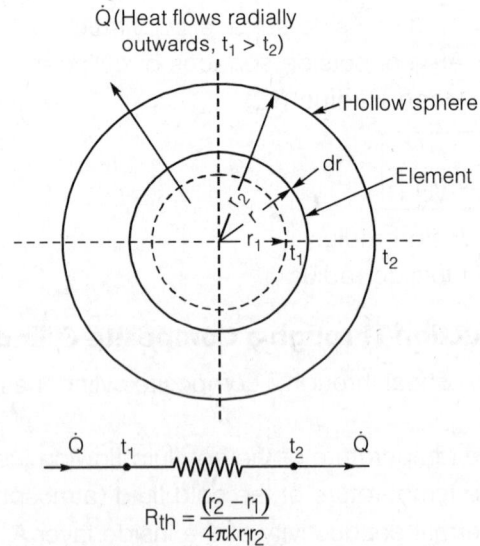

Let, r_1, r_2 = Inner and outer radii.

t_1, t_2 = Temperatures of inner and outer surfaces

k = Constant thermal conductivity of the material with the given temperature range.

• **Temperature Distribution**

$$\frac{t - t_1}{t_2 - t_1} = \frac{1/r - 1/r_1}{1/r_2 - 1/r_1}$$

$$\dot{Q} = \frac{t_1 - t_2}{\left[\dfrac{(r_2 - r_1)}{4\pi k r_1 r_2}\right]} \qquad [t_1 > t_2]$$

• **Resistance offered by sphere :**

Remember

$$R_{sphere} = \frac{r_2 - r_1}{4\pi K r_1 r_2}$$

16. Logarithmic Mean Area for the Hollow Sphere

$$A_m = \sqrt{A_i A_o}$$

Here A_i = Area of inner surface

A_o = Area of outer surface

Mean Radius of Sphere

$$r_m = \sqrt{r_1 r_2}$$

17. Heat Conduction Through a Composite Sphere

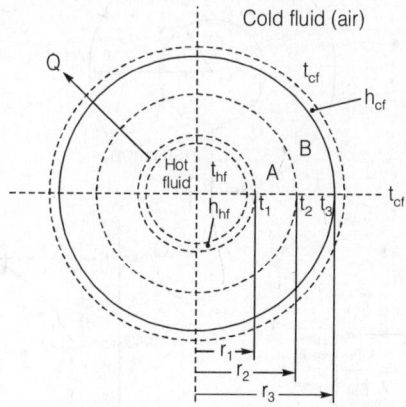

$$\dot{Q} = \frac{4\pi(t_{hf} - t_{cf})}{\dfrac{1}{h_{hf}r_1^2} + \dfrac{r_2 - r_1}{k_A r_1 r_2} + \dfrac{r_3 - r_2}{k_B r_2 r_3} + \dfrac{1}{h_{cf} r_3^2}}$$

If there are n concentric spheres then the above equation can be written as follows:

$$\dot{Q} = \frac{4\pi(t_{hf} - t_{cf})}{\left[\dfrac{1}{h_{hf} \cdot r_1^2} + \displaystyle\sum_{n=1}^{n=n} \left\{ \dfrac{r_{(n+1)} - r_n}{k_n \cdot r_n \cdot r_{(n+1)}} \right\} + \dfrac{1}{h_{cf} \cdot r_{(n+1)}^2} \right]}$$

Other symbol has usual meaning.

18. Critical Thickness of Insulation

- Critical radius is defined as the radius for which the heat transfer rate is **maximum**, thus resistance is minimum at this point.
- Adding insulation to a cylindrical pipe or a spherical shell, **increases** the **conduction resistance** of the insulation layer but **decreases** the **convection resistance** of the surface because of the **increase in outer surface area for convection**. The heat transfer from the pipe may increase or decrease depending on which effect dominates.
- The rate of heat transfer from the cylinder increases with the addition of insulation for $(r_2 < r_{cr})$ reaches a maximum when $r_2 = r_{cr}$ and starts to decrease for $r_2 > r_{cr}$.
- The radius of electric wires may be smaller than the critical radius.

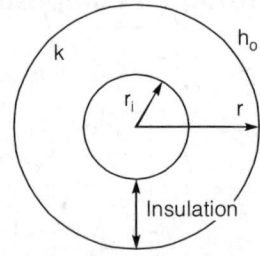

- **For Cylinder**

$$\boxed{r_c = \frac{k}{h_0}}$$

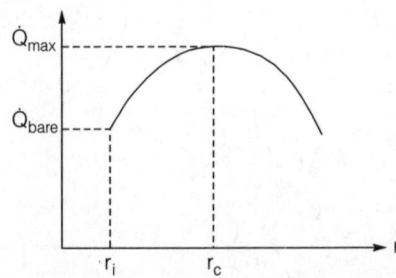

- Similarly for sphere

$$r_c = \frac{2k}{h_0}$$

Remember

- Upto critical radius convection dominates as increasing radius(r) convection resistance decreases thus heat transfer increases.

$$R_{convection} = \left[\frac{1}{h(Area)}\right] = \left[\frac{1}{hA}\right]$$

19. Plane Wall with Uniform Heat Generation

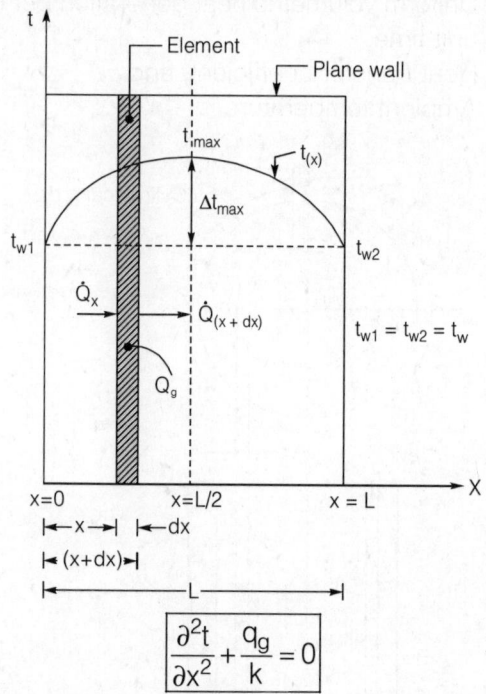

$$\frac{\partial^2 t}{\partial x^2} + \frac{q_g}{k} = 0$$

First and second integration of above equation will give

$$t = -\frac{q_g}{2k} \cdot x^2 + C_1 x + C_2$$

C_1 and C_2 can be calculated from initial boundary conditions.

$$t_{max} - t_w = \frac{q_g L^2}{2k}$$

Here,

t_{max} = Maximum temperature at centre
t_w = Wall temperature

Remember

- In case of plane wall at same Tw at both side with uniform heat generation, maximum temperature occurs at centre of wall.

20. Cylinder with Uniform Heat Generation

Let,   R = Radius of the rod
L = Length of the rod
k = Thermal conductivity (uniform)
q_g = Uniform volumetric heat generation per unit volume per unit time,
h = Heat transfer coefficient, and
t_d = Ambient temperature

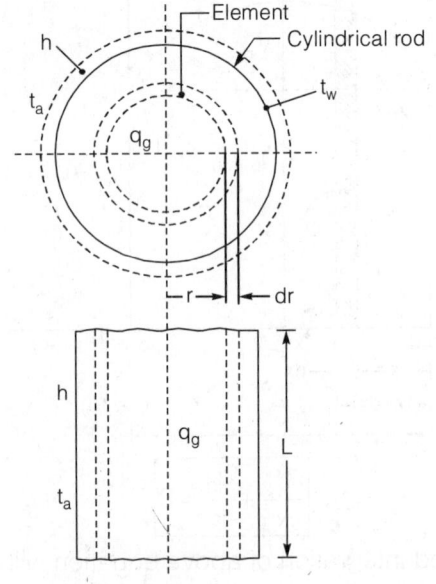

$$t = -\frac{q_g}{k} \cdot \frac{r^2}{4} + C_1 \ln r + C_2$$

C_1 and C_2 are constant of integration.

$$t_c - t_s = \Delta t_{max} = \frac{q_g R^2}{4k}$$

Here, t_c = Maximum temperature at the centre

t_s = Surface temperature

21. Heat Transfer from Extended Surface (FINS)

Finned surfaces are manufactured by extruding, welding or wrapping a thin metal sheet on a surface. Fins enhance heat transfer from a surface by exposing a larger surface area to convection.

- **Assumption**
  - (i) The fin material is homogeneous and isotropic
  - (ii) The temperature at any cross-section of the fin is uniform i.e. t = t (x) only.
  - (iii) The heat transfer coefficient 'h' is uniform over the entire surface
  - (iv) There is no heat generation
  - (v) Contact thermal resistance is negligible.
  - (vi) Heat conduction is steady state.
  - (vii) Thermal conditions at the tip of the fin to be non-variant in z-direction.
  - (viii) One dimensional heat conduction exists.

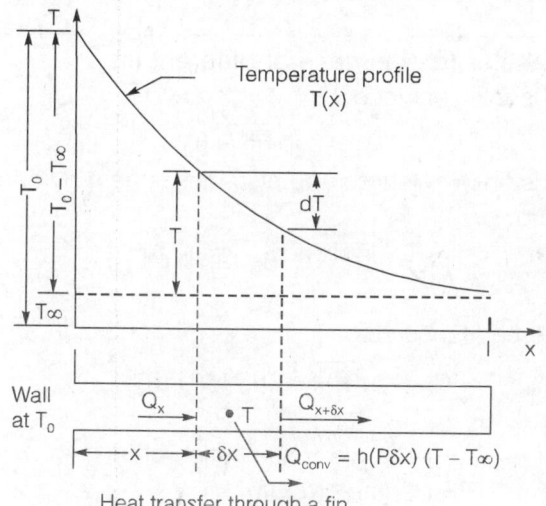

Heat transfer through a fin

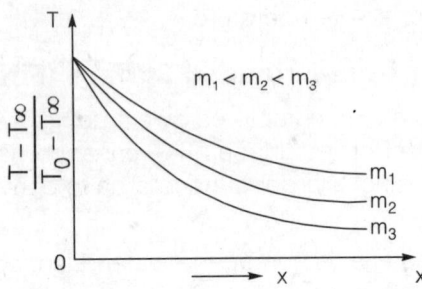

- **Slope of Heat Transfer line (m)**

$$m = \sqrt{\frac{hP}{kA}} \qquad \text{unit-m}^{-1}$$

Where, m = slope of heat transfer line
 h = convective heat transfer
 P = perimeter of cross-section
 A = area of cross-section
 K = Thermal conductivity of fin material

- Fin material with high thermal conductivity would have lesser slope of heat transfer line (m).

Remember

- **Heat transfer (q_{fin}) in case of different fin**
  (i) Fin is infinitely long.

$$\dot{q}_{fin} = \sqrt{hPKA}\ \theta_0$$

  (ii) Fin is finite in length and also looses heat by convection from its tip.

$$\dot{q}_{fin} = \sqrt{hPKA} \cdot \theta_0 \tan h(mL_C)$$

  (iii) Fin's tip is insulated

$$\dot{q}_{fin} = \sqrt{hPKA} \cdot \theta_0 \tan h(mL)$$

  Here $\theta_0 = T_o - T_\infty$
   A = Area of cross-section of fin
   P = Perimeter of fin
   L = Length of fin
   L_C = Corrected length

$$L_C = L + \frac{t}{2}, \text{ for rectangular fin of thickness t}$$

$$L_C = L + \frac{D}{4}, \text{ for circular fin of diameter D}$$

- **Fin efficiency for adiabatic tip (η_{fin})**

$$\eta_{fin} = \frac{q_{actual}}{q_{maximum}} = \frac{\tan h\,(mL)}{mL}$$

    q = heat transfer rate

- Fin efficiency for long fin (infinite)

$$\eta_{long\ fin} = \frac{q_{actual}}{q_{max}} = \frac{1}{mL}$$

- **Fin effectiveness (ϵ)**
  It is defined as ratio of the fin heat transfer rate to the heat transfer rate that would exist without the fin.
- **For adiabatic tip condition**

$$\epsilon_{fin} = \frac{\tan h\,(mL)}{\sqrt{\dfrac{hA}{KP}}}$$

- **For long fin (infinite)**

$$\epsilon_{fin} = \sqrt{\frac{kP}{hA}}$$

Remember

- Use of fin is justified if $\epsilon \geq 2$.
- Biot number of fin with good effectiveness should be less than 1.
- A fin with high (P/Ac) ratio would have $\epsilon > 1$ and such a fin is thin and closely spaced.
- Fin material should have convection resistance high than conduction resistance.
- $\dfrac{\eta_{fin}}{\varepsilon_{fin}} = \dfrac{\text{Cross section Area}(A_c)\text{of fin}}{\text{Surface Area }(A_s)\text{of fin}}$

22. Lumped Heat Capacitance Analysis

At any instant of time the body shall have uniform temperature throughout the body. The temperature of body does not change with position at any given time.

- Biot number < 0.1

$$\text{Biot Number} = \frac{\text{conduction resistance}}{\text{convective resistance}} = \frac{hL_c}{k}$$

$$L_C = \frac{\text{Volume of body}}{\text{Surface area of contact}}$$

For sphere, $\quad L_C = \dfrac{R}{3} \qquad\qquad$ (R = radius)

- The thermal conductivity in Biot Number refers to the conducting body, where as in Nusselt number, it refers to the conductivity of the convecting fluid.
- If

$$B_i = 1 \qquad \in\, = 1$$
$$B_i > 1 \qquad \in\, < 1$$
$$B_i < 1 \qquad \in\, > 1$$

- The use of fins is most effective in applications that involve low convection heat transfer coefficient.

23. Time Constant

- $\tau^* = \dfrac{\rho V C}{hA}$, τ^* is called *time constant.*

  Large time constant corresponds to slow system response, and a small constant refers to fast response. A low value of time constant can be achieved by
  (i) Decreasing wire diameter
  (ii) Using light metals of low density and low specific heat.
  (iii) Increasing the heat transfer coefficient.

24. Heisler Charts

Heisler Charts are extensively used to determine the temperature distribution and heat flow rate when both conduction and convection resistance are almost of equal importance. i.e. $B_i \approx 1$.

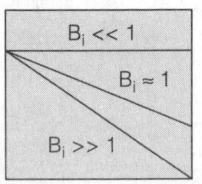

$B_i << 1$	$R_{conduction}$	$<< R_{convection}$
$B_i \approx 1$	$R_{convection}$	$\approx R_{conduction}$
$B_i >> 1$	$R_{conduction}$	$>> R_{convection}$

■■■■

Convection

2

1. Newton's Low of Cooling

The rate equation for the convective heat transfer between a surface and an adjacent fluid is prescribed by Newton's law of cooling.

$$\dot{Q} = hA(t_s - t_f)$$

Where, $\dot{Q}$ = Rate of conductive heat transfer

A = Area exposed to heat transfer

t_s = Surface temperature

t_f = Fluid temperature, and

h = Coefficient of convective heat transfer

The units of h are

$$h = \frac{\dot{Q}}{A(t_s - t_f)} = \frac{W}{m^2 \,^\circ C} \text{ or } W/m^2 \,^\circ C$$

2. Coefficient of Convective Heat Transfer (h)

The amount of heat transmitted for a unit temperature difference between the fluid and unit area of surface in unit time.

- **The value of 'h' depends on the following factors**
  - **(i)** Thermodynamic and transport properties (e.g. viscosity, density, specific heat etc.)
  - **(ii)** Nature of fluid flow
  - **(iii)** Geometry of the surface
  - **(iv)** Prevailing thermal conditions

3. Range of 'h' for Different State

States	Range of 'h' $(W/m^2 \text{-} k)$
Free convection in gas	3-25
Forced convection in gas	25-400
Free convection in liquid	50-350
Forced convection in liquid	350-3000
Condensation heat transfer	3000-25000
Boiling heat transfer	5000-50000

A. Forced convection
$$Nu = f(Re, Pr)$$

B. Free Convection
$$Nu = f(Gr, Pr)$$

Here,   Nu = Nusselt's number
Re = Reynold's number
Pr = Prandtl number
Gr = Grashoff number

4. Nusselt Number

The Nusselt number represents the enhancement of heat transfer through a fluid layer as a result of convection relative to conduction across the same fluid layer. The larger the Nusselt number, the more effective is the convection.

$$\text{Nusselt Number (Nu)} = \frac{q_{conv}}{q_{cond}} = \frac{h\delta}{k} \quad (\delta = \text{characteristic length})$$

- Nusselt number is to thermal boundary layer what the friction coefficient is to the velocity boundary layer.
- Nu >>> 1 then thermal *convection* >>> thermal *conduction*.

Remember
- Stanton number is called modified Nusselt.
- Stanton $(St) = \dfrac{h}{\rho V c_p} = \dfrac{Nu}{Re.Pr.} = \dfrac{Nu}{Pe}$

5. Thermal Boundary Layer

$$\delta_t = y \left(\text{for which } \frac{T_s - T_y}{T_s - T_\infty} = 0.99 \right)$$

Local convection coefficient

$$h = \frac{-K \left. \dfrac{\partial T}{\partial y} \right|_{y=0}}{T_s - T_\infty}$$

- Since $T_s - T_\infty$ is a constant, while δ_t increases with increasing x. Temperature gradients in the boundary layer must decrease with increasing x. According the magnitude of $\left.\dfrac{\partial T}{\partial y}\right|_{y=0}$ decreases with increasing x hence q_s'' and 'h' decrease with increasing x.

- The thermal boundary layer is best described by the dimensionless parameter Prandtl number (Pr)

$$Pr = \frac{\text{molecular diffusivity of momentum}}{\text{molecular diffusivity of heat}}$$

$$\boxed{Pr = \frac{\upsilon}{\alpha} = \frac{\mu c_p}{k} = \left(\frac{\delta}{\delta_t}\right)^3}$$

$$P_{randtl}\ (Pr) = \left[\frac{\upsilon/D}{\alpha/D}\right] = \left[\frac{\text{SchmditNo.(Sc)}}{\text{LewisNo.(Le)}}\right]$$

Here, δ = Velocity boundary layer thickness
           δ_t = Thermal boundary layer thickness

- The Pradntl numbers of fluids range from less than 0.01 for liquid metals to more than 100, 000 for heavy oils.
- Heat diffuses very quickly in **liquid metals** (Pr << 1) and very slowly in **oil** (Pr >> 1) relative to momentum.

6. Relationship between Local and Average Convection Coefficient

$$\boxed{\overline{h} = \frac{1}{A_s} \int_{A_s} h\, dA_s}$$

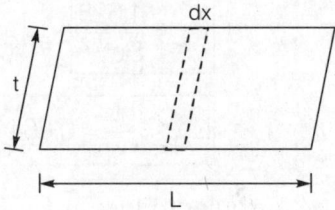

For flat plate

$$A_s = t \times L$$

$$dA_s = t \times dx$$

$$\therefore \quad \boxed{\overline{h} = \frac{1}{L} \int_0^L h\, dx}$$

$\overline{h}_L = 2h_x$ (Forced convection)

$\overline{h}_L = \dfrac{4}{3}h_x$ (Free convection)

7. Variation of 'h' with Nature of Flow for a Flat Plate

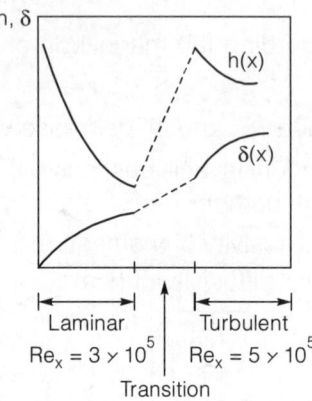

8. Friction Coefficient and Nusselt Number for Laminar and Turbulent Flow

- *Local friction coefficient* and the Nusselt number at location x over a flat plate.

$$C_{fx} = \frac{0.664}{Re_x^{1/2}}$$

$$Nu_x = \frac{h_x x}{k} = 0.332\, Re_x^{1/2}\, Pr^{1/3} \qquad (Pr \geq 0.6)$$

- Average Values

$$C_{fL} = \frac{1.328}{Re_L^{1/2}} \quad \text{and}$$

$$Nu_L = \frac{hL}{k} = 0.664\, Re_L^{1/2}\, P_r^{1/3}$$

- Turbulent Flow

The local friction coefficient and the Nusselt Number at location x over late plate are

$$c_{fx} = \frac{0.0592}{Re_x^{1/5}} \quad (5 \times 10^5 \leq R_{ex} \leq 10^7)$$

$$Nu_x = \frac{h_x x}{k}$$

$$= 0.0296 \, Re_x^{4/5} \, P_r^{1/3} (0.6 \le Pr \le 60)$$

- **Average Values**

$$C_{fL} = \frac{0.074}{Re_L^{1/5}}$$

and $$Nu_L = \frac{hL}{k} = 0.037 Re_L^{4/5} P_r^{1/3}$$

- For laminar flow $Nu_L = 2Nu_x$.

- For turbulent flow $\overline{Nu} = \dfrac{5}{4} Nu_x$.

9. Reynolds Number for Flow in Pipe

$$Re = \frac{\rho VL}{\mu}$$

Here, ρ = Density
$\quad\quad V$ = Velocity
$\quad\quad L$ = Characteristics length
$\quad\quad \mu$ = Viscosity

$\quad\quad$ Re < 2300 *Laminar flow*
2300 < Re < 4000 *Transition to turbulence*
$\quad\quad$ Re > 4000 *Turbulent flow*

10. Reynold's Analogy

It relates key engineering parameter of the viscosity, thermal and concentration boundary layer.

- **For Laminar boundary layer**

$$\frac{Nu_x}{Re_x \, Pr} = st_x = \frac{C_{fx}}{2} \left[st = \frac{h_x \Delta T}{\rho C_p U_\infty \Delta T} \right] \qquad \text{[when } Pr = 1]$$

11. Reynold's-colburn Analogy

$$St_x \cdot Pr^{2/3} = \frac{C_{fx}}{2}$$

St = Station number
C_{fx} = Local skin friction coefficient

12. Characteristics Dimension (CD)/Equivalent diameter of pipe (De)

Equivalent Diameter

$$De = CD = \frac{4 \times A}{P}$$

Here,   A = Area of cross-section
        P = Perimeter of cross-section

Rectangular Duct

Characteristic Dimension/Equivalent Diameter (De)

Remember

- De, would represent the same resistance against the flow and same heat transfer rate that takes place in actual pipe/flat surface.

13. Grashoff Number

The flow regime in *natural convection* is governed by another dimensionless number called the Grashoff Number, which represents the ratio of the buoyancy force to the viscous force acting on the fluid.

$$Gr_L = \frac{\text{buoyancy forces}}{\text{viscous force}}$$
$$= \frac{g\beta(T_s - T_\infty)L_C^3}{\nu^2}$$

Here,   g = Gravitational acceleration
            β = Coefficient of volume expansion
            T_s = Temperature of surface
            T_∞ = Temperature of fluid far from the surface, °C
            L_c = Characteristic length of the geometry
            ν = Kinematic viscosity of the fluid

- The Grashoff Number provides the main criteria in determining whether the fluid flow is laminar or turbulent in natural convection.
- For vertical plates, the *critical Grashoff Number* is 10^9. The flow regime on a vertical plate becomes turbulent at Grashoff Number greater than 10^9.
- Grashoff number plays same role in free convection as the Reynolds number in forced convection.

$$Re = \frac{\text{Inertia force}}{\text{Viscous force}}$$
$$Gr = \frac{\text{Buoyancy force}}{\text{Viscous force}}$$

If

(i) $\dfrac{Gr_L}{Re_L^2} = 1$ Effect of both forced convection should be considered.

(ii) $\dfrac{Gr_L}{Re_L^2} \gg 1$ forced convection effect neglected.

(iii) $\dfrac{Gr_L}{Re_L^2} \ll 1$ free convection effect neglected.

- **Vertical Plate**

(i) Hot

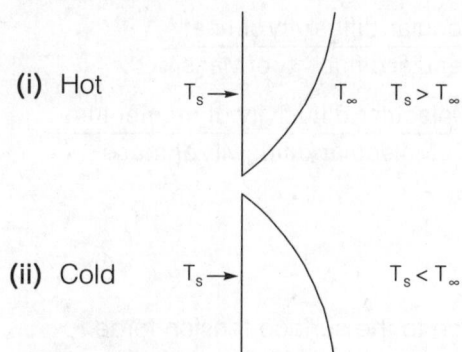

$T_s \rightarrow$ T_∞ $T_s > T_\infty$

(ii) Cold $T_s \rightarrow$ $T_s < T_\infty$

- **Horizontal Plate**

(i) Hot

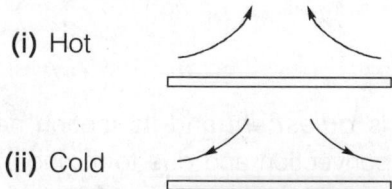

(ii) Cold

Upper surface of hot plate or lower surface of cold plate.

$$\overline{Nu}_L = 0.54\, Ra_L^{1/4} \ (\text{For la min ar})$$
$$\overline{Nu}_L = 0.15\, Ra_L^{1/3} \ (\text{For turbulent})$$

Where Rayleigh No. (Ra) = Gr · Pr

Forced Convection

	Laminar Flow	Turbulent Flow
Nu = 3.66	Constant surface temperature of pipe	Both constant heat flux & constant surface temperature we use Dittus Boelter equation
Nu = 4.364	Constant heat flux through pipe	Nu = 0.023 Re$^{4/5}$ (Pr)^n n = 0.4 heating n = 0.3 cooling

14. Convection Process of Boiling and Condensation

- Jacob Number

  It is the ratio of maximum sensible energy absorbed to the latent energy absorbed during liquid-vapour phase change.

- In many application the sensible energy is much less than the latent energy and 'Ja' has small numerical value.

$$Ja = \frac{C_p(T_s - T_{sat})}{h_{fg}}$$

- Lewis No. (Le) = $\dfrac{\text{Molecular diffusivity of heat}}{\text{Molecular diffusivity of Mass}}$

- Schmidt No. (Se) = $\dfrac{\text{Molecular diffusivity of momentum}}{\text{Molecular diffusivity of mass}}$

- $Pr = \left[\dfrac{Sc}{Le}\right] = \left[\dfrac{u/D}{\alpha/D}\right]$

- Bond Number

  Ratio of buoyancy force to the surface tension force.

$$Bo = \frac{g(\rho_1 - \rho_2)}{\sigma}$$

- Boiling

  - Pool boiling

    In pool boiling liquid is *quiescent* and its motion near the surface is due to free convection and due to mixing induced by bubble growth and detachment.

  - Forced Convection Boiling

    Here the fluid motion in induced by external means, as well as by free convention and bubble induced mixing.

    Boiling may also be classified according to weather it is subcooled or saturated.

  - Subcooled Boiling

    Here the temperature of the liquid is below the saturation temperature and the bubbles formed at the surface may condense in the liquid.

  - Saturated Boiling

    Here the temperature of the liquid slightly exceeds the saturation temperature, hence the bubbles formed at the surface are then propelled through the liquid by buoyancy forces, eventually escaping from the free surface.

- **Temperature Distribution in Saturated Pool Boiling**

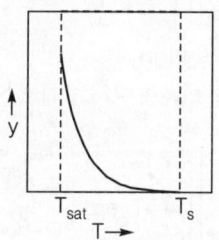

- **Boiling Curve**

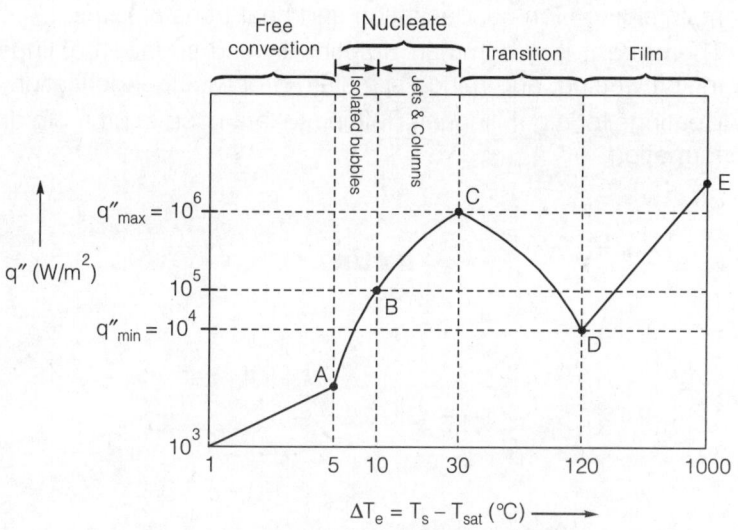

Typical boiling curve for water at 1 atm

Here, A = ONB (Onset of nucleate boiling)

C = q''_{max} = Critical heat flux

D = Leidenfrost point = q''_{min}

- **Surface Condensation**
  (i) Film wise condensation

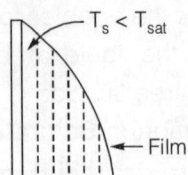

$T_s < T_{sat}$

Film

(ii) Dropwise condensation
90% of surface is covered by drops

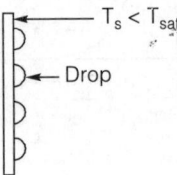

$T_s < T_{sat}$

Drop

Droplet formation is superior to film formation in terms of maintaining high condensation and heat transfer rates.

Therefore it is a common practice to use surface coatings that inhibit wetting, and hence stimulate dropwise condensation.

Coatings for e.g. Silicones, Teflon etc. are used to promote droplet formation.

■■■■

Heat Exchanger

3

A 'heat exchanger' may be defined as an equipment which transfers the energy from a hot fluid to a cold fluid, with maximum rate and minimum investment.

1. Type of Heat Exchanger

On the basis of nature of heat exchange process:

(i) Direct contact heat exchanger

(ii) Indirect contact heat exchanger.

It has two types

- ### Regenerator Type

  In a regenerator type of heat exchanger the hot and cold fluids pass alternately through a space containing solid particulars (matrix), these particles providing alternately a sink and a source for heat flow.

  ### Example:

  (i) IC engines and gas turbines

  (ii) Open hearth and glass melting furnaces

  (iii) Air heaters of blast furnaces

- ### Recuperators Type

  Recuperator is the most important type of heat exchanger in which the flowing fluids exchanging heat are on either side of dividing wall (in the form of pipes or tubes generally). These heat exchangers are used when two fluids cannot be allowed to mix i.e., when the mixing is undesirable.

  ### Examples:

  (i) Automobile radiators

  (ii) Oil coolers, intercoolers, air preheaters, economizers, superheaters, condensers and surface feed heaters of a steam power plant

  (iii) Milk chiller of pasteurizing plant

  (iv) Evaporator of an ice plant

Remember: ....

- Baffles are used in shell and tube type H.E. (Heat exchanger).

On the basis of relative direction of fluid motion:

- Parallel flow heat exchanger

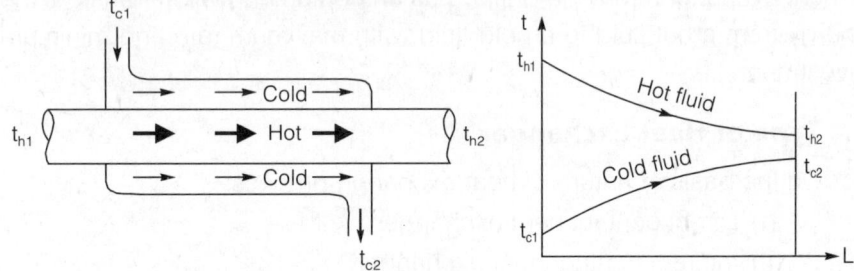

- Counter-flow heat exchanger

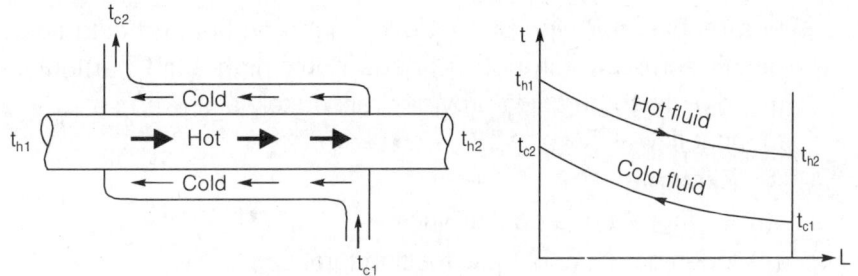

- Cross-flow heat exchanger

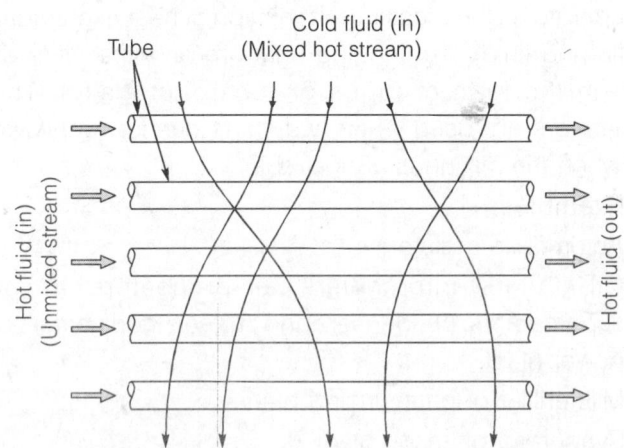

Remember

- Baffles are used on shell side to prevent stagnation of fluid and increases turbulence.
- Baffles increase the heat transfer coefficient of shell side fluid. and prevent fouling of Tubes.

2. Heat Exchanger Analysis

Let, $\dot{m}$ = Mass flow rate, kg/s,

   c_p = Specific heat of fluid at constant pressure, J/kg°C

   t = Temperature of fluid, °C, and

   Δt = Temperature drop or rise of a fluid across the heat exchanger.

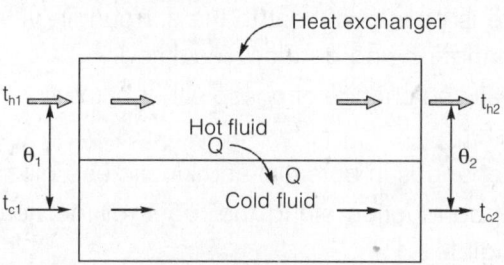

Subscripts h and c refer to the hot and cold fluids respectively; subscripts 1 and 2 correspond to the inlet and outlet conditions respectively.

Assuming that there is no heat loss to the surroundings and potential and kinetic energy changes are negligible, from the energy balance in a heat exchanger, we have

Heat given up by the hot fluid

$$Q = m_h c_{ph}(t_{h1} - t_{h2})$$

Heat picked up by the cold fluid

$$Q = m_c c_{pc}(t_{c2} - t_{c1})$$

Total heat transfer rate in the heat exchanger

$$Q = UA\theta_m$$

Where   U = Overall heat transfer coefficient between the two fluids

   A = Effective heat transfer area, and

   θ_m = Appropriate mean value of temperature difference or **Logarithmic Mean Temperature Difference** (LMTD)

3. Logarithmic Mean Temperature Difference (LMTD)

Logarithmic mean temperature difference (LMTD) is defined as the temperature difference which, if constant, would give the same rate of heat transfer as actually occurs under variable conditions of temperature difference.

Assumption:

(i) The overall heat transfer coefficient U is constant.

(ii) The flow conditions are steady.

(iii) The specific heats and mass flow rates of both fluids are constant.

(iv) There is no loss of heat to the surroundings, due to the heat exchanger being perfectly insulated.

(v) There is no change of phase either of the fluid during the heat transfer.

(vi) The changes in potential and kinetic energies are negligible.

(vii) Axial conduction along the tubes of the heat exchanger is negligible.

- **LMTD for parallel flow heat exchanger**

$$\text{LMTD} = \frac{\theta_1 - \theta_2}{\ln(\theta_1/\theta_2)}$$

> θ_1 = Temperature difference between hot and cold fluid at inlet.
>
> θ_2 = Temperature difference between hot and cold fluid at outlet.

- For parallel heat exchanger

$$\theta_1 = t_{h1} - t_{c1}$$
$$\theta_2 = t_{h2} - t_{c2}$$

- For parallel heat exchanger

$$\theta_1 = t_{h1} - t_{c2}$$
$$\theta_2 = t_{h2} - t_{c1}$$

Remember

- LMTD for cross flow and multiplies heat exchanger = (Correction factor) × LMTD (counter flow)
- NTU is measure of size of H.E. i.e. higher NTU means large heat transfer area of heat exchanger.

4. Number of Transfer Unit (NTU)

$$NTU = \frac{UA}{(\dot{m} C_p)_{small}}$$

U = Overall heat transfer coefficient

$\dot{m}$ = Mass flow rate

$\dot{m} C_p$ = Heat capacity

5. Effectiveness of Heat Exchanger

The heat exchanger effectiveness (ε) is defined as the ratio of actual heat transfer to the maximum possible heat transfer. Thus

$$\varepsilon = \frac{\text{Actual heat transfer}}{\text{Maximum possible heat transfer}}$$

$$= \frac{Q}{Q_{max}}$$

$$\varepsilon = \frac{C_h\left(t_{h_1} - t_{h_2}\right)}{C_{min}\left(t_{h_1} - t_{c_1}\right)} = \frac{C_c\left(t_{c_2} - t_{c_1}\right)}{C_{min}\left(t_{h_1} - t_{c_1}\right)}$$

Here, $C_h = \dot{m}_h C_{ph}$

$C_c = \dot{m}_c C_{pc}$

$C_{min} = Min\{C_h, C_c\}$

6. Effectiveness ($\in$) of parallel/counter flow heat exchanger

$$\epsilon_{parallel} = \frac{1 - e^{-NTU(1+C)}}{1 + C}$$

$$\epsilon_{counter} = \frac{1 - e^{-NTU(1-C)}}{1 - C \cdot e^{-NTU(1-C)}}$$

Here, C = Heat capacity ratio = $\dfrac{C_{min}}{C_{max}}$

Special cases

1. For Gas turbine and recuperator $C = 1$

$$\epsilon_{parallel} = \frac{1 - exp^{(-2NTU)}}{2}$$

$$\epsilon_{counter} = \frac{NTU}{1+NTU}$$

2. For Boiling and condensation C = 0

$$\epsilon = 1 - \exp^{(-NTU)}$$

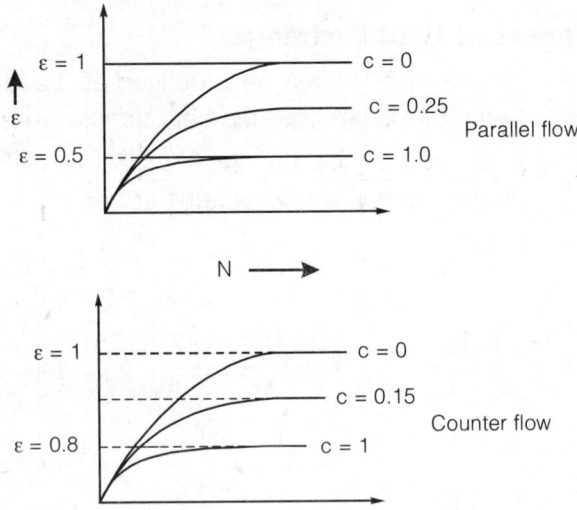

Radiation

4

Radiation is the energy emitted by matter in the form of ***electromagnetic waves*** as a result of changes in the electronic configurations of the atoms and molecules. In heat transfer study, we are interested in thermal radiation. In radiation, the internal energy of object decreases.

1. Electromagnetic Spectrum/Wave

All types of electromagnetic waves are classified in terms of wavelength and are propagated at the speed of light (c) i.e. 3×10^8 m/s. The electromagnetic spectrum is shown in figure. The distinction between one form of radiation and another lies only in its frequency (f) and wavelength (λ) which are related by

$$c = \lambda \times f$$

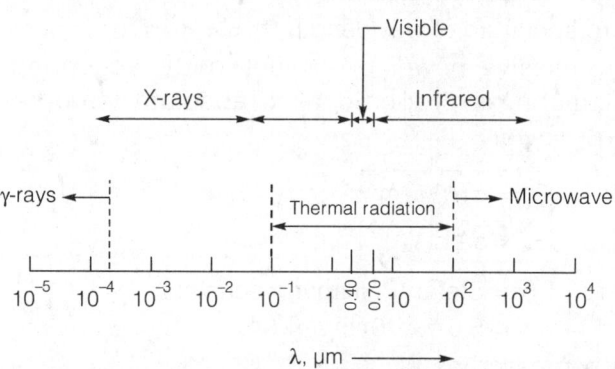

The emission of thermal radiation (range lies between wavelength of 10^{-7} μm and 10^4 μm) depends upon the nature, temperature and state of the emitting surface; however, with gases the dependence is also upon the thickness of the emitting layer and the gas pressure.

Remember: ....

- **Infrared** has wavelength range lie in **both** thermal radiation and microwave and its range is **0.8 μm** to **800 μm**.
- Thermal radiation exhibit characteristics similar to those of visible light.
- Thermal radiation can be reflected; refracted and are subject to scattering and absorption when they pass through media.

2. Properties of Surface Emission

The rate of emission of radiation by a body depends upon the following factors:

(i) The temperature of the surface.

(ii) The nature of the surface, and

(iii) The wavelength or frequency of radiation.

Properties:

- **Total emissive power (E)/Stefan-Boltzmann law**

  The "emissive power" is defined as the total amount of radiation (all wavelength range radiations) emitted by a body per unit area and time.

Remember: ....

- $$E_b = \int_{0}^{\infty} E_{b\lambda} d\lambda = \sigma T^4 \ w/m^2$$

- Entire spectrum of wavelength ($0 < \lambda < \infty$)

  The emissive power of a black body, according to Stefan-Boltzmann, is proportional to absolute temperature to the fourth power.

  $$E_b = \sigma T^4 \ W/m^2$$
  $$E_b = \sigma A T^4 \ W$$

  where σ = Stefan-Boltzmann constant

  $= 5.67 \times 10^{-8} \ W/m^2 K^4$

  It is expressed as W/m^2.

- **Emissitivity (ε)**

  It is defined as the ability of the surface of a body to radiate heat. It is also defined as the ratio of the Emissive power of any body to the Emissive power of a black body of equal temperature

  $\left(i.e., \varepsilon = \dfrac{E}{E_b} \right)$. Its values varies for different substances ranging

  from 0 to 1.

  Emissive power of a real surface

  $E = \varepsilon \sigma A t^4$ Watt

Remember: ....

- For black body $\varepsilon = 1$, for white body $\varepsilon = 0$
- For gray body it varies between '0' and '1'
- Emissitivity may vary with temperature or wavelength.
- A surface is said to be diffuse if its properties are independent of direction.
- A surface is said to be gray if its properties are independent of wavelength.

- **Irradiation (G)**

  Total incident radiation on a surface from all directions per unit time and per unit area of surface.

- **Radiosity (J)**

  It refers to all of the radiant energy leaving a surface per unit time, per unit area of surface.

  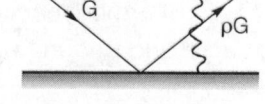

$$\boxed{J = E + \rho G}$$

Here,   E = Emissitivity

ρ = Reflectively

G = Irradiation

3. Absorptivity (α), Reflexivity (ρ), Transmissivity (τ)

(i) **Absorptivity:** Fraction of incident radiation absorbed.

(ii) **Reflexivity:** Fraction of incident radiation reflected.

(iii) **Transmittivity:** Fraction of incident radiation transmitted.

For black body

$\alpha = 1, \rho = 0, \tau = 0$

For opaque body

$\tau = 0, \alpha + \rho = 1$

For white body

$\rho = 1, \alpha = 0, \tau = 0$

4. Kirchhoff's Law

The law states that at any temperature the ratio of total emissive power E to the total absorptivity is a constant for all substances which are in thermal equilibrium with environment.

In other words: The emissivity of a body is equal to its absorptivity when the body remains in thermal equilibrium with its surroundings.

$\varepsilon = \alpha$ (Kirchhoff's Law)

5. Planck's Law

According to this

$$(E_\lambda)_b = \frac{2\pi c^2 h \lambda^{-5}}{\exp\left(\dfrac{ch}{\lambda kT}\right) - 1}$$

where, $(E_\lambda)_b$ = Monochromatic (single wavelength) emissive power of a black body,

c = Velocity of light in vacuum $\simeq 3 \times 10^8$ m/s

h = Planck's constant = 6.625×10^{-34} j.s

λ = Wavelength, μm

k = Boltzmann constant = 1.3805×10^{-23} J/K,

t = Absolute temperature, K

- **Monochromatic Emissive Power of Black Body**

  It is the amount of radiation energy emitted by a black body at a thermodynamic temperature T per unit time, per unit surface area and per unit wavelength, about the wavelength λ.

$$E_{b\lambda} = \frac{c_1 \lambda^{-5}}{\left[e^{c_2/\lambda T} - 1\right]} \quad c_1 = 0.374 \times 10^{-15} \text{ (W-m}^2) \ \& \ c_2 = 1.4385 \times 10^{-2} \text{ (m-K)}$$

6. Wien's Displacement Law

It gives a relationship between the temperature of a black body and the wavelength at which the maximum value of monochromatic emissive power occurs. A peak monochromatic emissive power occurs at a particular wavelength. Wein's displacement law states that the product of λ_{max} and t is constant.

∴ $\lambda_{max}t$ = Constant = 2900 μm.k

For $E_{b\lambda}$ to be maximum

$$\frac{d(E_{b\lambda})}{d\lambda} = 0$$

$$\lambda_{max}.T = \text{constant} = 2.898 \times 10^{-3} \text{ m}-\text{k}$$

Remember

- Stefan Boltzman law and weins displacement law is derived from max. Plank's law of spectral emissive power of blockbody.

7. Lambert's Cosine Law

The law states that the total emissive power E_θ from a radiating plane surface in any direction is directly proportional to the cosine of the angle of emission.

$$E_\theta = E_n \cos\theta$$

Here, E_n = Total emissive power in the direction of its normal
        θ = Angle subtended by normal to the radiating surface and direction vectors of emission of the receiving surface.

Remember: ....

- Equation is only true for a radiation surface whose radiation intensity is constant.
- Stefon -Boltzmann law and Wein's displacement law are derived from monochromatic emissive power (E_{bl}).
- Intensity of radiation (I) in a given direction from a surface per unit solid angle per unit of the projection of the surface on a plane normal to direction of radiation.
- Solid angle subtended by sphere at its centre 4π solid angle subended by hemisphere at centre 2π.
- Emissive power of black body is π times the intensity of radiation.

$$E_b = \pi I$$

8. Shape Factor

The shape factor may be defined as "The fraction of radiative energy that is diffused from one surface element and strikes the other surface *directly* with no intervening reflections.

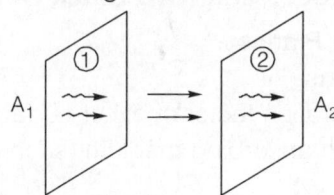

$$F_{1-2} = \frac{\dot{Q}_{1-2}}{\dot{Q}_1}$$

Here, F_{1-2} = Fraction of energy leaving surface '1' and reaching surface '2'.

$\dot{Q}_1$ = Rate of total energy radiated by A_1

$\dot{Q}_{1-2}$ = Fraction of rate of energy leaving surface '1' reaching surface '2'.

Reciprocity Theorem

$$A_1F_{1-2} = A_2F_{2-1}$$

- **Shape Factor of Some Important Surface**

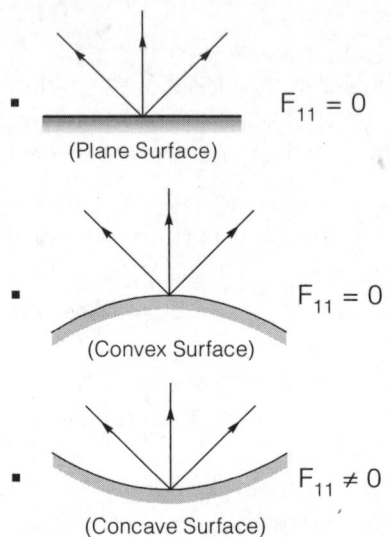

$F_{11} = 0$

(Plane Surface)

$F_{11} = 0$

(Convex Surface)

$F_{11} \neq 0$

(Concave Surface)

Remember

- Shape factor depends only on geometry of body.
- In general, $0 \le F_{mn} \le 1$.
- When one body is completely enclosed inside the another body, then shape factor of inner body with respect to outer body is equal to 1.

9. Heat Exchange between Non-black Bodies

- **Infinite Parallel Planes**
  Assumptions:
  (i) The configuration factor of either surface is unity.
  (ii) There is non-absorbing medium (such as air) in between the surfaces.
  (iii) The emissive and reflective properties are constant over all the surfaces.

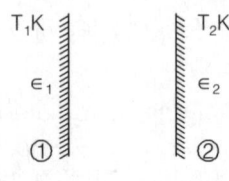

Here, $T_1 > T_2$

Net radiation heat exchange (q_{1-2})

$$q_{1-2(net)} = \frac{\sigma\left(T_1^4 - T_2^4\right)}{\dfrac{1}{\epsilon_1} + \dfrac{1}{\epsilon_2} - 1} \; W/m^2$$

If surface are black, $\epsilon_1 = 1, \epsilon_2 = 1$.

$$q_{1-2 \,(net)} = \sigma\left(T_1^4 - T_2^4\right) W/m^2$$

- **Equivalent Emissivity**

$$\bar{\epsilon} = \frac{1}{\dfrac{1}{\epsilon_1} + \dfrac{1}{\epsilon_2} - 1}$$

- **Infinite long concentric cylinder**

   Consider two concentric cylinders as shown in figure of areas A_1 and A_2, emissivities ε_1 and ε_2 and their surfaces maintained at temperatures T_1 and T_2 respectively.

Now, $\boxed{A_1 F_{1-2} = A_2 F_{2-1}}$ (By reciprocity theorem)

But, $\boxed{F_{1-2} = 1}$

(Since all the radiations emitted by the inner cylinder are intercepted by the outer cylinder)

∴ $\boxed{F_{2-1} = \dfrac{A_1}{A_2}}$

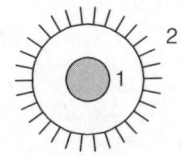

$\boxed{F_{21} + F_{22} = 1}$

$\boxed{F_{22} = 1 - \dfrac{A_1}{A_2} = 1 - \dfrac{D_1}{D_2}}$

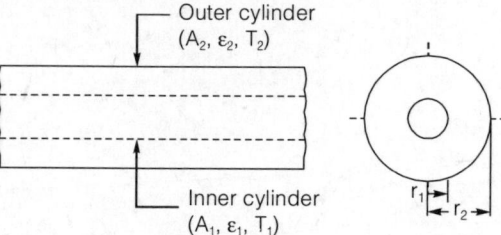

Outer cylinder (A_2, ε_2, T_2)

Inner cylinder (A_1, ε_1, T_1)

Net radiation heat exchange between surface is

$$q_{1-2} = \frac{A_1 \cdot \sigma\left(T_1^4 - T_2^4\right)}{\dfrac{1}{\epsilon_1} + \dfrac{A_1}{A_2}\left(\dfrac{1}{\epsilon_2} - 1\right)} \text{ Watt}$$

- **Radiation Shield Concept**

  For N shields between 1 and 2

$$q_{12}(\text{shield}) = \frac{A\,\sigma(T_1^4 - T_2^4)}{\left(\dfrac{1}{\epsilon_1} + \dfrac{1}{\epsilon_2} - 1\right) + \sum_{j=1}^{N}\left(\dfrac{1}{\epsilon_{j,1}} + \dfrac{1}{\epsilon_{j,2}} - 1\right)}$$

Remember

- If the shields have the same emissivities as the bodies then

  (i) If one shield is used 50% reduction

  (ii) If nine shields are used 90% reduction.

■■■■

A Handbook on
Mechanical Engineering

Basic Thermodynamics

CONTENTS

Basic Concepts

1. System and Surrounding

- A *system* is a matter or region on which analysis is done. System is separated from the surrounding by boundary. Everything external to the system is called *surroundings*. System and surrounding together is called a *universe*.
- Boundary
  - It separate system and surroundings
  - It can be fixed or movable
  - Fixed boundary e.g. rigid box containing gas
  - Movable boundary e.g. cylinder with piston

2. Type of System

Types of system	Mass Transfer	Energy transfer	Example
Closed	No	Yes	Piston cylinder *without* valves
Open	Yes	Yes	Turbine, pump, compressor
Isolated	No	No	Universe, hot coffee in a perfectly insulated thermos

3. Thermodynamics Equilibrium

A body is said to be in thermodynamic equilibrium if it is in
- Thermal equilibrium: Equality of *temperature*
- Mechanical equilibrium: Equality of *forces* and *couples*
- Chemical equilibrium: Equality of chemical potentials.

4. Property of System

Properties are *point* function and are *exact* or perfect differentials e.g. internal energy, enthalpy, entropy

- Intensive properties
  These properties are *independent* of mass. e.g. pressure, temperature density, specific volume, Specific heat (c_p, c_v), specific enthalpy.

- **Extensive properties**
  These properties are *dependent* on mass. e.g. volume, energy, Heat capacity (C_V, C_p), enthalpy. All specific properties are intensive properties.

Point functions	Does not depend on path history (T, P, V)
Path functions	Depend on path history (work, heat)

5. Process

Change of state is known as process.

- **Reversible process**
  - A process when reversed in direction follows the *same path* as that of the forward path without leaving any effect on *system* and *surroundings*.

- **Irreversible process**
  - The process which is not reversible is known as irreversible.
  - All *actual* process are irreversible process.

Remember

- A system is said to have undergone a cycle when the initial and final points are same.
- For a cycle change in property is zero.
- A *frictionless* Quasi static process is called a reversible process.

---

6. Gibs Phase Rule

$$P + F = C + 2$$

Here

  P = Number of phases (solid, liquid, gas)

  F = Minimum number of *independent* intensive variables required to fix the state (Degree of freedom)

  C = Number of components (e.g. If O_2 and N_2, then C = 2)

If in a system water vapour and liquid water coexist together then only one intensive property required i.e. either pressure or Temperature (DOF = 1).

7. Zeroth Law of Thermodynamics/Temperature Measurement

- When a body A is in thermal equilibrium with a body B & also separately with a body C then body B & C will be in thermal equilibrium with each other.

- Zeroth law of thermodynamics is the basis of *temperature* measurement.

8. Type of Thermometer

Type of thermometer	Principle	Thermometric property
Resistance	Wheat store bridge	Resistance
Thermocouple	See back effect	E.M.F (voltage)
Constant volume gas thermometer	Ideal gas equation	Pressure
Constant pressure gas thermometer	Ideal gas equation	Volume

- Ideal gas thermometer are *independent* of material of construction.
- Thermocouple uses copper-constantan, platinum-rhodium combinations.

9. Conversion of Temperature Unit/Triple point of Water

- Conversion of temperature unit:

$$\frac{°C}{5} = \frac{°F - 32}{9} = \frac{T - 273.15}{5}$$

Here,

$$°C = \text{Temperature in degree Celsius}$$
$$°F = \text{Temperature in degree Fahrenheit}$$
$$T = \text{Temperature in Kelvin}$$

- **Triple point of water**
  It is the point where all the phases coexist. For water its value is 273.16 k = 0.01 °C.

10. Method of Temperature Measurement

- Used before 1954

$$T = \frac{100(P - P_i)}{(P_s - P_i)}$$

Here

     T = Temperature

     P = Pressure corresponding to temperature 'T'

     P_i = Pressure corresponding to ice point (0°C)

     P_s = Pressure corresponding to steam point (100°C)

- **Used after 1954**

$$T = 273.16\left(\frac{P}{P_{tp}}\right)$$

Here

     T = Temperature

     P = Property corresponding to temperature 'T'

     P_{tp} = Property corresponding to triple point of water (273.16 k)

■■■■

Energy Interaction

2

1. Work

Work is said to be done by the system if the sole effect on the things external to the system can be reduced in raising of weight (weight may not actually be raised).

- **Sign convention**
  - Work done **by** the system is **positive**
  - Word done **on** the system is **negative**

Remember: ....

- Work is a **path** function and inexact or imperfect differential. Work is not a property.

- For quasistatic (reversible) process work done is calculated by $\int p dV$.

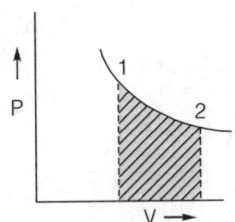

Conditions for applying above equation

(i) System must be closed system

(ii) Process must be **reversible** process

(iii) Work should cross the boundary

- If process becomes **Irreversible** then, **Irreversibility** may be due to following reason :

  (a) Quasi equilibrium with friction

  (b) Non-equilibrium process

  then $\int dw = \int P_{external} dv$

- For Reversible process $P_{external} = P_{Internal}$

Remember

2. Representation of Various Process on P-V diagram

$$Pv^K = Constant$$

Process	K
P = const.	K = 0
V = const.	K = ∞
T = const.	K = 1
Adiabatic	K = γ
Polytropic	K = n

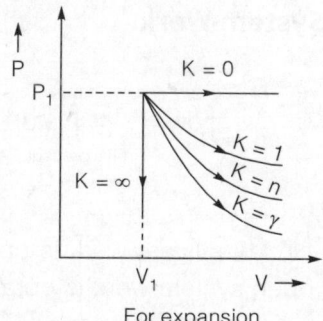

For expansion

3. Closed System Work in Various Process

Process	Work done
Constant pressure (isobaric)	$w_{1-2} = p(v_2 - v_1) = mR(T_2 - T_1)$
Constant volume (isochoric)	$w_{1-2} = 0$
Constant Temperature (isothermal)	$w_{1-2} = p_1 v_1 \ln\left(\dfrac{p_1}{p_2}\right)$ $= mRT \ \ln\left(\dfrac{p_1}{p_2}\right) = mRT \ \ln\left(\dfrac{v_2}{v_1}\right)$
Adiabatic (isentropic)	$w_{1-2} = \dfrac{(p_1 v_1 - p_2 v_2)}{(\gamma - 1)} = \dfrac{mR(T_1 - T_2)}{\gamma - 1}$
Polytropic	$w_{1-2} = \dfrac{(p_1 v_1 - p_2 v_2)}{(n - 1)} = \dfrac{mR(T_1 - T_2)}{n - 1}$

Remember

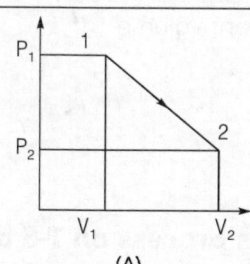

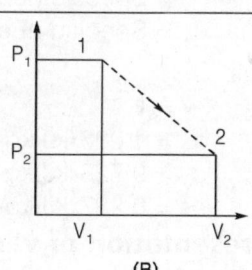

(A)   (B)

(A) Reversible closed system process

$$W = \int P dv$$

$$P_{external} = P_{Internal} = P$$

(B) Irreversible closed system process

$$W = \int P_{external} dv$$

$$P_{external} \neq P_{Internal}$$

4. Open System Work

$$\int dw = - \int_{\substack{\text{final pressure} \\ \text{initial pressure}}} v.dp$$

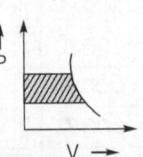

Remember: ....
- The closed system work is obtain by plotting it on *volume axis* and open system work is obtain by plotting it on *pressure axis*.

5. Heat, Specific Heat Ratio

- Heat

$$Q = mC \Delta t$$

Here, Q = Heat

       m = mass

       C = Specific heat

Heat flow *out* of a system is taken as *negative* while heat flow *into* a system is taken as *positive*.

- Heat like work is also a *path* function so is *inexact* or imperfect differential.

6. Specific Heat Ratio (γ)

$$\gamma = \frac{c_p}{c_v}$$

Here, c_p = Specific heat at constant pressure

       c_v = Specific at constant volume

For air

$$\gamma = 1.4$$
$$c_p = 1.008 \text{ kJ/kg-K}$$
$$c_v = 0.728 \text{ kJ/kg-K}$$
$$R = 0.287 \text{ kJ/kg-K}$$

7. Representation of various process on T-S diagram

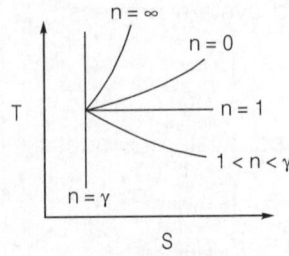

■■■■

First Law of Thermodynamics

3

1. First Law of Thermodynamics

- For a cycle (Closed System)

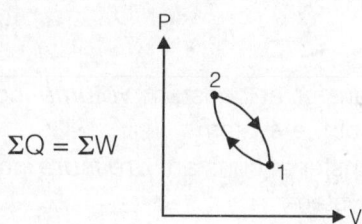

$$\Sigma Q = \Sigma W$$

- For a process

$$\delta Q = dE + \delta W$$
$$\delta Q = dU + \delta W \text{ (If KE and PE are negligible)}$$

Remember: ....

- For *isolated* system energy (E) always constant.
- Energy is a *point* function and a property of the system. Energy is an extensive property while *specific* energy is an intensive property.
- The internal energy depends only on temperature, for an ideal gas.

2. Heat Transfer for Closed System

- Heat transfer at constant volume

$$Q_V = (\Delta u) = \int_{T_1}^{T_2} C_V dT$$

C_v = specific heat at constant volume

- Heat transfer at constant pressure

$$Q_P = (\Delta h) = \int_{T_1}^{T_2} C_p dT$$

C_p = Specific heat at constant pressure

- For isothermal process

$$dU = 0$$

∵ U is function of temperature only, for ideal gas

$$dQ = dW$$

- In adiabatic process

Heat transfer = 0

dQ = 0
- In polytropic process

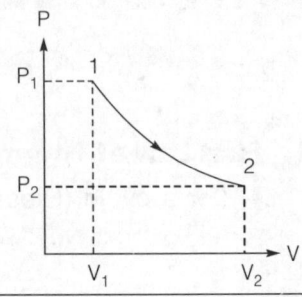

$$dQ = \frac{P_1 V_1 - P_2 V_2}{n-1}\left(\frac{\gamma-n}{\gamma-1}\right)$$

$$= \text{(Polytropic work)}\left(\frac{\gamma-n}{\gamma-1}\right)$$

Remember

- Heat transfer at constant *volume* increases the *internal* energy of the system.
- Heat transfer at constant *pressure* increases the *enthalpy* of the system.
- In polytropic process, heat transfer is not zero.

3. Enthalpy (H)

It is the heat content of a body. Its unit is joule.
$$H = U + PV$$

4. Polytropic Specific Heat

$$C_{polytropic} = \frac{\gamma-n}{1-n}C_v$$

Here,   n = Polytropic index

$$\gamma = \frac{C_p}{C_v}$$

Remember: ....
- Polytropic specific heat is always negative.

$$\int dQ = \Delta Q = \int mC_v dT\left(\frac{\gamma-n}{1-n}\right)$$

$$\Delta Q = \int dQ = \int mC_{polytropic}dT$$

5. Perpetual Motion Machine of First Kind (PMM1)

There can be no machine which would continuously supply mechanical work without some other form of energy disappearing simultaneously. Such a fictitious machine is called perpetual motion machine of first kind.

■■■■

First Law Applied to Flow Process

4

1. Steady Flow Process/Steady Flow Energy Equation

In a flow if thermodynamic properties **do not** change with **time** at different locations, the process is steady flow process.

In steady flow process there is no accumulation of mass or energy in control volume i.e. conservation of mass and energy occurs.

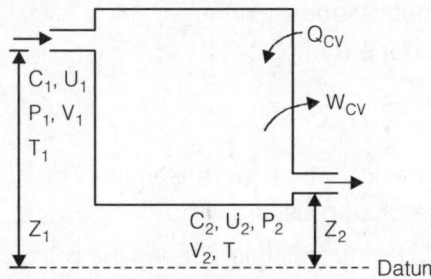

- **Steady Flow Energy Equation (S.F.E.E.)**

$$H_1 + \frac{1}{2}mC_1^2 + mgz_1 + Q_{cv} = H_2 + \frac{1}{2}mC_2^2 + mgz_2 + W_{CV}$$

For Per Unit Mass

$$h_1 + \frac{1}{2}C_1^2 + gz_1 + q_{cv} = h_2 + \frac{1}{2}C_2^2 + gz_2 + w_{cv}$$

Here, H = Enthalpy
  C = Velocity
  Z = Height
  Q_{cv} = Heat given
  W_{cv} = Work done
  m = Mass flow rate
  h = Specific enthalpy

- **SFEE can also be written as:**

$$(u_1 + P_1 v_1) + gz_1 + q_{cv} + \frac{C_1^2}{2} = (u_2 + P_2 v_2)\frac{C_2^2}{2} + gz_2 + w_{cv}$$

$$u_1 + \frac{c_1^2}{2} + gz_1 + q_{cv} = u_2 + \frac{c_2^2}{2} + gz_2 + w_{Total}$$

where, $w_{(Total)}$ = [Displacement work done on system + displacement work done by system + w_{cv}]

$$w_{(Total)} = -P_1 v_1 + P_2 v_2 + w_{cv}$$

2. Application of S.F.E.E is Steady Flow Process

- Nozzle

$$C_2 = \sqrt{2(h_1 - h_2)} = \sqrt{2C_p(T_1 - T_2)}$$

Here, T = Temperature
h = Enthalpy
C_2 = Velocity at outlet

Conditions:

(i) Nozzle is perfectly insulated

(ii) Neglect potential energy change

(iii) No work is done by nozzle

(iv) $C_1 < < < < C_2$

Remember: ....

- A *nozzle* is a device which increases the velocity or K.E. of fluid at the expense of its pressure drop.

- A *diffuser* is a device which increases the pressure of fluid at the expense of K.E.

- Turbine

$$W_{CV} = h_1 - h_2$$

Here, h_1 = Enthalpy at inlet of turbine
h_2 = Enthalpy at outlet of turbine

Conditions:

(i) Turbine is perfectly insulated

(ii) Change in potential energy and kinetic energy neglected

(iii) $C_1 \approx C_2$

- Compressor

$$W = h_2 - h_1$$

Here

h_1 = Enthalpy at inlet of compressor
h_2 = Enthalpy at outlet of compressor

Conditions:

(i) Compressor is perfectly insulated

(ii) Change in potential energy and kinetic energy neglected

(iii) $C_1 \approx C_2$

- Throttling device

$$h_1 = h_2$$

Condition
(i) No heat transfer
(ii) No work transfer
(iii) Constant Enthalpy
(iv) Irreversible process

- When a fluid flow through a narrow passage like an orifice, partially opened valve, there is an **appreciable drop in pressure**. The process is throttling process.

3. Free Expansion

Expansion against **vacuum** is called free expansion. For free expansion, internal energy is same and if it is an ideal gas then $T_1 = T_2$.

     T_1 = Initial temperature

     T_2 = Final temperature

Remember: ....

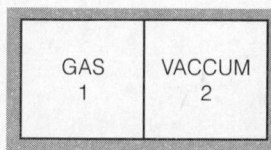

Insulated system

$$\boxed{\Delta Q = 0}$$
$$\boxed{\Delta w = 0}$$
$$\boxed{\Delta U = 0}$$

- Both chamber 1 and 2 are considered as system in free expansion. Thus $u_1 = u_2$

4. Unsteady State Transient

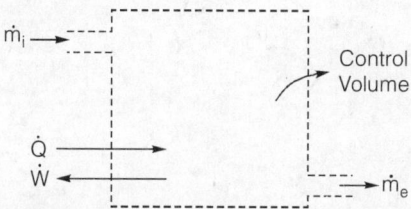

Here $\dot{m}_i$ = Mass flow/sec at inlet

     $\dot{m}_e$ = Mass flow/sec at outlet

From mass conservation

- $\left(\dfrac{dm}{dt}\right)_{c.v.} = \dot{m}_i - \dot{m}_e$
- From energy balance equation we have

$$\left(\frac{dU}{dt}\right)_{cv} = \dot{m}_i h_i - \dot{m}_e \dot{h}_e + \dot{Q} - \dot{W}$$

Here

$\qquad U$ = Total internal energy

$\qquad CV$ = Control volume

$\qquad \dot{Q}$ = Heat input in the CV

$\qquad \dot{W}$ = Work done by CV

■■■■

Second Law of Thermodynamics 5

Work is said to be *high grade energy* and heat is a low grade energy. The complete conversion of **low grade energy** into **high grade energy** in a cycle is impossible while the complete conversion of high grade energy into low grade energy is possible. Heat energy is a low grade energy.

1. Second Law of Thermodynamics

- ### Kelvin Plank Statement
  It is impossible for a heat engine to produce net work in a complete cycle if it exchanges heat only with bodies at a single fixed temperature.

  Two **reversible adiabatic** paths **cannot** intersect each other which violates the Kelvin-Plank's statement.

  The machine which violate Kelvin plank statement is called **PMM2**.

- ### Clauses statement
  It is impossible to construct a device which operating in a cycle transfers heat from cooler body to a hotter body without any work input.

2. Thermal Energy Reservoir (TER)/Heat Engine/Heat Pump/ Refrigerator

- ### Thermal energy reservoir
  A thermal energy reservoir (TER) is defined as a large body of infinite heat capacity which is capable of absorbing or rejecting an unlimited quantity of heat without any appreciable changes in its thermodynamic properties.

- ### Heat Engine
  A heat engine works on a thermodynamic cycle in which there is net heat transfer to the system and a net work transfer from the system.

  Thermal Efficiency

  $$\eta_{th} = \frac{W_{net}}{Q_1} = \frac{Q_{net}}{Q_1} = 1 - \frac{Q_2}{Q_1} \quad ...(i)$$

  $$\eta_{th\,(rev.cyc)} = 1 - \frac{T_2}{T_1}$$

- Heat pump

$$C.O.P = \frac{Q_1}{Q_1 - Q_2} \quad ...(ii)$$

$$C.O.P_{(rev.cyc)} = \frac{T_1}{T_1 - T_2}$$

- Refrigerator

$$C.O.P = \frac{Q_2}{Q_1 - Q_2} \quad ...(iii)$$

$$C.O.P_{(rev.cyc)} = \frac{T_2}{T_1 - T_2}$$

- $(C.O.P.)_{H.P} = 1 + (C.O.P.)_R$
- The above formulae (i) (ii) and (iii) are applicable for both **reversible** and **irreversible** cycles.

If 2-Reversible engine operating between thermal reservoir at T_1, T_2 and T_3 ($T_1 > T_2 > T_3$) as shown below:

- **Case-I :** If the efficiency is same for engine.

$$T_2 = \sqrt{T_1 T_3}$$

- **Case-II:** If work output of both engine same.

$$T_2 = \frac{T_1 + T_3}{2}$$

---

3. Perpetual Motion Machine of Third Kind (PMM-3)

Continual motion of a movable device in complete absence of friction is known as PMM-3.

4. Carnot Theorem

It states that all heat engines operating between a given constant temperature source and given constant temperature sink, none has higher efficiency than a reversible engine.

Remember: ....

- The efficiency of all reversible heat engines operating between the **same temperature** levels is the same.

- The effect of *decreasing* T_2 to increase thermal efficiency of Carnot cycle is *more than* the effect of increasing T_1 by *same amount*.
- First law of thermodynamics gave origin to *internal energy* and second law leads to *entropy*.

5. Third law of thermodynamics

It is impossible by any procedure, no matter how idealized, to reduce any system to absolute zero of temperature in a finite number of operations.

■■■■

Entropy

6

Second law of thermodynamics leads to entropy.

1. Clausus Theorem

- The cyclic integral of $\dfrac{dQ}{T}$ for a reversible cycle is equal to zero.

$$\oint_R \frac{dQ}{T} = 0$$

- The cyclic integral of $\dfrac{dQ}{T}$ for a irreversible cycle is less than zero.

$$\oint_R \frac{dQ}{T} < 0$$

2. Entropy/Entropy Change

- $ds \geq \dfrac{dQ}{T}$

- $ds = \left(\dfrac{dQ}{T}\right)_{rev.}$

- $(ds)_{universe} \geq 0$
  Here

  ds = Entropy change
  dQ = Change in heat
  T = Temperature

Remember: ....

- Entropy change in **reversible** process is a point function and exact differential.
- It is an **extensive** property.
- Reversible adiabatic process is an isentropic process but reverse is **not always** true.
- When the system is at equilibrium, any conceivable change in entropy would be zero.

3. Inequality of Clausus

- Inequality of Clausus - It provides the criterion of irreversibility of cycle.

$$\oint \frac{dQ}{T} \leq 0$$

$$\oint \frac{dQ}{T} = 0, \text{ the cycle is reversible}$$

$$\oint \frac{dQ}{T} < 0, \text{ the cycle is irreversible and possible}$$

$$\oint \frac{dQ}{T} > 0, \text{ the cycle is impossible}$$

4. Change in Entropy of System

- Change in entropy of a system is due to heat transfer and internal irreversibility (entropy generation).

$$S_2 - S_1 = \underbrace{\int_1^2 \frac{dQ}{T}}_{\text{due to heat transfer}} + \underbrace{S_{generation}}_{\substack{\text{due to internal} \\ \text{irreversibility}}}$$

Remember: ....

- Entropy *generation* is a path function.
- Entropy *generation* is always a positive value.

5. Twin Equation of Entropy

- $T.ds = du + P.dv$
- $T.ds = dh - v.dP$

Above equations are applicable for *reversible* and *irreversible* process and for open and closed system both.

6. Entropy Change (ΔS) For Finite Body

$$\Delta S = mc\ln\frac{T_f}{T_i}$$

Here, C = Specific heat

       T_f = Final temperature

       T_i = Initial temperature

       m = Mass of body

7. Entropy Change (ΔS) During Phase Change

$$\Delta s = \frac{ml}{T}$$

Here, l = Latent heat of phase change (vaporization, condensation, fusion etc.)

        T = Saturation temperature (freezing point, boiling point)

8. Entropy Change For an Ideal Gas Between State '1' and State '2'.

- $S_2 - S_1 = C_p \, ln\frac{V_2}{V_1} + C_v \, ln\frac{P_2}{P_1}$

- $S_2 - S_1 = C_v \cdot ln\frac{T_2}{T_1} + R \cdot ln\frac{V_2}{V_1}$

- $S_2 - S_1 = C_p \, ln\frac{T_2}{T_1} - R \, ln\frac{P_2}{P_1}$

Here, S = Entropy

         P = Pressure

         V = Volume

         T = Temperature

         R = Characteristic gas constant

Remember: ....

- Change in specific entropy of a system undergoing Isothermal process by ideal gas given by:

$$\Delta S = (C_P - C_V) \, ln\frac{V_2}{V_1}$$

- Also for Isothermal process change in specific energy given by:

$$\Delta S = -(C_P - C_V) \, ln\frac{P_2}{P_1}$$

9. Slope of Constant Volume and Constant Pressure Line

$$\left(\frac{dT}{dS}\right)_v = \frac{T}{C_v}$$

$$\left(\frac{dT}{dS}\right)_P = \frac{T}{C_P}$$

- Slope of constant volume line = g(slope of constant pressure line).

■■■■

Available Energy, Availability and Irreversibility

7

The part of low grade energy which is available for conversion to high grade energy is referred to as available energy, which is also known as *exergy* while the part of low grade energy which, according to second law, must be rejected is called unavailable energy. Unavailable energy is known as *anergy*.

1. Available Energy (Cycle)

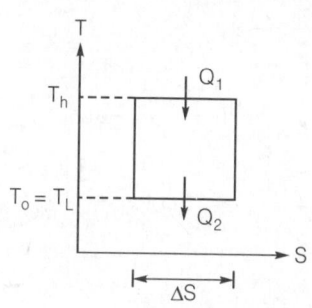 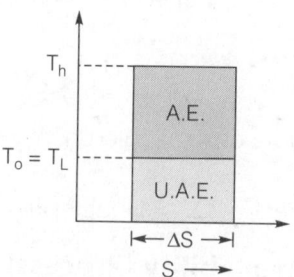

$$T_o = \text{Ambient temperature}$$

$$W_{max} = Q_1 \cdot \eta_{max}$$

$$\Rightarrow \qquad = Q_1 \left(1 - \frac{T_o}{T_H}\right)$$

$$W_{max} = Q_1 - Q_2 = Q_1 - (\Delta S) T_o$$

Here

$$(\Delta S) T_o = \text{Unavailable energy}$$

- The maximum work can be obtain when the lower temperature would be ambient temperature.

Remember: ....

- Decrease in available energy take place due to irreversible process when heat is transferred through a finite temperature difference.

- Same amount of heat has more available energy when it is transferred from higher temperature body than when it is transferred from lower temperature body.

- The degradation is more for energy loss at a higher temperature than that at a lower temperature.

- The first law states that the energy is always conserved *quantity wise* while the second law emphasizes that energy always degrades *quality wise*.

2. Loss of Available Energy in a Cycle

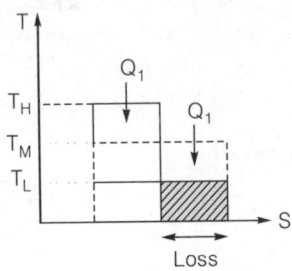

Loss of Available energy $= Q_1 \left[\dfrac{T_L}{T_M} - \dfrac{T_L}{T_H} \right]$

Here Q_1 = Heat input

3. Availability (Process)

Availability (A) of a given system is defined as the maximum useful work that is obtainable in a process in which the system comes to *equilibrium* with its surroundings.

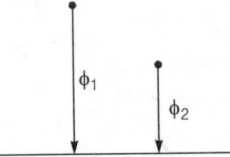

Availability $= \phi_1 - \phi_2$

- Availability of a closed system

$$= \phi_1 - \phi_2 \qquad (\phi = U - T_o S + P_o V)$$
$$W_{max.useful} = (U_1 - U_2) - T_o(S_1 - S_2) + P_o(V_1 - V_2)$$

Here

     U = Internal energy

     T_o = Surrounding temperature

Other symbol has usual meaning.

- Availability of open system

$$= \phi_1 - \phi_2 \qquad (\phi = h - T_o S)$$
$$W_{(max. useful)} = (h_1 - h_2) - T_o(S_1 - S_2)$$

4. Irreversibility (I)

- Irreversibility is the loss of available energy due to dissipation of energy or entropy generation.

$$\text{Irreversibility} = W_{max} - W_{act}$$
$$= T_0 (\Delta s)_{univ}$$

5. Gouy-Stodula theorem

$$I \propto (\dot{\delta}s)_{gen}$$

6. Gibbs function (G)

$$G = H - TS$$

- It is applicable for **open** system.

Remember

- The change in availability of a open system is equal to change in Gibbs function of system at constant temperature.
- Gibbs function is useful in evaluating the availability of systems in which chemical reaction occurs.
- The change in availability of closed system is **NOT** equal to change in Helmholtz function at given temperature.

7. Helmholtz function (F)

$$F = U - TS$$

- It is applicable for **closed** system.

■■■■

Properties of Pure Substance 8

A pure substance is a substance of **constant** chemical composition throughout its mass. It is one component system. It may exist in one or more phases.

1. Saturation State/Critical Temperature/Critical Pressure/ Triple Point

- **Saturation state**
  A saturation state is a state from where a change of phase may occur without a change of pressure or temperature.
- **Critical temperature**
  At critical temperature a liquid completely changes to vapour and vice-versa. Also above critical temperature a vapour cannot be liquefied by any amount of pressure.
- **Critical pressure**
  At critical temperature the corresponding minimum pressure required to transform a vapour to liquid is called critical pressure.
- **Triple point**
  Triple point is the fixed point (fixed temperature and pressure) at which solid, liquid and vapour phases co-exist in equilibrium.

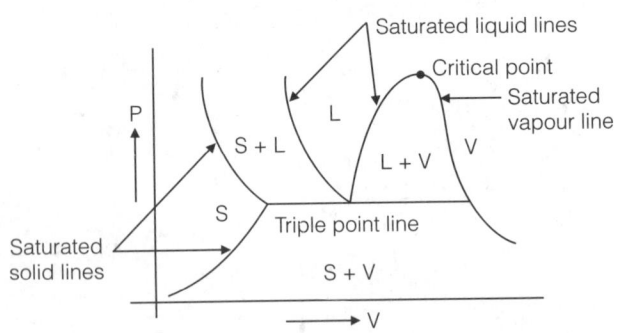

Remember: ....

- Transformation of solid to vapour directly is called *sublimation*.
- Transformation of vapour to solid directly is called *ablimation*.
- Enthalpy of vaporization at critical point is zero.

2. Degree of subcooling

$$T_{saturation} - T_{actual}$$

where, $T_{sat.} = T_{saturated}$

$T_{sub.} = T_{subcooled}$

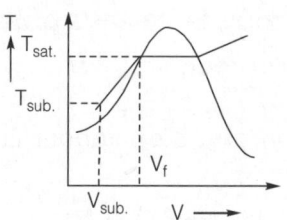

Remember: ....

- A subcooled liquid has specific volume less than $V_{saturated}$ at liquid state(V_f) thus sometimes also called compressed liquid.

3. Degree of superheating

$$= T_{(actual)} - T_{(saturation\ vapour)}$$

4. Dryness fraction

$$x = \frac{m_v}{m_v + m_l}$$

Here, m_v = mass of vapour

      m_l = mass of liquid

- Various properties of pure substance based on dynes fraction.

$$v = v_f + x(v_g - v_f)$$

Similarly

$$h = h_f + x(h_g - h_f)$$
$$s = s_f + x(s_g - s_f)$$

Here

v, v_f, v_g = specific volume of moist vapour, liquid, vapour

h, h_f, h_g = specific enthalpy of moist vapour, liquid, vapour

s, s_f, s_g = specific entropy of moist vapour, liquid, vapour

5. Enthalpy of Liquid at Various Point

- Wet region
$$h_1 = h_f + x(h_g - h_f)$$
- On-vapour line
$$h_1 = h_g = h_f + \text{Latent heat}$$
$$[h_g - h_f = \text{L.H.}]$$
- On superheated region
$$h_1 = h_g + c_p(T_1 - T_s)$$

Here

$$c_p = \text{Specific heat of vapour}$$

6. Entropy at various point

- Wet region

$$s_1 = s_f + x(s_g - s_f)$$

- On saturated vapour line

$$s_1 = s_f + \frac{L.H}{T} \qquad \left(s_g - s_f = \frac{h_g - h_f}{T} \right)$$

- In super heated region

$$s_1 = s_g + \Delta s = s_g + c_p \, In \frac{T_1}{T_{saturation}}$$

7. Some Important Points to Remember

- A gas or a *pure* substance require *two* known properties (P, V, T, H etc.) to describe it completely i.e. it have two degrees of freedom.

- A liquid and vapour in equilibrium state (*saturated state*) has *one degree* of freedom.

- A liquid, vapour and solid in equilibrium (*triple point*) has *zero degree* of freedom.

- Throttling calorimeter is used to measure *dryness fraction* of *pure* substance.

- On P-T diagram, triple point and critical point both are points.

- On P-V diagram, *triple point is a line* and critical point is a point.

- On T.V. diagram for water we have density maximum at 4°C.

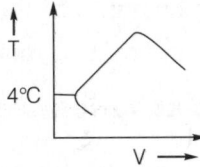

- Thus in container full of ice. When ice starts melting it level decreases upto 4°C of water.

- $C_P = C_V$ at 4°C as $\left(\dfrac{\partial V}{\partial T} \right)_P = 0$

- $C_P - C_V = -T \left[\left(\dfrac{\partial V}{\partial T} \right)_P \right]^2 \left(\dfrac{\partial P}{\partial V} \right)_T$

■■■■

Properties of Gases and Gas Mixture

A hypothetical gas which obeys ideal gas equation at all pressures and temperature is called an ideal gas.

At *very low pressure and high temperature*, real gas approaches the ideal gas behavior.

1. Ideal Gas Equation

$$PV = n\bar{R}T$$
$$PV = mRT$$

$$\text{Number of mole (n)} = \frac{\text{Mass (m)}}{\text{Molecular weight}}$$

$$\bar{R} = RM$$

Here

$\bar{R}$ = Universal gas constant = 8.314 kJ/kgK
R = Characteristics gas constant
M = Molecular weight

2. Mole Fraction

$$x_1 = \frac{n_1}{\Sigma n}$$

$$x_1 = \frac{P_1}{P}$$

Here, P_1 = Partial pressure of gas 1

3. Equivalent Gas Constant

$$R_e = \frac{m_1R_1 + m_2R_2 + m_3R_3 + \cdots}{m_1 + m_2 + m_3}$$

Here,

m = Mass
R = Gas constant of individual gas

4. Equivalent Molecular Weight (Me)

$$M_e = x_1 M_1 + x_2 M_2 + x_3 M_3 + \cdots + x_n M_n$$

Here, x = Mole fraction of different gas

M = Molecular weight of different gas

5. Value of 'γ' For Different Gases

Gases	Value of γ
Monatomic gases	5/3
Diatomic gases	7/5
Polyatomic gases	4/3
Air	1.41

6. Vender Waals Equation For Real Gas

Volume of
gas molecules

$$\left\{P + \frac{a}{v^2}\right\}(V - b) = RT$$

Force of cohesion
among gas molecules

- Relation among a, b, R, V_c, T_c, P_c

$$a = 3P_c V_c^2$$

$$b = \frac{1}{3}V_c$$

$$\therefore \quad P_c = \frac{a}{27b^2}, \quad V_c = 3b$$

$$T_c = \frac{8a}{27bR}$$

$$R = \frac{8a}{9T_c V_c} = \frac{8}{3}\frac{P_c V_c}{T_c}$$

Here

a, b = vander Waals constant

P_c, V_c, T_c = critical pressure volume and temperature of the gas

- Boyle's temperature (T_B)

$$T_B = \frac{a}{bR}$$

■■■■

Thermodynamic Relations Equilibrium and Stability

1. Theorem 1

If $\quad dz = M.\,dx + N.dy$, thus z-is exact when

$$\left(\frac{\partial M}{\partial y}\right)_x = \left(\frac{\partial N}{\partial x}\right)_y$$

2. Theorem 2

If 'f' is a function of (x, y, z) i.e.
$$f = \phi(x,\ y,\ z)$$
and if there is some relationship between x, y, z then

$$\left(\frac{\partial x}{\partial y}\right)_f \cdot \left(\frac{\partial y}{\partial z}\right)_f \cdot \left(\frac{\partial z}{\partial x}\right)_f = 1$$

3. If $z = f(x, y)$ then

$$\left(\frac{\partial x}{\partial y}\right)_z \cdot \left(\frac{\partial y}{\partial z}\right)_x \cdot \left(\frac{\partial z}{\partial x}\right)_y 0 = -1$$

4. Maxwell's Equation

$$\left(\frac{\partial T}{\partial V}\right)_s = \left(\frac{-\partial P}{\partial S}\right)_V$$

$$\left(\frac{\partial T}{\partial P}\right)_s = \left(\frac{\partial V}{\partial S}\right)_P$$

$$\left(\frac{\partial P}{\partial T}\right)_V = \left(\frac{\partial S}{\partial V}\right)_T$$

$$\left(\frac{\partial V}{\partial T}\right)_P = \left(\frac{-\partial S}{\partial P}\right)_T$$

5. Tds Relations

- First T. ds equation

$$T.\,ds = C_V \cdot dT + T.\left(\frac{\partial P}{\partial T}\right)_V \cdot dv$$
$$= dU + P.dV$$

- Second T. ds equation

$$T.\ ds = C_p.dT - T\left(\frac{\partial V}{\partial T}\right)_P \cdot dP$$

6. Difference in Specific Heat

$$C_p - C_v = -T\left[\left(\frac{\partial V}{\partial T}\right)_P\right]^2 \left(\frac{\partial P}{\partial V}\right)_T$$

- $\left(\frac{\partial P}{\partial V}\right)_T$ always negative.

  Therefore $C_p > C_v$

Special case - At 4°C density of water is maximum

$$\left(\frac{\partial P}{\partial V}\right)_T = 0$$

Thus $C_p = C_v$

7. Energy equation

$$dU = C_v \cdot dT + \left[T\left(\frac{\partial P}{\partial T}\right)_v - p\right]dV$$

8. Clapeyron's equation

$$\frac{dP}{dT} = \frac{h_g - h_f}{T(V_g - V_f)}$$

9. Clausus Clapeyron equation

$$\frac{dP}{dT} = \frac{P(\text{Latent Heat})}{RT^2}$$

It apply to change of liquid to gas only.

10. Joule-Kelvin Effect

When a gas is throttled then first its temperature increases (heating) as the pressure decreases but after a particular pressure, temperature decreases (cooling) as pressure decreases. At different initial temperatures different such pressures exist.

- Joule-Kelvin constant

$$\left(\frac{dT}{dP}\right)_h = \frac{1}{c_p}\left[T\left(\frac{\partial V}{\partial T}\right)_P - V\right]$$

Its value is zero, for ideal gas

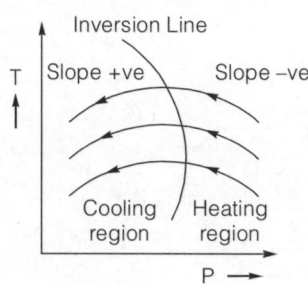

- In cooling reign, Joule-Kelvin constant is negative and in heating region, it is positive.
- First Tds equation

$$Tds = c_V dT + T\left(\frac{\partial P}{\partial T}\right)_V dV = c_V dT + \left(\frac{T\beta}{K_T}\right)dV$$

- Second Tds equation

$$Tds = c_p dT - T\left(\frac{\partial V}{\partial T}\right)_P dP = c_p dT - (T\beta V)dP$$

Remember

- Refrigeration effect can be obtained for certain gases by throttling and value of Joule Thompson coefficient (μ) is greater than 1 ($\mu > 1$).

11. Coefficient of volume expansibility (β)

$$\beta = \frac{1}{V}\left(\frac{\partial V}{\partial T}\right)_P$$

12. Isothermal compressibility (k_T) (opposite to bulk modulus)

$$k_T = -\frac{1}{V}\left(\frac{\partial V}{\partial P}\right)_T$$

$$C_p - C_v = \frac{TV\beta^2}{k_T}$$

Remember: ....

- Compressibility factor $= \dfrac{PV}{RT}$

- Compressibility factor for an *ideal gas* $= 1$

- $\left(\dfrac{\partial P}{\partial V}\right)_{T_{cr}} = 0$, $T_{Cr} =$ Critical temperature

- $\left(\dfrac{\partial^2 P}{\partial V^2}\right)_{T_{cr}} = 0$

- $\left(\dfrac{\partial^3 P}{\partial V^3}\right)_{T_{cr}} < 0$

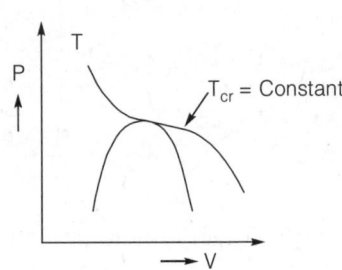

13. Entropy Change in Diffusion

O_2	N_2
p_1	p_2
T_1	T_2

$$\Delta S = -m_1 R_1 \ln\frac{p_1}{P} - m_2 R_2 \ln\frac{p_2}{P}$$

Where

    p_1, $p_2 =$ Partial pressure of O_2 and N_2

      $P =$ Final pressure of mixture

    R_1, $R_2 =$ Characteristic gas constant for O_2 and N_2

    m_1, $m_2 =$ Mass of O_2 and N_2

■■■■

A Handbook on
Mechanical Engineering

Power Plant Engineering

CONTENTS

◀ ☐ ▶

Power Plant Economics

1. Load Curve

A curve showing the load demand (variation) of consumer with respect to time is known as load curve. This curve may be for daily, weekly, monthly and on yearly basis. This a graph between *load and duration*.

The energy consumption of consumer is given by an

$$kwh = \int_{t_1}^{t_2} (kw) \cdot dt$$

2. Connected Load

It is the sum of ratings in kilowatt of equipments installed in the consumer House/Area.

3. Demand Factor

It is expressed as the ratio of maximum demand to the connected load. Its value *can not be* more than unity.

4. Average Load

It is the ratio of area under load curve (kwh) and time period for which average load is calculated.

5. Load Factor

It is the ratio of average load for a given period to the peak load during the same period. Its value will be *always less* than one.

6. Plant Use Factor (Plant Load Factor)

It is the ratio of energy produced in a given time to maximum possible energy that could be produced during actual number of hours of operation.

$$\text{Plant use factors} = \frac{\text{Annual production of energy}}{\text{Operational hour in year} \times \text{capacity of plant}}$$

■■■■

1. Type of Coal

- **Anthracite**

  Anthracite contains *more than* 86% fixed carbon and less volatile matter, volatile matter helps in the ignition of coal. So it is often difficult to burn anthracite.

- **Bituminous**

  It contain **46-86%** of fixed carbon and 20-40% of volatile matter. *Lower* the volatility *higher* the heating value.

- **Lignite**

  It is the lowest grade of coal containing moisture as high as **30%** and high volatile matter.

- **Peat**

  Peat contains up to 90% moisture and is not attractive as a utility fuel.

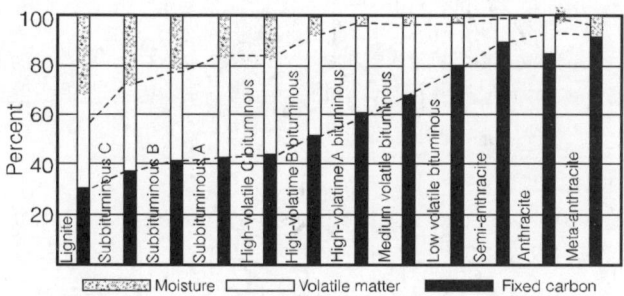

2. Coal Analysis

- **Proximate analysis**

  The proximate analysis indicates the behavior of coal when it is heated.

  - **Moisture content of coal**

    When 1g sample of coal is subjected to a temperature of about 105°C for a period of 1 hour, the loss in weight of the sample gives the moisture content of the coal.

  - **Volatile matter**

    When 1g sample of coal is placed in a covered platinum crucible and heated to 950°C and maintained at that temperature for about 7 min, there is a loss in weight due to the elimination of *moisture and volatile matter*. Volatile matter consists of hydrogen and certain hydrogen - carbon compounds which can be removed from the coal simply by heating it.

- **Ash content**

  By subjecting 1g sample of coal in an uncovered crucible to a temperature of about 720°C until the coal is completely burned, a constant weight is reached, which indicates that there is only ash remaining in the crucible.

- **Fixed carbon**

  Fixed carbon is the difference between 100% and the sum of the percentages of moisture ash and volatile matter.

Thus proximate analysis of coal gives

$$FC + VM + M + A = 100\% \text{ by mass}$$

where, FC = Fixed carbon

VM = Volatile matter

M = Moisture

A = Ash

- More volatile a coal more it will smoke but volatility helps in ignition.
- Lower rank coals are characterized by a *greater* oxygen content, that *aids* ignition and enhances combustibility and flame stability.

- **Ultimate analysis**

  The ultimate analysis gives the chemical elements that comprises the coal substance, together with ash and moisture.

  $$C + H + O + N + S + M + A = 100\% \text{ by } mass$$

Here,

C = Carbon

H = Hydrogen

O = Oxygen

N = Nitrogen

S = Sulphur

M = Moisture

A = Ash

- Coal that does not cake is called free burning coal.
- A free burning coal has a *high* value of Swelling Index; which indicates that it somewhat expands in volume during combustion.

3. Heating Value

It is the heat generated when the products of complete combustion of a sample of fuel are cooled to initial temperature of fuel-air. It is determined by *bomb-calorimeter*. It has two type.

- **Higher heating value (HHV)**
  In this assumption is that the water vapour in the product *condenses* and thus include the *latent heat* of vaporization of water vapour formed during combustion.

- **Lower heating value (LHV)**
  In this assumption is that the water vapour forms by combustion *leaves* as vapour itself.

$$\boxed{L.H.V. = H.H.V. - m_w h_{fg}}$$

Here,

h_{fg} = Latent heat of vaporization

m_w = Mass of water vapour

$m_w = M + 9H + \gamma_A W_A.$

Where, M and H are the mass fractions of moisture and hydrogen in the coal, γ_A is the *specific humidity* of atmospheric air and W_A is the actual amount of air supplied per kg of coal.

4. Dulong and Petit Formula

If the *ultimate analysis* is known, the HHV of anthracite and bituminous coals can be determined approximately by using Dulong and Petit formula as:

$$\boxed{HHV = 33.83\,C + 144.45\left(H - \frac{O}{8}\right) + 9.38\,s \text{ in MJ/kg.}}$$

Here,

C = Carbon

H = Hydrogen

O = Oxygen

S = Sulphur

All content by *mass* fraction.

5. Orsat Gas Analysis

Orsat gas analyzer measures the *volume* or **mole fraction** of CO_2, CO and O_2 in the dry flue gas. An orsat analyzer contains three pipettes containing chemical solutions.

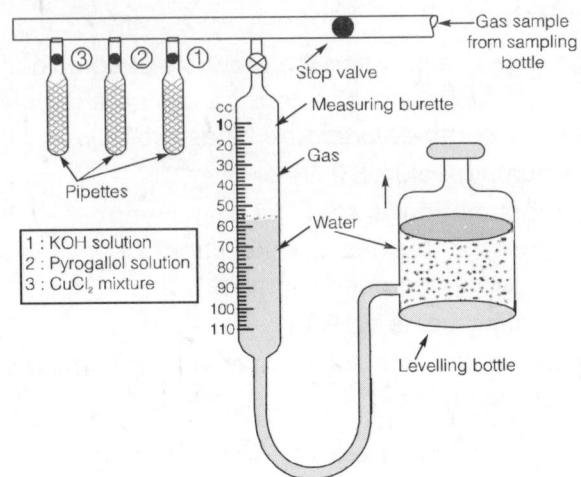

1 : KOH solution
2 : Pyrogallol solution
3 : CuCl₂ mixture

Remember

- **KOH** absorbs CO_2, **Pyrogallol solution** absorbs O_2 and $CuCl_2$ absorbs CO.

- The remaining unabsorbed gas is nitrogen. Since the sample is collected over water, any water vapour in the flue gas would have condensed during the collection process. The SO_2 gas will react with water in the container. So

  $CO_2 + CO + O_2 + N_2 = 100\%$ by *volume*

6. Height of Chimney

$$P = 353\,H\left(\frac{1}{T_1} - \frac{m+1}{m \cdot T_2}\right) \text{ mm of water}$$

Here,   P = Drought pressure in mm of water
   H = Height of chimney above the fire grate in metres
   m = Mass of air used for per kg of fuel
   T_1 = Absolute temperature of air outside chimney (Kelvin)
   T_2 = Absolute temperature of air inside chimney (Kelvin)

7. Condenser Vacuum Efficiency ($\eta\varpi$)

$$\eta_v = \frac{\text{Actual vaccum}}{\text{Ideal vaccum}}$$

8. Condenser Efficiency of Surface Condenser (η_c)

$$\eta_c = \frac{\text{Actual temperature rise of water}}{\text{Maximum possible temperature rise}}$$

■ ■ ■ ■

1. Classification of Boiler

- **Fire tube boiler**

  In the fire tube boilers; the hot *gases are inside* the tubes and a water surrounds the tubes. Example *Cochran, Lancashire, Locomotive* etc.

- **Water tube boiler**

  In the water tube boilers, the *water is inside* the tubes and hot gas surround them Example *Babcock and Wilcox, stirling, yarrow boiler* etc.

- **Externally fired boiler**

  In the externally fired boiler *fire is outside* the shell. Example *Babcock and Wilcox boiler, stirling boiler* etc.

- **Internally fired boiler**

  In case of internally fired boilers, the furnace is located inside the boiler shell Example *Cochran, Lancashire boiler* etc.

- **Forced circulation boiler**

  In forced circulation, circulation of water is done by *a pump* Example *Velox, Lamont, Benson* etc.

- **Natural circulation boiler**

  In natural circulation type of boiler water in boilers takes place due to *natural convection* currents Example *Lancashire, Babcock & Wilcox* etc.

- **Super critical boiler**

  The boiler which operating parameter is above critical point which is $P = 221.2$ *bar* and $T = 373°C$, is called super-critical boiler.

Remember

Super-critical boiler *must be* without drum. But subcritical boiler may be *with* or *without* drum.

A supercritical boiler with $P > 221.2$ Bar and temperature $T_{max.} = 540°C$ is called **once through boiler**.

---

2. Boiler Mounting

The item which is used for *safety* of boiler is called boiler mounting. The items such as stop valve, safety valves, water level, gauges, *fusible plug,* blow of cock, pressure gauges, water level indicator.

Remember: ....

- The function of a fusible plug is to protect the boiler against damage due to overheating for low water level. It is fitted on fire box crown plate or over the combustion chamber. *Gun metal* is used for fusible plug.

3. Boiler Accessories

The item which is used to *increase the efficiency* of boiler is called boiler accessories.

The items such as superheaters, economizers feed pumps etc. are termed as accessories, they form integral part of boiler.

4. Circulation Ratio

$$\frac{\text{Flow } rate \text{ of saturated water in downcomer}}{\text{Flow } rate \text{ of steam released from the drum}}$$

Circulation ratio of super critical boiler is 1 and for sub critical boiler it will be *more than* 1.

In order to avoid overheating range of circulation ratio:

Remember $6 \leq CR < 25$

5. Analysis of Steam/Rankine Cycle with Superheating

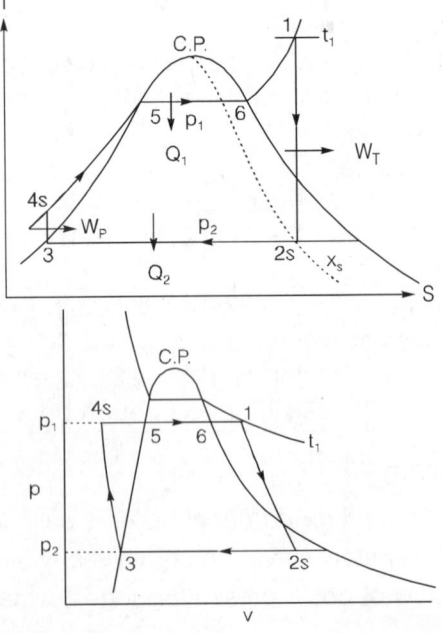

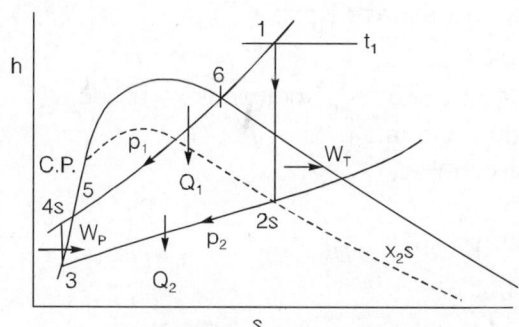

- Process 4s-1 : Reversible constant pressure heating of water. (Q_1) in boiler
- Process 1-2s : Reversible adiabatic expansion of steam in turbine. (W_T)
- Process 2s-3 : Reversible constant pressure heat rejection. (Q_2) in condenser
- Process 3-4s : Reversible adiabatic compression. (W_P) in compressor

For 1 kg of fluid, the steady flow energy equation to each processes:

For boiler $\quad\quad Q_1 = h_1 - h_{4s}$

For turbine $\quad\quad W_T = h_1 - h_{2s}$

For condenser $\quad Q_2 = h_{2s} - h_3$

For pump $\quad\quad W_P = h_{4s} - h_3$

- **Efficiency of Rankine cycle (Ideal)**

$$\eta = \frac{W_{net}}{Q_1} = \frac{(h_1 - h_{2s}) - (h_{4s} - h_3)}{h_1 - h_{4s}}$$

Remember: ....

- Pump work is very small as compare to turbine work so often neglected.
- Only sensible heat will be extracted by condenser.
- In Rankine cycle rw (work ratio) is approximately unity (rw ≃ 1)

- $$r_\omega = \frac{W_T - W_P}{W_T} \quad \text{work ratio}$$

- $$r_{Bw} = \frac{W_P}{W_T} \quad \text{back work ratio}$$

6. Heat Rate and Steam Rate

- **Heat rate**

  The cycle efficiency is sometimes expressed alternatively as heat rate which is the rate of heat input (kJ/s) required to produce unit shaft output (1 kW).

  $$\text{Heat rate (H.R.)} = \frac{Q_1}{W_T - W_P} = \frac{1}{\eta} \frac{kJ}{kWs}.$$

- **Steam rate S.S.C.**

  The capacity of a steam plant is often expressed in terms of steam rate or **specific steam consumption**. It is defined as the rate of steam flow (kg/s) required to produce unit shaft output (1 kW).

  $$\boxed{\text{S.S.C} = \text{Stream rate} = \frac{1}{W_{net}}\, kg/kWs}$$

- Larger the value of net output the lesser will be the value of S.S.C. and smaller will be plant size.

7. Steam Cycle/Mean Temperature of Heat Addition

In the Rankine cycle, heat is added reversible at a constant pressure but at infinite temperatures. If T_m is the *mean temperature of heat addition* then

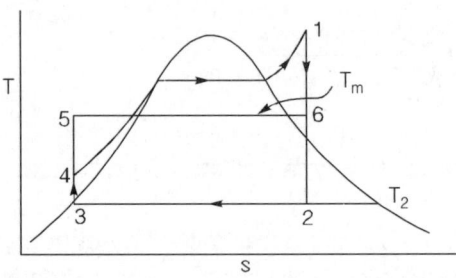

Heat added is $Q_1 = h_1 - h_4 = T_m (S_1 - S_4)$

$\therefore \quad T_m = \dfrac{h_1 - h_4}{S_1 - S_4}$

Heat rejected $Q_2 = h_2 - h_3 = T_2 (S_2 - S_3)$

$$\eta_{Rankine} = 1 - \frac{Q_2}{Q_1} = 1 - \frac{T_2}{T_m}$$

Remember

- Lower is the condenser Pressure, the higher will be the Efficiency of the Rankine cycle. Since it is fixed so $\eta_{Rankine} = f(T_m)$ only.
- The higher the mean temperature of heat addition, the higher will be the cycle efficiency.
- An increase in the superheat at constant pressure increases the mean temperature of heat addition and hence cycle efficiency. Moreover, with increase in superheat, the expansion line of steam in the turbine shifts to the right, as a result of which the quality of steam at turbine shifts to the right, as a result of which the quality of steam at turbine exhaust increases and performance of the turbine improves.
- It is desirable that most of the turbine expansion exhaust is not allowed to exceed 12% or the quality of steam to fall below 88%.

8. Reheating of Steam

If a steam pressure higher than $(p_1)_{max}$ is used in order to limit the quality to 0.88 at turbine exhaust, reheating of steam has to be adopted. In reheat cycle the expansion of steam from the initial state to the condenser pressure is carried out in **two or more steps** depending upon the number of reheats used.

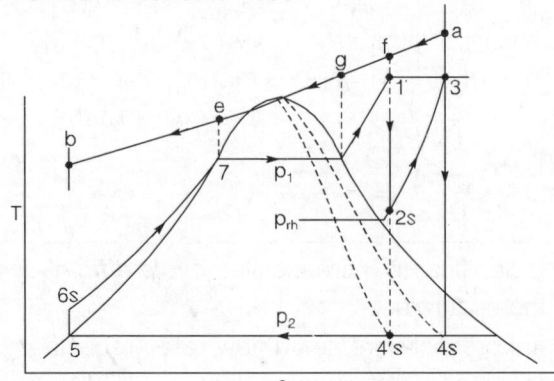

Heat addition $(Q_1) = h_1 - h_{6s} + h_3 - h_{2s}$

Heat rejected $(Q_2) = h_{4s} - h_5$

$$W_T = h_1 - h_{2s} + h_3 - h_{4s}$$

$$W_p = h_{6s} - h_5$$

$$\eta = \frac{(h_1 - h_{2s} + h_3 - h_{4s}) - (h_{6s} - h_5)}{h_1 - h_{6s} + h_3 - h_{2s}}$$

- The **net work output** of the plant increases with reheat, and hence the steam rate decreases.
- Reheating also improves the quality of steam at turbine exhaust.
- The **optimum** reheat pressure for most of the modern power plant is 0.2 to 0.25 of the initial steam pressure. If reheating is done in this pressure range then efficiency increases.

9. Regenerative Feed Water Heating

In practical regenerative cycle the feed water enters the boiler at high temperature as compare to the corresponding temperature in case of simple Rankine cycle and it is heated by steam extracted or bled from intermediate stages of the turbine.

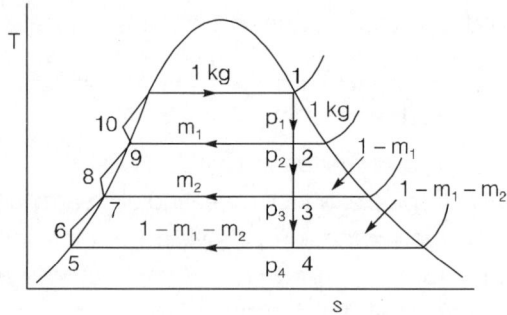

$$W_T = 1\,(h_1 - h_2) + (1 - m_1)\,(h_2 - h_3) + (1 - m_1 - m_2)\,(h_3 - h_4)$$
$$W_P = (1 - m_1 - m_2)\,(h_6 - h_5) + (1 - m_1)\,(h_8 - h_7) + (h_{10} - h_9)$$
$$Q_1 = 1\,(h_1 - h_{10});\quad Q_2 = 1\,(1 - m_1 - m_2)\,(h_4 - h_5)$$

$$\eta = \frac{Q_1 - Q_2}{Q_1} = \frac{W_T - W_P}{Q_1}$$

- It significantly increases the **cycle efficiency** and reduces the heat rate.
- It increases the steam flow rate, (requiring bigger boiler) for same turbine work output.
- If there is no change in boiler output, turbine output drops.
- It reduces the steam flow to the condenser (need smaller condenser)
- *Infinite* or *a series of steam extraction* will lead to Ideal regenerative cycle called **carnotization of Rankine cycle.**

10. Efficiencies of a Steam Power Plant

- Overall efficiency

$$\eta_{overall} = \frac{\text{Power available at the generator terminals}}{\text{Rate of energy release by the combustion of fuel}}$$

$$= \frac{MW_e \times 10^3}{W_f \times C.V.}, \ W_f = \text{Fuel burning rate.}$$

- The boiler efficiency

$$\eta_{boiler} = \frac{\text{Rate of energy absorption by water from steam}}{\text{Rate of energy release by the combustion of fuel}}$$

$$= \frac{W_s(h_1 - h_4)}{W_f \times C.V.},$$

W_s = steam generation rate.

- Mechanical efficiency

$$\eta_{turbine(mech)} = \frac{\text{Brake output of the turbine}}{\text{Internal output of the turbine}}$$

$$= \frac{\text{brake output}}{w_s(h_1 - h_2)}$$

- Generator Efficiency

$$\eta_{generator} = \frac{\text{Electrical output at generator terminals}}{\text{Brake output of the turbine}}$$

- $$\eta_{overall} = \eta_{boiler} \times \eta_{cycle} \times \eta_{turbine} \times \eta_{generator}$$

■ ■ ■ ■

Steam Nozzle and Turbine

4

1. Velocity of Pressure Pulse in a Fluid (C) For Iso-Entropic Process

$$C = \sqrt{\left(\frac{\partial P}{\partial \rho}\right)_s}$$

Here,   P = Pressure

      ρ = Density

      s = Entropy

2. Velocity of Sound in Ideal Gas (C)

$$C = \sqrt{\gamma R T}$$

Here,   R = Gas constant

      T = Temperature

      γ = Specific heat ratio

3. Mach Number (M)

$$M = \frac{\text{Actual velocity}}{\text{Sonic Velocity}}$$

M > 1, flow is supersonic

M < 1, flow is subsonic

M = 1, flow is sonic

4. Stagnation Properties

It is defined as the state a fluid in motion would reach if it is brought to rest *isentropically* in steady flow and zero work output device.

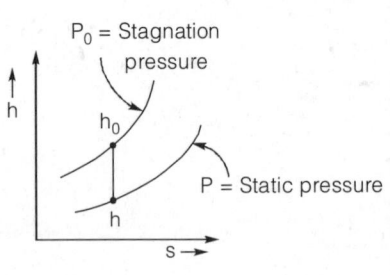

$$h_o = h + \frac{V^2}{2}$$

- Stagnation temperature ratio (T_0/T)

$$\frac{T_0}{T} = 1 + \frac{\gamma - 1}{2} \cdot M^2$$

Here,   T = Static temperature

γ = Ratio of specific heat

M = Mach number

- **Stagnation Pressure (P_0)**

$$\frac{P_0}{P} = \left(\frac{T_0}{T}\right)^{\frac{\gamma}{\gamma-1}}$$

Here,   P = Static pressure

γ = Ratio of specific heat

5. Nozzle/Diffuser Identification

$$\frac{dA}{A} = (M^2 - 1)\frac{dV}{V}$$

Here,   A = Area

V = Velocity

If M < 1, i.e. inlet velocity is subsonic, as flow area 'A' decreases, the pressure decreases and velocity increases and vice-versa.

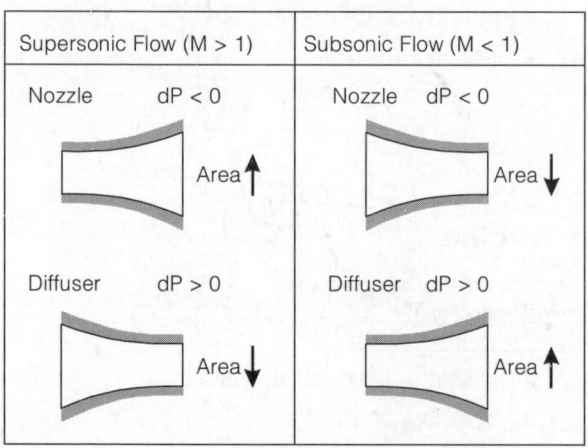

Supersonic Flow (M > 1)	Subsonic Flow (M < 1)
Nozzle dP < 0 Area ↑	Nozzle dP < 0 Area ↓
Diffuser dP > 0 Area ↓	Diffuser dP > 0 Area ↑

6. Critical Pressure Ratio and Choked Flow

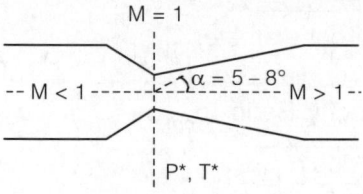

Critical pressure ratio (P*)

$$\frac{P^*}{P_0} = \left(\frac{2}{\gamma+1}\right)^{\frac{\gamma}{\gamma-1}}$$

Here,

P_0 = Inlet pressure

γ = Ratio of specific heat

- Critical temperature ratio (T*)

$$\frac{T^*}{T_0} = \frac{2}{\gamma+1}$$

Here, T_0 = Inlet temperature

- The mass flow rate is maximum for choked flow.

7. Flow in Steam Nozzle

$$V = 44.72\,(h_0 - h_1)^{1/2}$$

Here,

V = Velocity

h = Enthalpy of different state

- For dry saturation steam

$$V = 1.03\,(P_0\,V_0)^{1/2}$$

- For super-heated steam

$$V = 1.06\,(P_0\,V_0)^{1/2}$$

8. Impulse Turbine

- $$\text{Blade velocity } (V_b) = \frac{\pi D N}{60}$$

Here, D = Mean diameter of blade

N = Speed in RPM

- **The area of Flow**

$$A_b = \frac{\pi}{4}\,(D_2^2 - D_1^2)$$

Here, D_1 = Root diameter

D_2 = Tip diameter

- Optimum Velocity Ratio ($\rho_{opt.}$)

$$\rho_{opt.} = \frac{\cos\alpha}{2z} = \frac{u}{V}$$

Here, α = Nozzle angle

z = Number of stage

- **Maximum efficiency (η_{max})**

$$\eta_{max} = \frac{1+k_b}{2} \cdot \cos^2 \alpha$$

Here, k_b = Blade friction factor

α = Nozzle angle

- **Graph of pressure compounding**

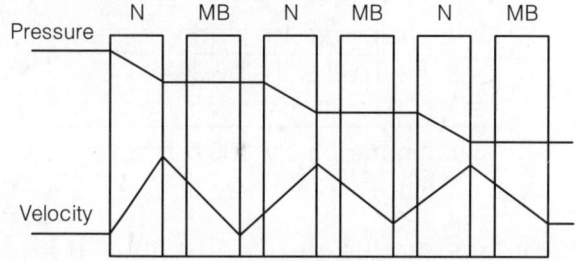

Where, N = Nozzle, MB = Moving blade

The pressure drop occur only in nozzle. There is no pressure drop while steam flows through the blades.

- **Graph of velocity compounding**

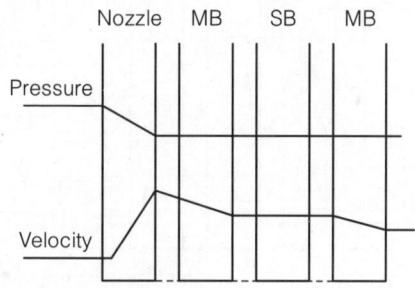

Where, SB = Stationary blade

All the pressure drop and enthalpy drop of steam takes place in a single row of nozzles.

- $(\Delta H)_{stage} = \dfrac{(\Delta H)_{total}}{\text{Total Number of stage}}$

- De-level turbine is a single stage impulse turbine.
- Blade are **symmetrical** if inlet blade angle will be equal to outlet blade angle.

Remember

9. Reaction Turbine

- Optimum velocity ratio (ρ)

$$\rho = \cos \alpha$$

α = nozzle angle

- Maximum efficiency (η_{max})

$$\eta_{max} = \frac{2\cos^2 \alpha}{1 + \cos^2 \alpha}$$

- Degree of reaction (R)

$$R = \frac{\text{Enthalpy drop in moving blade}}{\text{Total enthalpy drop}}$$

Remember

- Hero's Turbine is purely 100% reaction turbine because $\Delta h_{(FB)} = 0$ thus R = 1 for hero turbine.

- Parsons turbine $\Delta h_{(MB)} = \Delta h_{FB}$ thus $R = \frac{1}{2}$ called 50% reaction turbine.

- Graph of reaction turbine

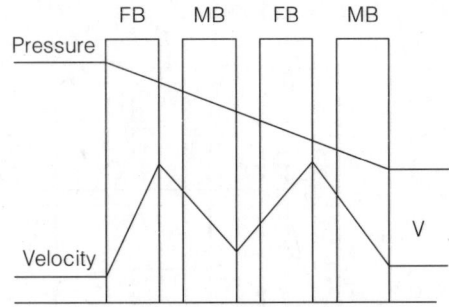

- $$\boxed{(\text{Power})_{n^{th}\,stage} = \frac{\text{Total power}}{(n_{th}^2)}}$$

Here,

n_{th} = Number of particular stage.

■■■■

5

Combined Cycle Plant

The gas turbine obtained its power by utilizing the energy of burnt gases and air which are at high temperature and pressure by expanding through the several rings of fixed and moving blades.

1. Simple Open Cycle Gas Turbine/Air Standard Brayton (Joule) Cycle

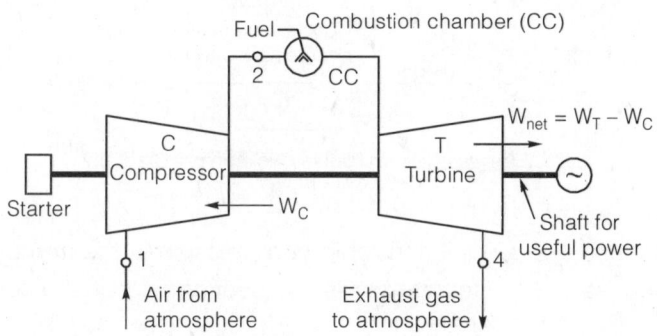

Simple Open Cycle Gas Turbine
Showing Blade Rows

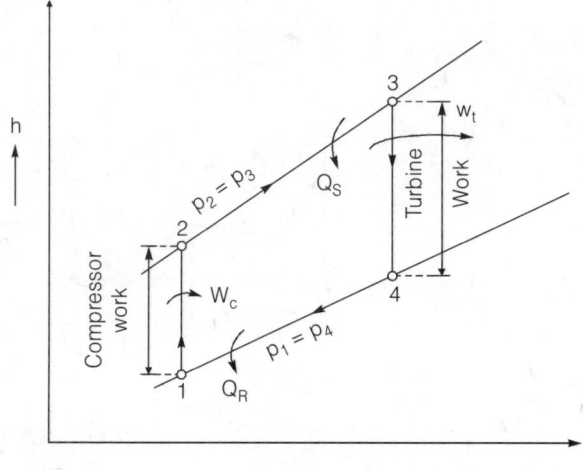

Representation of Joule cycle on h-s or (T-s) Diagram

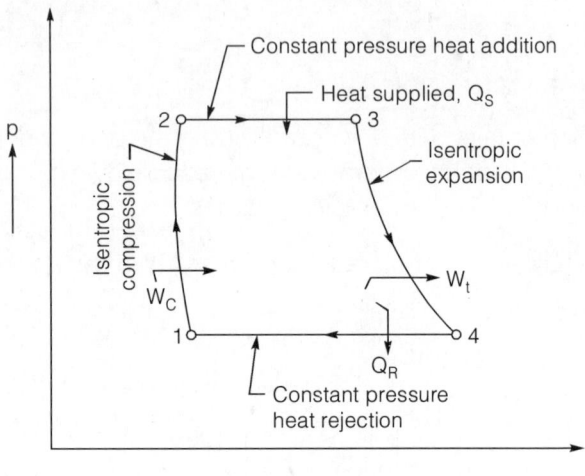

Representation of Joule Cycle on p-v Diagram

- Process 1-2 : The isentropic compression in the compressor.
- Process 2-3 : Constant pressure heat addition in the combustion chamber.
- Process 3-4 : Isentropic expansion in the turbine.
- Process 4-1 : Constant pressure heat rejection.

Therefore, the thermal efficiency

$$\eta_{th} = 1 - \frac{1}{r_p^{(\gamma-1)/\gamma}}$$

Here   Pressure ratio $(r_p) = \dfrac{P_2}{P_1}$

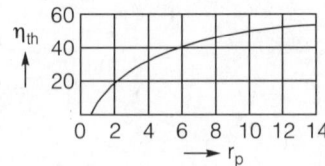

Remember: ....

- *Optimum* pressure ratio for maximum specific output in simple gas turbine cycle.

$$\left(r_p\right)_{optimum} = \left(\frac{P_2}{P_2}\right)_{optimum} = \left(\frac{T_{max}}{T_{min}}\right)^{\gamma/2(\gamma-1)}$$

For w_{max}. i.e. maximum specific output.

$$W_{net(max.)} = c_p\left[\sqrt{T_3} - \sqrt{T_1}\right]^2$$

$$\eta_{optimum} = 1 - \sqrt{\frac{T_{min.}}{T_{max.}}}$$

2. The Cycle Air Rate, Work Ratio and Specific Fuel Consumption

- **Air rate**

  Air rate, AR, is defined as the air flow required per kWh output.

  $$AR = \frac{3600}{w_{net}} \frac{kg\,of\,air}{kWh}$$

  The reciprocal of air rate is termed as specific power. Air rate is actually the criterion of the *size of the plant* i.e. the lower the air rate the smaller the plant.

- **Cycle work ratio**

  The work ratio, WR, is defined as the ratio of net work to the turbine work. Thus

  $$WR = \frac{w_{net}}{w_T}$$

  Here, $w_{net} = W_T - W_C$

  For gas turbine the work ratio *should be high*. The work ratio may be also used as a guide in the determination of the size of the gas turbine and as indicator of the sensitivity of the plant to the decreases of the component efficiency.

- **Specific fuel consumption**

  It is the amount of fuel required for producing unit kWh power.

  $$sfc = \frac{3600}{W_{net}} \frac{kg\,of\,fuel}{kWh}$$

3. Effect of Modification on Performance of Simple Gas Turbine Cycle

Optimum modification to cycle	Work output	Thermal Efficiency
Regeneration	No effect	Increases
Intercooling	Increases	Decreases
Reheat	Increases	Decreases
Reheat + Regeneration	Increases	Increases
Intercooling + Regeneration	Increases	Increases
Reheat + Intercooling	Increases	Decreases
Reheat + Intercooling + Regeneration	Increases	Increases

Remember

- Efficiency of Brayton cycle with regeneration :

$$\eta = 1 - \frac{T_1}{T_3}(r_p)^{\gamma-1/\gamma}$$

4. Gas Turbine-Steam Turbine Power Plant (Combined Cycle)

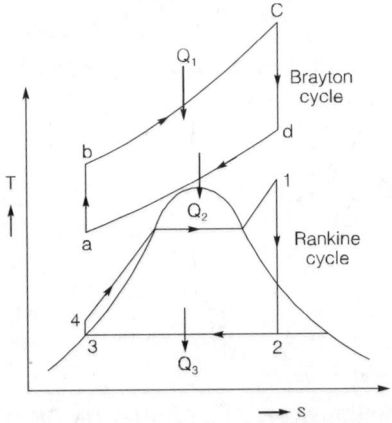

The air standard cycle for a gas turbine power plant is the Brayton cycle. It consist of two reversible adiabatic and two reversible isobars, but unlike Rankine cycle the working fluid **does not undergo** phase change.

A gas turbine may be used in conjunction with a steam turbine plant in an utility based load station, to offer the utilities the gas turbine advantages of quick starting and stopping and permit **flexible operation** of the combined plant over a wide range of loads.

- Overall efficiency of combined plant

$$\eta = \eta_1 + \eta_2 - \eta_1\eta_2$$

- Advantage of gas turbine
  - Less installation cost
  - Less installation time
  - Quick starting and stopping
  - Fast response to load changes.
- Disadvantage of gas turbine
  - Large compressor work input.
  - Large Exhaust loss; since exhaust gas temperature is quite high and also the mass flow rate of gas is large due to high air-fuel ratio used.
  - Machine inefficiencies (compressor and turbines)
  - Low cycle efficiency, due to large exhaust loss, large compressor work and machine inefficiencies.

Remember

- The *overall* efficiency of combined cycle power plant is nearly 50%.
- Nearly **66%** of power developed by turbine is consumed *by compressor*.
- **Cogeneration plant** is plant that uses both **electrical power** and **process heat** simultaneously.
- A plant where **power generation** is utmost uses turbine called **pass out turbine** or **Extraction turbine**.
- A plant where **process heat** is utmost uses turbine called **back pressure turbine**.

■■■■

Compressors

Function of compressor is to take a definite quantity of fluid (*gas or air*) and deliver it at a required pressure.

1. Type of Compressor

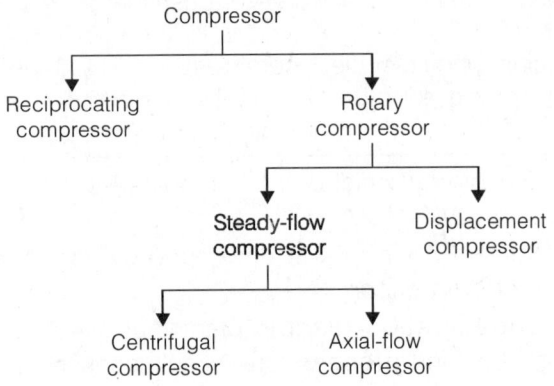

- **Steady-flow compressor**
  Those compressor in which compression occur by transfer of kinetic energy from a rotor.

Remember: ....

- *For low pressure ratios* (4 : 1) the centrifugal compressor is lighter and is able to operate effectively over a *wide range of mass flow* at any one speed.
- For *larger* unit and *high mass flow rate* the axial flow compressor is more efficient.
- Pressure rise *per stage* in axial flow compressor is *low* as compare to *per stage* in centrifugal compressor *but high* pressure ratio can be obtained in axial flow compressor by using *more number of stages*.
- In aircraft and gas turbine axial flow compressor is used. Since it has low frontal area.

2. Centrifugal Compressor

A centrifugal compressor consists of three components.

(i) A stationary casing

(ii) A rotating impeller

(iii) A diffuser

It consist of impeller with a series of curved radial vanes. The static pressure of the air increase from the eye to the tip of impeller in order to provide the centrifugal force on the air. As air leaves the impeller tip, it is passed through diffuser passage which convert most of kinetic energy into increase in enthalpy and hence the pressure of air is further increased.

- **Pressure, velocity variation of air**

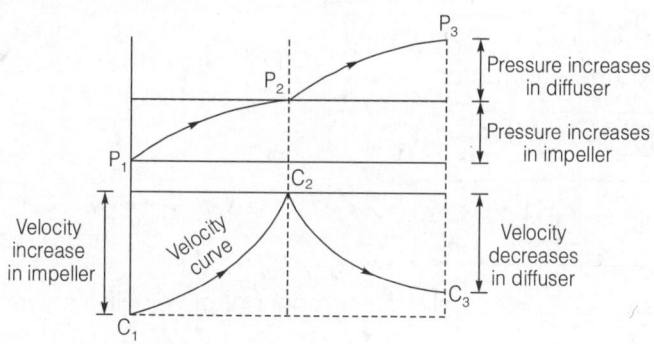

- **Work done (Euler's work)**

Work per k.g. of air entering and leaving the impeller in a unit time.

$$\boxed{W = C_{w2}\, C_{b/2}}$$

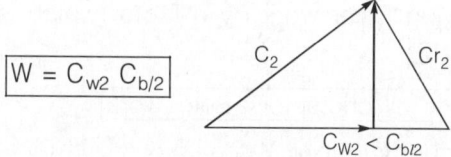

Here

$\qquad C_{w2}$ = Velocity of whirl at outlet

$\qquad C_{b/2}$ = Mean blade velocity at exit

For ideal condition, radial blade,

$\qquad C_{w2} = C_{b/2}$

$\therefore \qquad \boxed{W = C_2^2} \qquad$ (C_2 = Absolute velocity at outlet of rotor)

Since air can not leave at a velocity greater than impeller velocity.

- **Power required per impeller**

$$P = \dot{m} \, C_{w2} \, C_{bl2} \, \text{Watt} \qquad (\dot{m} = \text{Mass flow rate})$$

Remember: ....

- In practice, nearly *half* of the total pressure is achieved in *impeller* and remaining half in *diffuser*.
- For pressure ratio **4 : 1** *single stage* compressor is used and for pressure ratio **12 : 1** *multi stage* compressor is used.
- As compare to reciprocating compressor the velocity in centrifugal compressor are very large.
- **Isentropic efficiency**

$$\eta_{isentropic} = \frac{\text{Isentropic work}}{\text{Actual work}}$$

- **Slip and slip factor**

$$Slip = C_{bl_2} - C_{w_2}$$

$$Slip \ factor \, (\phi_s) = \frac{C_{w_2}}{C_{bl_2}}$$

Here, C_{w_2} = Actual whirl component of velocity

C_{bl_2} = Ideal whirl component of velocity

- **Work factor (ϕ_w)**

  The actual work per kg of air by compressor is *always greater than* Euler work due to friction losses. Then actual work is obtained by multiplying Euler work by a factor which is known as work factor.

$$\text{Actual work} = \text{Work factor} \times \text{Euler work}$$

- **Pressure coefficient (ϕ_p)**

$$\phi_p = \frac{\text{Isentropic work}}{\text{Euler work}}$$

$$\phi_p = \text{Work factor} \times \text{Slip factor} \times \eta_{isentropic}$$

Value of work factor generally lies in the range (1-1.03)

Remember: ....
Prewhirl is provided in centrifugal compressor to avoid shock wave formation because of high speed of turbine blade.
Thus mach number (M) < 1 is desired.
Generally a **prewhirl** of 6° is provided

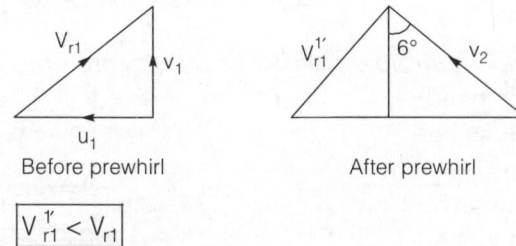

Before prewhirl   After prewhirl

$$\boxed{V_{r1}^{1'} < V_{r1}}$$

3. Axial Flow Compressor

It consists of row of *moving blade and fixed blade* arranged around the rotor and stator respectively. There is usually large number of stage to maintain a constant work input per stage. Generally **5-14 *stage*** have been used.

• **Work done per kg**

$$W = C_{bl}(C_{w_2} - C_{w_1}) \text{ N-m}$$

Here, C_{w_1} = Whirl component of velocity at inlet

C_{w_2} = Whirl component of velocity at outlet

C_{bl} = Mean velocity at entrance or exit

• C_{w_1}, at inlet is not 'zero' because air flows axially *not radially*.

• **Degree of reaction of axial compressor (R_c)**

$$R_c = \frac{\text{Pressure rise in rotor blade}}{\text{Pressure rise per stage}}$$

Remember: ....
• Arrangement is such that equal temperature rise in will occur in moving and fixed blade.
• Design of blade is based on *'Aero dynamic theory'*.
• Shape of blade — *Aerofoil shape*.
• With 50% reaction blading it has symmetrical, blade and with this type losses reduced significantly.

- The *radial blades* are used exclusively in *turbojet* and *space craft* applications, because centrifugal effect of the *curved blades* create bending moment, hence stress which restricts its maximum speed.

4. Compressor Characteristics (Choking and Surging)

- Choking

  In this *maximum flow rate* occur at *minimum pressure ratio*.

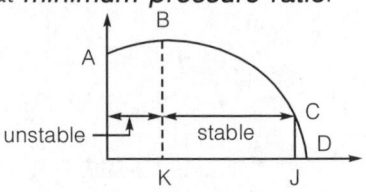

A = Shut off head

BK = Surging line

CJ = Chock line

AB = Unstable flow

BC = Stable flow

In AB, any small disturbance causing a check in mass flow cause a fall in pressure ratio and flow may reverse at some point.

After temporary disturbance, small disturbance result in flow to oscillate rapidly.

This oscillation is noisy and if continue, can damage compressor, this is called *'surge'*.

Point B, marks the limit of useful operation. Efficiency of compressor will be highest near point *adjacent* to 'B'.

- Surging occurs at point 'B' due to high pressure at down stream than pressure at compressor delivery.

5. Efficiency Curve of Axial and Centrifugal Compressor

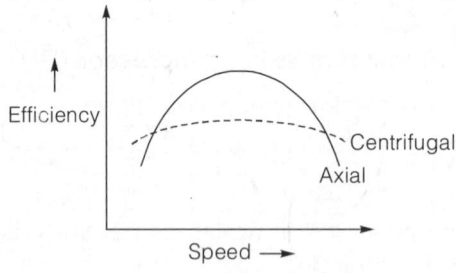

■■■■

A Handbook on
Mechanical Engineering

Refrigeration & Air-Conditioning

CONTENTS

Refrigeration

1

It is the process of producing lower temperature compare to surrounding.

1. Refrigeration Machine

A refrigerating machine is a device which will either cool or maintain a body at a temperature below that of the surroundings. Heat must be made to flow from a body at low temperature to the surroundings at high temperature.

2. Refrigeration Effect (R.E)

It is the amount of heat, which is to be extracted from storage space in order to maintain lower temperature.

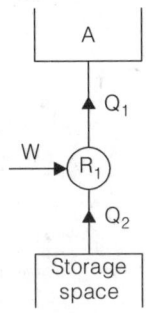

$$C.O.P. = \frac{Q_2}{Q_1 - Q_2} = \frac{R.E.}{Q_1 - Q_2} = \frac{R.E}{W_{in}}$$

Q_1 = Heat rejected to surrounding

W = Work done

Remember: ....

- There are two statements of the second law of thermodynamics, the Kelvin Planck statement, and the Clausius statement. The *Kelvin-Planck* statement related to *heat engines* and the *Clausius* statement pertains to *refrigerators* and *heat pumps*.
- Above two statements are for a *cycle not* for a *process*.

Remember

- COP reflect about running cost.
- Higher the COP the lower will be running cost.

3. Refrigeration Capacity (R.C)

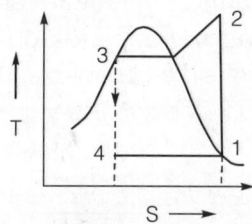

R.C. = Refrigeration effect × mass flow rate of refrigerant

$$R.C. = (h_1 - h_4) \times \dot{m}_{Ref.}$$

∴
$$C.O.P. = \frac{R.E.}{W} \times \frac{\dot{m}}{\dot{m}} = \frac{R.C.}{P_{input}}$$

P_{input} = Power input

4. Unit of Refrigeration

It is amount of heat, which is to be extracted from **1 tonne** of **water** at 0°C in order to convert it into ice at 0°c in **24 hours**.

$$1\,TR = 3.5\,kJ/sec = 210\,kJ/Min$$

5. Ideal Refrigeration Cycle/The Carnot Principle

Reverse Carnot cycle

$$C.O.P. = \frac{Q_2}{Q_1 - Q_2} = \frac{T_l}{T_h - T_l}$$

T_h = temperature of atmosphere
T_l = Temperature of storage space

Remember
- All reversible cycles have the same C.O.P. operating between *same* temperature range.
- C.O.P. *does not* depends upon *working substance* in case of reversible Carnot cycle.
- Carnot C.O.P. for *cooling* varies from 0 to ∞.
- Carnot C.O.P. for *heating* varies from 1 to ∞.
- $C.O.P_{heat\ pump} = 1 + C.O.P_{refrigerator}$

6. Volumetric Efficiency of Reciprocating Compressor ($\eta_{vol}.$)

$$\eta_{vol} = \frac{\text{Actual volume of refrigerant}}{\text{swept volume}} = \frac{\dot{m}v_1}{\frac{\pi}{4}\cdot D^2 \cdot L.N.K.}$$

Where, $\dot{m}$ = Actual mass flow of refrigeration (kg/min)

D = Diameter of piston

v_1 = Specific volume at inlet of compressor

N = Speed (rpm)

K = Number of cylinder

$$\eta_{vol} = 1 + C - C\left(\frac{P_2}{P_1}\right)^{\frac{1}{n}}$$

Where, C = Clearance ratio = $\dfrac{\text{clearance volume}}{\text{swept volume}}$

P_2 = Final pressure

P_1 = Initial pressure

η = Polytropic index of expansion

7. Vapour Compression Refrigeration Cycle

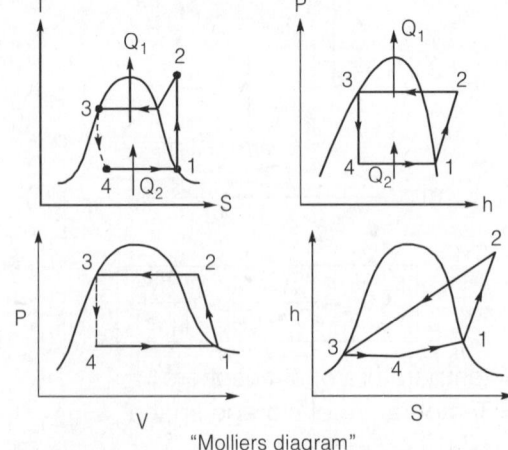

"Molliers diagram"

- Process 1-2 : isentropic compression
- Process 2-3 : constant pressure heat rejection
- Process 3-4 : iso-enthalpic expansion
- Process 4-1 : constant pressure heat absorption

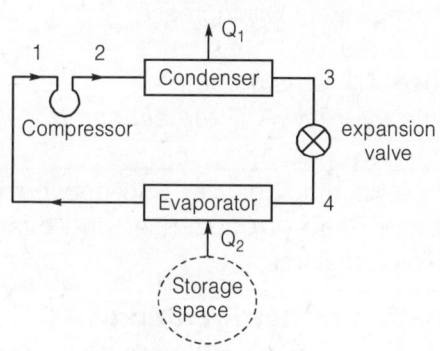

$$C.O.P = \frac{Q_2}{W} = \frac{h_1 - h_4}{h_2 - h_1} = \frac{h_1 - h_3}{h_2 - h_1} \qquad [\because h_4 = h_3]$$

8. Effect of Various Properties on Performance of Vapour Compression Cycle

- Effect of decrease in evaporator pressure
  - Refrigerating effect will decrease
  - COP will decrease
  - Volumetric efficiency will decrease
  - Work done will increase

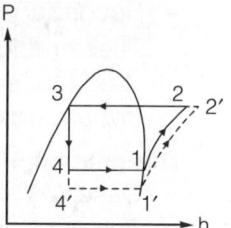

- Effect of increase in condenser pressure
  - Refrigerating effect will decrease
  - COP will decrease
  - Volumetric efficiency will decrease
  - Work done will increase

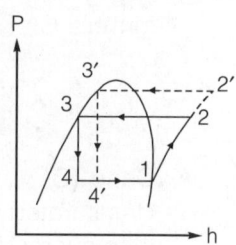

- Effect of Super Heating
  - Refrigerating effect will increase
  - Work done will be directly proportional to inlet temperature.
  - COP will depend on refrigerating substance. For *R-12* it will *increase*

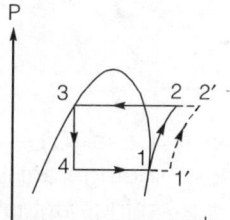

and for NH_3 it will *decrease.*

- Volumetric efficiency will remain almost constant.

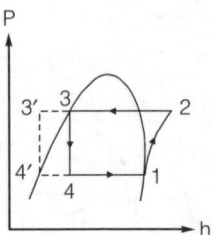

- Effect of Sub-Cooling
  - Refrigerating effect will increase
  - COP will increase
  - Work done will *remain same*
  - Volumetric efficiency will *remain same*

Remember

- Volumetric efficiency in Sub-cooling remain constant because Ratio of condenser and evaporator pressure remains constant.

---

9. Type/Designation of Refrigeration

Type of Refrigerants

- **Primary refrigerant**

  Primary refrigerant are those refrigerant which absorb heat directly from storage space by undergoing in a cycle. It absorbs *latent heat.* e.g. NH_3, R-11, R-12 and R-22.

- **Secondary refrigerant**

  These refrigerant is first cooled by primary refrigerant and then used for cooling at the required place. Secondary refrigerant *will not undergo* in a cycle and it absorb *sensible heat.*

Designation of Refrigerants

- **Case (A)**

  Refrigerant is saturated hydrocarbon (R-11, R-12, R-134) chemical formula-$C_m H_n F_p Cl_q$, then such refrigerant is designed as

  $$R-(m-1)(n+1)P$$

  $$\boxed{n + P + q = 2m + 2}$$

- **Case (B)**

  Unsaturated hydrocarbon (R-1150) chemical formula-$C_m H_n F_p Cl_q$, then such refrigerant is designed as

  $$R-1(m-1)(n+1)P$$

  $$\boxed{n + P + q = 2m}$$

- **Case (C)**

  Refrigerant is inorganic compound (NH_3, H_2O, Air) then such refrigerant is designed as

$$R-(700 + \text{Molecular weight of Refrigerant})$$

10. Desirable Properties of Refrigerant

- Thermodynamics Properties
  - Critical temperature should be high.
  - Specific heat of refrigerant in *liquid* phase must be *low* and in *vapour* phase must be *high*.
  - Enthalpy of vaporization should be high.
  - Thermal heat conducitivity should be high.
  - Freezing point should be low.
  - Specific volume of refrigerant at the inlet of compressor should be low.
  - Temperature of refrigerant at the outlet of compressor should be low.
- Chemical Properties
  - It should not be toxic and non flammable.
  - Refrigerant should not be misicible with oil or *completely* immisicible with oil.
  - It should not attack material of construction of equipment.
- Physical Properties
  - Its viscosity should be low.
  - There is no leakage tendency.

11. Cascade Refrigeration:

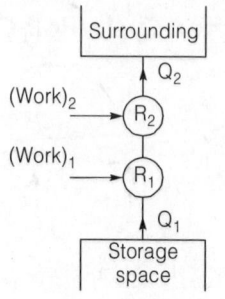

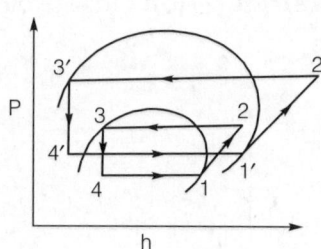

$$(COP)_{overall} = \frac{(COP)_1\,(COP)_2}{1+(COP)_1+(COP)_2}$$

- It is used to obtain very low temperature.
- Cascade refrigeration save considerable compression work.

Remember

- Multistage compression also uses for obtaining very low temperature and saves considerable compression work.

12. Vapour-Absorbation Cycle (V_A)

It is heat operating device that is low grade energy. It is used where *waste heat available*. In this liquid is pumped thus, work is small.

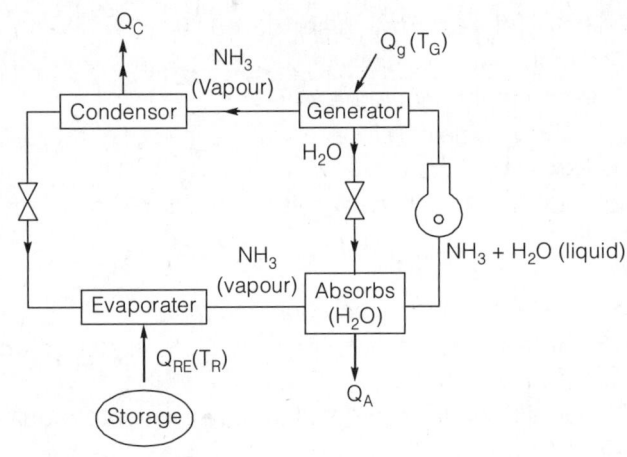

$$COP_{max} = \left(\frac{T_G - T_O}{T_G}\right)\left(\frac{T_R}{T_O - T_R}\right)$$

T_o = Atmospheric temperature

13. Gas Refrigeration Cycle/Reverse Brayton Cycle/Bell-Colemn Cycle

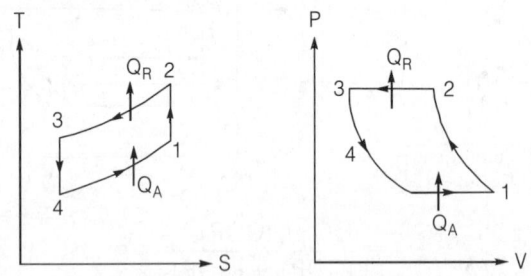

- Process 1-2 : isentropic compression
- Process 2-3 : constant pressure heat rejection
- Process 3-4 : isentropic expansion
- Process 4-1 : constant pressure heat absorbtion

$$C.O.P. = \frac{1}{\left(\dfrac{T_2}{T_1}\right) - 1} = \frac{1}{(r_p)^{\frac{\gamma-1}{\gamma}} - 1}$$

$$r_p = \text{Pressure ratio} = \frac{P_2}{P_1}$$

$$\gamma = \frac{C_p}{C_v}, \qquad \frac{T_2}{T_1} = \frac{T_3}{T_4}$$

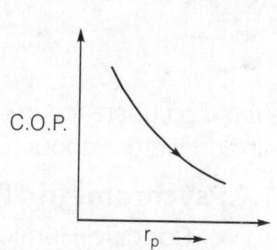

C.O.P.

$r_p \longrightarrow$

Remember

- Its COP is *low* but used in *aircraft* because of low weight of refrigerator *per tonne* of refrigeration.

14. Application of Different Refrigerants

Refrigerant	Application
NH_3	Cold storage, ice plants, refrigerators
H_2O	Water, Li Br absorbtion system
CO_2	Used as dry ice in transport
R-11	Central air conditioning
R-12	Domestic refrigerator
R-22	Window type air conditioner
R-113	Air conditioning

■■■■

Working substance in air conditioning is *moist air* which comprise of dry air and water vapour. *Degree of freedom* of moist air is *3*.

1. Psychrometric Properties/Properties of Moist Air

- Specific Humidity or Humidity Ratio (ω)

$$\omega = \frac{\text{mass of vapour}}{\text{mass of air.}} = \frac{m_v}{m_a}$$

$$\omega = 0.622 \frac{P_v}{P_a} = 0.622 \frac{P_v}{P_t - P_v}$$

$$P_t = P_v + P_a$$

Here,

m_a = mass of dry air

P_v = Partial pressure of vapour

P_a = Partial pressure of air

P_t = Total pressure (Atmospheric)

- Relative Humidity (ϕ)

$$\boxed{\phi = \frac{m_v}{m_{vs}} = \frac{P_v}{P_{vs}}}$$

Given at same volume (V) and temperature (T)

Here, $V_{vapour} = V_{vapor\ saturated}$

$T_{vapour} = T_{vapor\ saturated}$

m_v = mass of vapour

m_{vs} = mass of vapour in saturated condition at *same volume* and *temperature*

P_v = Partial pressure of vapour

P_{vs} = Partial pressure of vapour in saturated condition

- Wet Bulb Depression

= Dry bulb temperature – Wet Bulb Temperature

- Degree of Saturation (μ)

$$\mu = \frac{\text{Actual specific humidity } (\omega)}{\text{Saturated specific humidity } (\omega_s)}$$

$$\Rightarrow \qquad \mu = \frac{\omega}{\omega_s} = \phi\left(\frac{P_t - P_{vs}}{P_t - P_v}\right)$$

- **Enthalpy of Moist Air (H)**

$$H = H_a + H_v$$
$$H = m_a\,h_a + m_v\,h_v$$

$$h = c_{p_a}t + \omega[2500 + 1.88t]\ \text{kJ/kg dry air}$$

H_a = Enthlapy of air in kJ

H_v = Enthalpy of vapour in kJ

Here,　m_a = mass of air in kg

m_v = mass of vapour in kg

t　 = Dry bulb temperature in °C

C_{p_a} = Specific heat of air at constant pressure

Remember

- Relative humidity of **saturated air** will be **100%**.
- Wet bulb temperature (wbt) is always less than dry bulb temperature (dbt) **except** when the **air is saturated**.
- Wet bulb temperature is an indirect measure of the dryness of the moist air.
- For saturated air, DPT = WBT = DBT
- WBT is not thermodynamic property as its temperature measurement involves heat and mass transfer.
- WBT is temperature of air measured by thermometer whose bulb is rapped with wet cloth over the bulb.

2. Psychromatic Chart

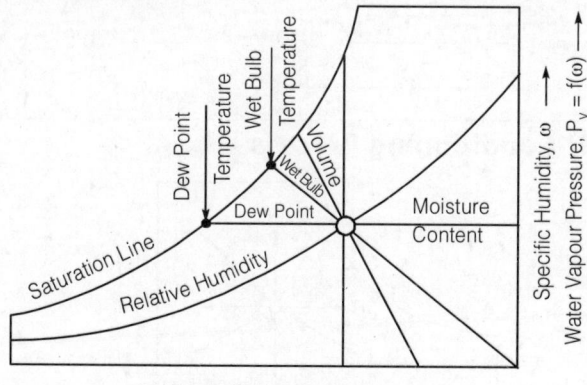

Dry-bulb Temperature, t ⟶

- Constant enthalpy lines are along the constant wet bulb temperature.

3. Psychrometry of Air-Conditioning Process

- **Mixing Process**
  - Specific humidity of the mixture (ω)

  $$\omega = \frac{m_{a_1}\,\omega_1 + m_{a_2}\,\omega_2}{m_{a_1} + m_{a_2}}$$

  - Enthalpy of the mixture (h)

  $$h = \frac{m_{a_1}\,h_1 + m_{a_2}\,h_2}{m_{a_1} + m_{a_2}}$$

  - Temperature of the mixture (t)

  $$t = \frac{m_{a_1}\,t_1 + m_{a_2}\,t_2}{m_{a_1} + m_{a_2}}$$

  where for the two moist air stream

  m_a = Mass of the dry air

  ω = Specific humidity

  h = Specific enthalpy

  t = Temperature (in °C)

Remember

- Dew point temperature is the temperature of air starts condensing at partial pressure (P_v) i.e. saturation temperature corresponding to partial pressure (P_v).

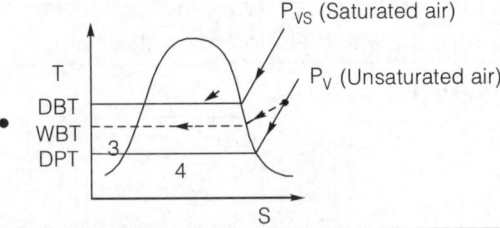

4. Basic Air-Conditioning Process

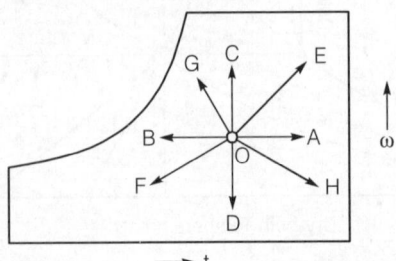

Process in diagram	Type of Air - Conditioning
OA	Sensible Heating
OB	Sensible Cooling
OC	Humidification
OD	Dehumidification
OE	Heating and Humidification
OF	Cooling and Dehumidification
OG	Cooling and Humidification
OH	Heating and Dehumidification

5. Sensible Heat factor (S.H.F.)

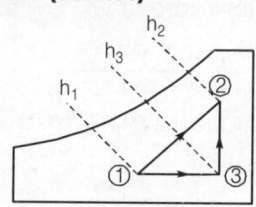

$$SHF = \frac{\text{Sensible heat}}{\text{Latent heat}} = \frac{h_3 - h_1}{h_2 - h_1}$$

h_1 = enthalpy at point 1

h_2 = enthalpy at point 2

h_3 = enthalpy at point 3

Remember
- SHF will be 1 for sensible heating or sensible cooling.
- SHF will be 0 for humidification or dehumidification.

6. By-Pass factor (BPF)

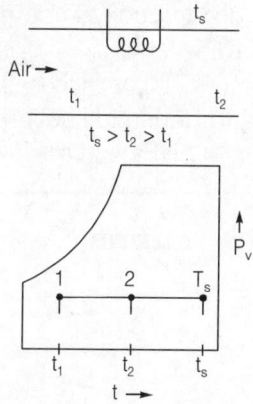

$$BPF = \frac{Actual\,loss}{Ideal\,gain}$$

$$BPF = \frac{t_s - t_2}{t_s - t_1}$$

Here   't' = temperature

- **Efficiency of coil**

$$\eta_{coil} = \frac{t_2 - t_1}{t_s - t_1} = \frac{Actual\,gain}{Ideal\,gain}$$

$$\eta + BPF = 1$$

- By pass factor represents how much effective the coil is.
- Contact factor = 1 – BPF

7. By Pass Factor (x) of Coil with More Than 1 Row (n coil)

$$x_{equivalent} = x^n$$

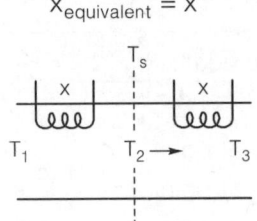

n = number of coil

- Above equation is applicable only if all coils are similar.

- For sensible heating coil temperature (T_c) is greater than DBT of air.
- For sensible cooling coil temperature lesser than DBT and greater than DPT.

$$DPT < T_{coil} < DBT$$

- For cooling and dehumidification (summer air conditioner)

$$T_{coil} < DPT < DBT$$

■■■■

A Handbook on
Mechanical Engineering

6

I.C. Engine

CONTENTS

◀ ☐ ▶

Cycle

1. Assumption of Air-Standard Cycle

- Air is a working fluid.
- Air behave as an *ideal* gas.
- Mass of air remains constant i.e. *closed* system analysis.
- Specific heat of air remains constant.
- The working fluid *does not* undergoes any chemical change.
- All process are *reversible*.

2. Different Parameter Associated With Piston Cylinder Arrangement

- Piston-Displacement Volume/Swept Volume

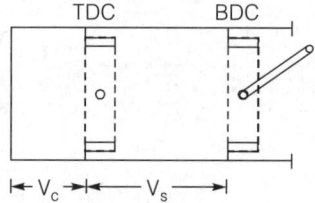

TDC = Top dead centre
BDC = Bottom dead centre
 D = Nominal Inner diameter of cylinder called piston diameter (In general)
 L = Length travelled by piston between TDC & BDC

$$V_s = \frac{\pi}{4}D^2 \times L$$

- Clearance volume (V_c)
  It is the volume, when piston is at top dead centre
  $$V_{total} = V_c + V_s$$
- Compression ratio (r)

  $$r = \frac{\text{Volume before compression}}{\text{Volume after compression}}$$

  $$r = \frac{V_{total}}{V_c} = \frac{V_s + V_c}{V_c} = 1 + \frac{V_s}{V_c}$$

3. Otto Cycle

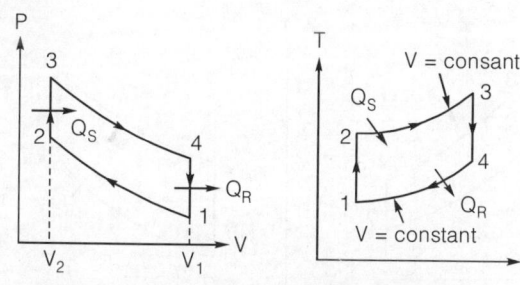

- Process 1-2 : Reversible adiabatic compression
- Process 2-3 : Constant volume heat addition
- Process 3-4 : Reversible adiabatic expansion
- Process 4-1 : Constant volume heat rejection
- **Efficiency**

$$\eta = 1 - \frac{T_1}{T_2} = 1 - \frac{1}{(r)^{\gamma-1}}$$

Here, $r = \dfrac{V_1}{V_2}$

- As the compression ratio r increases, efficiency increases.
- Limit of compression ratio is subject to **knocking** phenomenon in SI engine.

Remember

- An isentropic expansion, could do more work in **Otto Cycle** if allowed to continue to **lowest cycle pressure** then otto cycle becomes **Atkinson Cycle** (with same temperature limit).

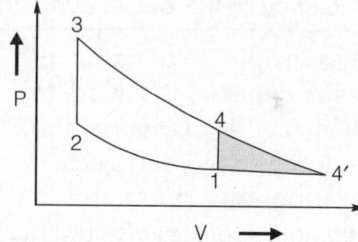

Otto cycle 1 – 2 – 3 – 4

Atkinson cycle 1 – 2 – 3 – 4

- Shaded area shown is 1 – 4 – 4' is additional work obtained in Atkinson cycle.
- For same temperature limit, Atkinson cycle will have less heat rejection than otto cycle as the heat rejection in Atkinson is at constant pressure.

4. Diesel Cycle

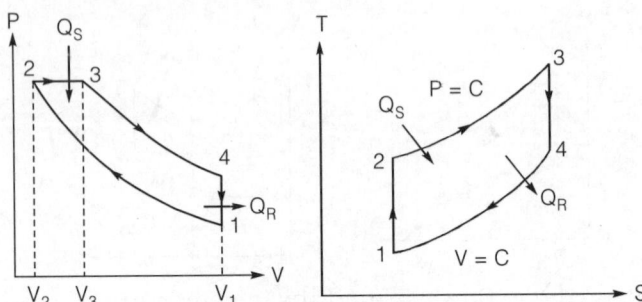

- Process 1-2 : Reversible adiabatic compression
- Process 2-3 : Constant *pressure* heat addition
- Process 3-4 : Reversible adiabatic expansion
- Process 4-1 : Constant volume heat rejection

- Cut-off ratio (r_c) = $\dfrac{V_3}{V_2}$

- Expansion ratio (r_e) = $\dfrac{V_4}{V_3}$

- Compression ratio (r) = $r_c \times r_e$

- Efficiency (η) = $1 - \dfrac{1}{\gamma \cdot (r)^{\gamma-1}} \cdot \dfrac{r_c^\gamma - 1}{r_c - 1}$

- As the compression ratio *increases* or cut-off ratio *decreases*, thermal efficiency of the diesel cycle *increases*.

Remember

- If Isentropic expansion of diesel cycle is allowed to lowest cycle pressure and then followed by heat rejection at constant pressure it becomes *Joule cycle*. (With same temperature limit)

  Cycle 1 – 2 – 3 – 4   **Diesel Cycle**

  Cycle 1 – 2 – 3 – 4′   **Joule Cycle**

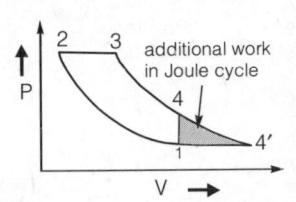

- For same temperature limit Joule cycle will have less heat rejection than diesel because heat rejection in Joule cycle is at constant pressure.

5. Dual Cycle

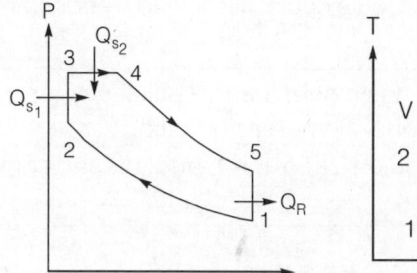 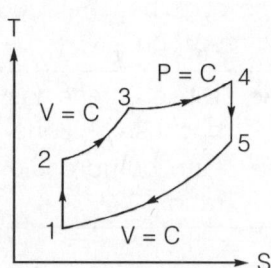

- Process 1-2 : Reversible adiabatic compression
- Process 2-3 : Constant volume heat addition
- Process 3-4 : Constant pressure heat addition
- Process 4-5 : Reversible adiabatic expansion
- Process 5-1 : Constant volume heat rejection

- Compression ratio $(r) = \dfrac{V_1}{V_2}$

- Pressure ratio $(\alpha) = \dfrac{P_3}{P_2}$

- Thermal efficiency $\eta_{th} = 1 - \dfrac{1}{r^{\gamma-1}}\left[\dfrac{\alpha r_c{}^{\gamma} - 1}{(\alpha - 1) + \alpha\gamma(r_c - 1)}\right]$

Remember: ....
- If pressure ratio = 1, dual cycle becomes Otto cycle.
- If $r_c = 1$, dual cycle become diesel cycle.

6. Comparison of Cycle

- For same *compression ratio* and *heat addition*

  $\eta_{otto} > \eta_{dual} > \eta_{diesel}$
- For same *compression ratio* and *heat rejection*

  $\eta_{otto} > \eta_{dual} > \eta_{diesel}$
- For same *maximum temperature* and *heat rejection*

  $\eta_{diesel} > \eta_{dual} > \eta_{otto}$
- Same *maximum pressure* and *heat input*

  $\eta_{diesel} > \eta_{dual} > \eta_{otto}$
- For same *maximum pressure* and *output*

  $\eta_{diesel} > \eta_{dual} > \eta_{otto}$

- For same temperature limit if isentropic expansion is allowed to lowest cycle pressure. Otto cycle becomes Atkinson cycle and Diesel cycle becomes Joule cycle, if followed by constant pressure heat resection.

Remember

- Diesel engine has higher thermal efficiency than otto cycle because of higher compression ratio.
- Otto compression ratio 6-11 and diesel compression ratio 13-22.

■■■■

Combustion of Fuel/Knocking

1. Equivalence Ratio (ϕ)

$$\phi = \left(\frac{F}{A}\right)_{Actual} \bigg/ \left(\frac{F}{A}\right)_{stoichiometric}$$

$\phi > 1$-Rich mixture

$\phi = 1$-Stoichiometric mixture

$\phi < 1$-Lean mixture

Remember: ....

- S.I. engine A/F ratio 10-17
- C.I. engine A/F ratio 18-100

S.I. engine	C.I. engine
A/F ratio not optimum in multi cylinder engines	Excellent distribution of fuel in multicylinder engiens thus better efficiency and balance in C.I. engine
Specific fuel consumption (s.f.c) Full load - low Part load & Idling - worse	Specific fuel consumption (s.f.c) Full load - Better Part load - Mucht better Thus C.I. engine wide application

2. Minimum Air Required for Combustion of Fuel (1 kg)

$$\left(\frac{A}{F}\right)_{stoichiometric} = \left[8(H_2) + \frac{8}{3}(c) + S - O_2\right] \times \frac{100}{23}$$

Here H_2 — Hydrogen percentage

C — Carbon percentage

S — Sulphur percentage

O_2 — Oxygen percentage

Remember: ....

- Oxygen (O_2) content in air is 23% *by mass*.

3. Factor Tending to Reduce Knocking in SI Engine

- Self ignition temperature of fuel   — High
- Compression ratio   — Low

- Inlet temperature and pressure   — Low
- Super charging   — No
- Spark advance   — No
- Flame travel distance   — Small
- Engine size   — Small
- Turbulence   — High
- Engine speed   — High
- Octane rating   — High

4. Factor Tending to Reduce Knocking in CI Engine

- Compression ratio   — High
- Self ignition temperature   — Low
- Delay period   — Low
- Inlet temperature and pressure   — High
- Combustion chamber temperature — High
- Rotation per minute (RPM)   — Low
- Engine size   — Small
- Cetane rating   — High
- Super charging   — Yes
- Injection angle advance   — No

5. Carburetion

Carburetion is the process of formation of a combustible fuel/air mixture by mixing the proper amount of fuel and air before admission into engine cylinder, and the device which performs this function is known as a carburettor.

- **Simple Carburetor**

  A simple carburetor consists of a choke tube also known as venturi, a float which acts as reservoir for fuel, a throttle valve, and a fuel discharge nozzle.

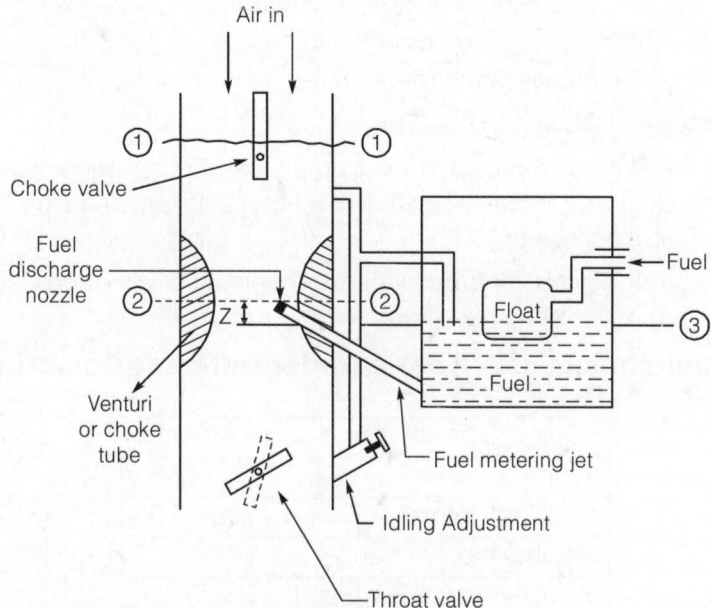

- **A/F ratio provided by a simple carburettor**
  - (i) Considering air as *compressible*

$$\text{A/F ratio} = \frac{(\dot{m}_a)_{actual}}{(\dot{m}_f)_{actual}}$$

$$= \frac{(C_{da})\dfrac{P_1 A_2}{R\sqrt{T_1}}\sqrt{2C_p\left[\left(\dfrac{P_2}{P_1}\right)^{2/\gamma} - \left(\dfrac{P_2}{P_1}\right)^{\frac{\gamma+1}{\gamma}}\right]}}{C_{df}\,A_f\sqrt{2\rho_f(P_1 - P_2 - \rho_f gz)}}$$

  - (ii) Considering air as *incompressible*

$$(\text{A/F})\text{ratio} = \frac{C_{da}\,A_2\sqrt{\rho_a(P_1 - P_2)}}{C_{d_f}\,A_f\sqrt{\rho_f(P_1 - P_2 - \rho_f\,gz)}}$$

Because, at higher air flow rates $(P_1 - P_2) >> gz\,\rho_f$

$$\therefore \quad (\text{A/F})_{ratio} = \frac{C_{d_a}\,A_2}{C_{d_f}\,A_f}\sqrt{\frac{\rho_a}{\rho_f}}$$

A_f = Area of cross-section of nozzle

A_2 = Area of air inlet at throat

Other symbol has usual meaning.

'Z' generally in the range of 4-6 mm.

Remember: ....

- Air fuel mixture provided by a simple carburettor becomes riches at higher air flow rate i.e. when $P_1 - P_2 = \Delta P$ is *large* and the density of air *decreases*.

- Also *at higher altitudes* when the density of air *decreases* the A/F mixture becomes progressively *richer*.

6. Requirements of A/F Ratio Under Different Operations

Operations	A/F ratio
Idling	12-12.5
Cruising/Normal	16-16.5
Maximum power range	12-13
Transient operation/Starting	3-5

7. Injection

The purpose of carburetion and fuel injection is the same i.e. preparation of the combustible charge. But in case of carburetion fuel is atomized by *air speed greater than the fuel speed* at the fuel nozzle. Whereas in fuel injection, *fuel speed at the point of delivery is greater than the air speed* to atomize the fuel.

- Basic fuel injection equations
  - Volume of fuel injected per cycle
    $$= C_d \times A \times V_f \times t$$

Here,

C_d = Coefficient of discharge

A = Area of orifice

v = Velocity of fuel

t = Time for injection

  - Velocity of injection (V_f) of fuel

$$V_f = \sqrt{\frac{2(P_{inj} - P_{cylinder})}{\rho_f}}$$

Here

P_{inj} = Injection pressure

$P_{cylinder}$ = Cylinder pressure

ρ_f = Density of fuel

- Time for one injection

$$\therefore \quad \boxed{\text{Time} = \frac{(\theta)_{rad}}{\omega} = \frac{\pi\theta}{180} \times \frac{60}{2\pi N}}$$

$$\boxed{T = \frac{\theta}{6N} \, \sec onds}$$

Here

N = Number of rotation Per minute

θ = Angle of crank travel over which injection takes place (radian)

8. Super Charging

- **Supercharging of SI engines**

This is *not done commonly* but it is used *only for racing car* engines and for *aircraft engines*. This is because increase in supercharging pressure increase the tendency to detonate and preignite. Hence SI engines using supercharging make use of lower compression ratio (r_k). But a lower compression ratio leads to a lower thermal efficiency which increase the fuel consumption. Therefore the supercharged SI engines have a higher fuel consumption.

- **Supercharging of CI engines**

Supercharging is normally preferred for CI engines because

(i) It improves combustion. This is because increased pressure and temperature reduces ignition delay and therefore decreases the knocking tendency.

(ii) The improved combustion allows a relatively poor quality of fuel to be used in CI engines.

(iii) Supercharging leads to reduction in smoke.

Remember

- Supercharging in S.I. is limited due to detonation while supercharging in C.I. is limited due to mechanical loading and thermal loading.

9. Effects of Supercharging on Engine Performance

- Power output increase because of increased amount of air inducted per cycle, recovery of some work as the gas exchange work and due to better scavenging.

- Only a small increase in mechanical efficiency is seen. This is because higher supercharging pressure increases the bearing load, friction. But the increase in brake mean effective pressure is higher than the effects of increase in friction load. Hence mechanical efficiency improves but only by a small amount.

Remember

- For supercharged *SI engine*, fuel consumption *increases* while *CI engines* it *decreases*.

- Betters and smooth combustion in supercharged C.I. reduces fuel consumption.

■■■■

Engine Performance Parameter

1. Indicated Power (IP)

$$IP = \frac{P_m \, LAN \, K}{60} \qquad \text{[For 2-S engine]}$$

$$IP = \frac{P_m LANK}{120} \qquad \text{[For 4-S engine]}$$

2. Brake Power (BP)

$$BP = \frac{2\pi \, NT}{60}$$

N = Number of rotation (RPM)
T = Torque

3. Frictional Power (FP)

F.P. = Indicated power – Brake power

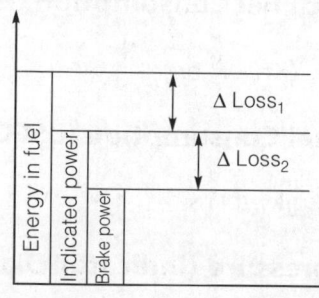

- Δ Loss$_1$ = Energy lost in Exhaust, coolant, radiation.
- Δ loss$_2$ = Energy lost due to friction pumping.

4. Mechanical Efficiency (η_m)

$$\eta_m = \frac{B.P.}{I.P.}$$

5. Volumetric Efficiency

$$\eta_v = \frac{\dot{m}_a}{\rho_a \, V_d}$$

Here,

$\dot{m}_a$ = Air entering (kg/sec)

V_d = Displaced volume

ρ_a = Density of air

6. Indicated thermal efficiency (η_{ith})

$$\eta_{ith} = \frac{I.P.}{\dot{m}_f \times C.V.}$$

m_f = mass flow rate (kg/sec)

C.V. = calorific value (J/kg)

7. Brake Thermal Efficiency (η_{bth})

$$\eta_{b.th.} = \frac{B.P.}{\dot{m}_f \times C.V.}$$

8. Mechanical Efficiency (η_M)

$$\eta_M = \frac{\eta_{bth}}{\eta_{ith}}$$

9. Indicated Specific Fuel Consumption (ISFC)

$$Isfc = \frac{\dot{m}_f}{I.P.} \ (Kg/kW\text{-}hr)$$

10. Brake Specific Fuel Consumption (BSFC)

$$bsfc = \frac{\dot{m}_f}{B.P.} \ (kg/kw\text{-}hr)$$

11. Mean-effective pressure (Indicator Diagram)

$$P_m = \frac{Ka_d}{l_d}$$

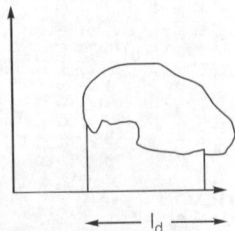

l_d

K = Spring constant

a_d = Area of indicator diagram

l_a = Length of indicator diagram

- Mean effective pressure is a measure of the effectiveness with which the displaced volume of the engine is used to produce work high value of m.e.p. is desirable.

12. Relative Efficiency

$$\eta_{relative} = \frac{\text{Actual thermal efficiency}}{\text{Air s tan dard efficiency}}$$

Fuel Quality

4

1. Constituents of Fuel

- Paraffins — Straight chain saturated compound
- Olefins — Straight chain unsaturated compound
- Naphthalene — Cyclic saturated compound
- Aromatic — Cyclic unsaturated compound.

2. Octane Number

It is percentage of iso-octane in the fuel containing iso-octane and n-Heptane that gives the same knocking intensity as that of the fuel whose octane number is calculated. Octane number 80 means the fuel is equivalent to 100% mixture of iso-octane and n-Heptane where iso-octane is 80% and n-Heptane is 20%.

Remember: ....

- Octane number of iso-octane-**100** and octane number of n-heptane-**zero**.
- Octane number of a fuel can be increased by adding *Tetra ethyl lead* (TEL).
- TEL causes spark plug fouling so **ethylene dibromide** is used to avoid lead deposits for spark plug fouling.

3. Cetane Number

It is the percentage by volume of cetane ($C_{16} H_{34}$) in a mixture of cetane and α-methyl naphthalene that has the same performance in the standard test engine as that of the fuel whose cetane number is calculated. So if a fuel is equivalent to 100% mixture of cetane and α-methyl naphthalene where cetane is 85% and α-methyl naphthalene is 15% then cetane number of the fuel is 85.

- Cetane numbers of cetane is 100 and cetane numbers of a-methyl naphthalene is zero.
- CN = 15 is for Isocetance or Heptamethyl Nonane (HMN), which is stable in nature.

Remember

4. Relation between octane number (O.N.) and Cetane number (C.N.)

$$C.N. = 60 - \frac{O.N.}{2}$$

5. Fuel Sensitivity/Anti Knock Index

Fuel sensitivity = Research octane number (RON)
                   – Motor octane number (MON)

$$\text{Anti knock index} = \frac{R.O.N. + M.O.N.}{2}$$

6. Important Pollution Level Graph Related to Patrol and Diesel Engine

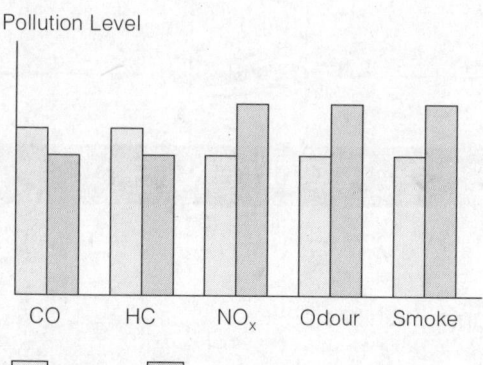

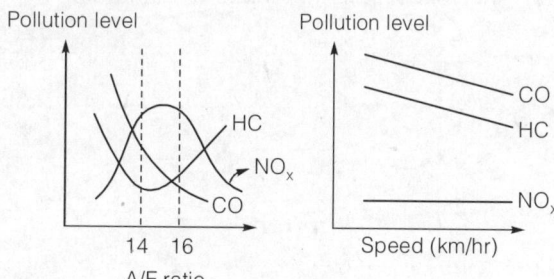

A Handbook on
Mechanical Engineering

7

Theory of Machines

CONTENTS

1. Mechanism

If a number of bodies assembled in such a way that the motion of one causes *constrained* and *predictable* motion to others, it is known as 'Mechanism'.

It *transmit* or *modify* a *motion* e.g. slider-crank mechanism, type writer, spring toys.

Machine

A machine is a mechanism or a combination of mechanism which, apart from imparting motion to the part, *also* transmits and modifies the available *mechanical energy* into some kind of desired work.

2. Type of Constrained Motion

- Completely constrained motion

  When motion between two elements of a pair is in a definite (*single*) direction irrespective of the direction of force applied. It is known as completely constrained motion.

- Successfully constrained motion

  When motion between two element of a pair is *possible* in *more than one* direction but is made to have motion *only in one* direction by using some *external* means, it is called successfully constrained motion e.g.

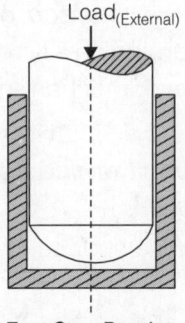

Foot Step Bearing

a piston in a cylinder of an internal combustion engine is made to have only reciprocating motion due to constrain of the piston pin (external), cam and follower, shaft in foot step bearing.

- **Incompletely constrained motion**

  When the motion between the elements of a pair is possible in *more than one direction* and depends upon the *direction of force applied*, it is known as incompletely constrained motion e.g. cylindrical shaft in round bearing.

3. Rigid, Resistant Body

- **Rigid body:** It does not suffer any distortion, under the action of force.
- **Resistant body:** Those body which are rigid for the purpose they have to serve for e.g. belt drive, where belt is rigid when subjected to tensile forces.

4. Link

A link is defined as a member of mechanism, connecting other member and having motion relative to them.

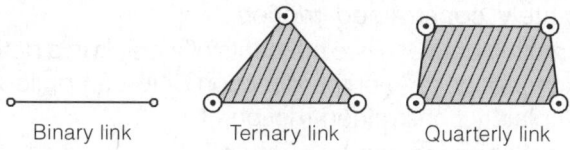

Binary link   Ternary link   Quarterly link

5. Classification of Kinematic Pair

- **According to nature of contact**
  - **Lower pair**

    A pair of links having *surface or area* contact between the member. e.g. nut turning on a screw, shaft rotating in a bearing, all pair of slider-crank mechanism.

  - **Higher pair**

    When a pair has *point or line contact* between the links, it is known as higher pair.

    e.g. wheel rolling on a surface, cam and follower pair.

Remember

- Higher pair involves two dissimilar surface contact.

- **According to Nature of Mechanical Constraint**
  - **Closed pair**

    When the element of pair held *mechanically*, it is known as closed pair.

    The contact between the two can be broken by only destruction of *at least one* of the members.

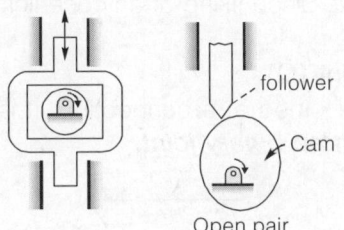

Open pair

Remember: ....

- All lower pairs are closed pair.
  - **Open (unclosed pair)**

    When two links of a pair are in contact either due to force of gravity or some spring action.

    e.g. cam follower.

- **According to Nature of Relative Motion**
  - **Sliding pair**

    If two links have sliding motion relative to each other, they form sliding pair e.g.

    A rectangular rod in a rectangular hole in a prism.

  - **Turning pair**

    When one link has a turning or revolving motion relative to the other, they constitute a turning or revolving pair. e.g. circular shaft revolving in a bearing.

  - **Rolling pair**

    When the link of pair have rolling motion relative to each other, they form a rolling pair, e.g.

    Rolling wheel on flat surface, ball and roller bearing.

  - **Screw pair (Helical)**

    It two mating link have turning as well as sliding motion between them, form a screw pair e.g. lead screw, nut of lathe.

  - **Spherical pair**

    When one link in the form of a sphere turns inside a fixed link, it is a spherical pair e.g. ball in socket.

6. Type of Joints

There are three types of joint:

*   Binary joint (B)

    If *two* links are joined at the same connection.

*   Ternary joint (T)

    If *three* link are joined at a connection. It is consider, equivalent to *two binary joint*. Since fixing of any one link constitutes two binary joints.

*   Quaternary joint (Q)

    If four links are joined at a connection. It is quaternary joint. It is equivalent to *three binary joint*.

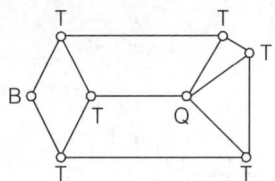

Remember: ....

*   If *n number* of links are connected at a joint, it is equivalent to *(n – 1) binary joints*.

7. Degree of freedom

The connection of a link with another imposes certain constrains on relative motion thus,

Degree of freedom = 6 – number of restraints

*   Number of restraints can never be zero (joint disconnected).
*   Number of restraints can never be 6 (joint become *rigid*).

8. Degree of Freedom of Space Mechanism (3-D)

$$F = 6(L - 1) - 5P_1 - 4P_2 - 3P_3 - 2P_4 - P_5$$

Here,

F = Degree of freedom (D.O.F)

L = Total number of links in mechanism

P_1 = Number of pair having one D.O.F

P_2 = Number of pair having two D.O.F

P_3 = Number of pair having three D.O.F

P_4 = Number of pair having four D.O.F

P_5 = Number of pair having five D.O.F

- Degree of freedom of plane (2D) mechanism (Grabbler Criterion)

$$F = 3(L - 1) - 2P_1 - P_2$$

Here,   L = Number of link in a mechanism

     P_1 = Number of pair having one degree of freedom

     P_2 = Number of pair having two degree of freedom

- Kutzback's equation

$$F = 3(L - 1) - 2j - h$$

Here,   L = Number of link

     j = Number of *binary joint*

     h = Number of *higher joint*

- Grubler's Equation

For those mechanism which have *single* degree of freedom and *zero* higher pair.

$$3l - 2j - 4 = 0$$

Here, l = Number of links

     j = Number of binary joints

- Degree of Freedom

     F = 0 (Frame)

     F < 0 (Redundant frame)

     F > 0 (constrained/unconstrained frame)

- All mechanism have *minimum 4* number of link.

9. Kinematic Chain

When all the links are connected in such a way that first link is joined with the last link, then the structure formed is known as *closed chain*. The closed chain will be a kinematic chain when the relative motion between the links is *either completely* constrained or *successfully* constrained.

- Condition for a kinematic chain

There are two conditions and both the conditions are equivalent.

     ▪ Relation between the number of links and number of pair.

$$l = 2P - 4$$

Where, l = Number of link

        P = Number of pair

- **Relation between number of binary joints and number of links**

$$\boxed{2J = 3l - 4}$$

$$2J = 3l - 4$$

Where,   J = Number of binary joint

      l = Number of link

(i) If   L.H.S > R.H.S

    Locked chain or frame or structure

(ii) If   L.H.S = R.H.S

                      Kinematic chain (constrained chain)

(iii) If   L.H.S < R.H.S

    unconstrained chain

- **Redundant chain**

It does not allow any motion of a link relative to other.

10. Frame/Structure

If one of the link of redundant chain is fixed. It is known as structure or a locked system.

Remember
- No relative motion.
- Capable of transmitting force only.
- Power/energy can not be transmitted.
- Degree of freedom of *structure* is *zero*.
- Degree of freedom of *super structure* is *less than zero*.

11. Simple Mechanism

All the mechanism having *4-links* are *simple* mechanism and the mechanism having *more than* 4 links are *compound* mechanism. There are three different simple mechanisms.

- **Four bar mechanism**

It consists of four link and four turning pair.

- **Slider crank mechanism**

It consists four link, three turning pair and one sliding pair.

- **Double slider crank mechanism**

It consists four link, two turning pair and two sliding pair.

12. Four-Bar Mechanism

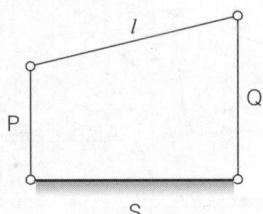

- **Grashof's law**

  $$\boxed{S + l \le P + Q}$$

  Here,

  S = Shortest link

  l = Longest link

  P, Q = Adjacent link to shortest link

Case-I

If $S + l < P + Q$

- S is fixed   – Double crank mechanism.
- P or Q fixed   – Crank-rocker mechanism
- l is fixed   – Rocker-rocker mechanism.

Case-II

If $S + l = P + Q$

- All link have different length then same as case-I.
- Parallelogram linkage-crank-crank mechanism.

  i.e. S = P, l = Q

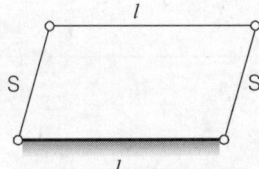

- S is fixed   – Double crank mechanism.
- l is fixed   – Double crank mechanism.
- Deltoid linkage

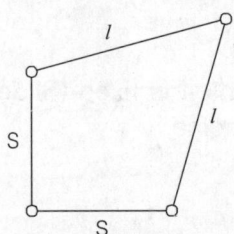

(a) S is fixed – Crank-crank mechanism.

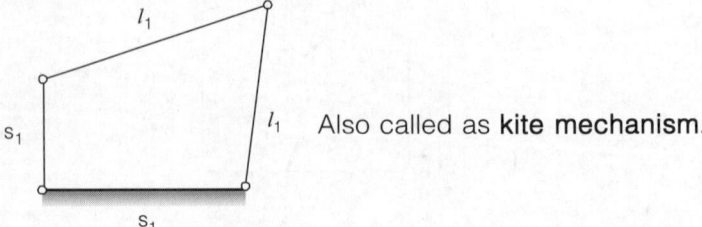

Also called as **kite mechanism**.

(b) l is fixed – Crank-rocker mechanism.

Case-III $s + l > P + Q$

Grashof's law is not satisfied and it will give *rocker-rocker* mechanism.

13. Inversion of 4-Bar Mechanism

- Coupling rod of locomotive – Crank-crank mechanism
- Beam engine   – Crank-rocker mechanism
- Watt's indicator   – Rocker-rocker mechanism

Remember

- Approximate straight line mechanism are *watts indicator*, *modified* Scott-Russel mechanism, Grass Hopper mechanism.
- Exact straight line mechanism are Peculiar mechanism, *Hart mechanism*, Scott-Russel mechanism.

14. Inversion of Slider-Crank Mechanism

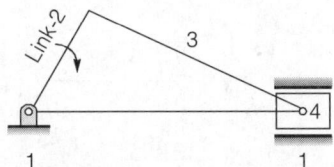

- **First Inversion-link 1 is fixed**
  Reciprocating engine/compressor.
- **Second Inversion-link 2 is fixed (Crank)**
  Whitworth quick return mechanism, rotary (radial) engine.
- **Third Inversion-link 3 is fixed (Connecting Rod)**
  Crank and slotted lever mechanism, oscillating cylinder mechanism.
- **Fourth Inversion-link 4 is fixed (Slider)**
  Hand pump, bull engine

15. Inversion of Double Slider Crank Mechanism

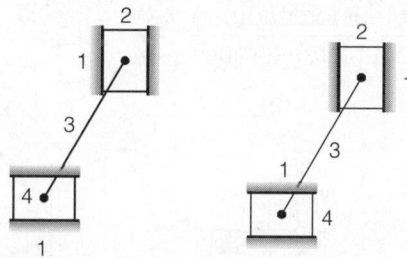

- **First Inversion-link 1 is fixed**
  Elliptical trammel.
    - Turning pair
    - Sliding pair

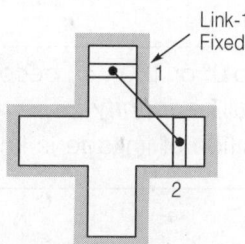

- **Second Inversion-slider 2 is fixed**
  Scotch yoke mechanism.

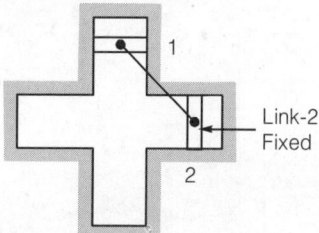

- **Third Inversion-link 3 is fixed**
  Oldham coupling.

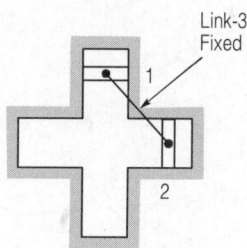

Oldham coupling is used to connect two shaft which has **lateral misalignment.**

16. Mechanical Advantage

$$\text{M.A} = \frac{\text{Output force/torque}}{\text{Input force/torque}}$$

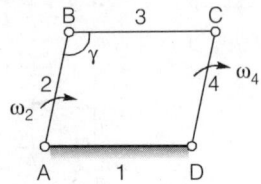

Power input = Power output

$$T_2\omega_2 = T_4\omega_4$$

$$\text{M.A} = \frac{T_4}{T_2} = \frac{\omega_2}{\omega_4}$$

Remember

- If γ is equal to 0° or 180°, ω_4 become zero thus mechanical advantage will be *infinity*.
- Extreme position of linkage is knows as *toggle position*.

■■■■

Velocity Analysis

2

Velocity in machines can be determined by either analytically or graphically. This chapter deals with graphical analysis.

1. Velocity Analysis

- Let V_{ao} = Velocity of A relative to O
  V_{ba} = Velocity of B relative to A
  V_{bo} = Velocity of B relative to O

$$V_{bo} = V_{ba} + V_{ao}$$

- Velocity of A relative to O = $\dfrac{\text{Arc } AA'}{\delta t}$

$$V_{ao} = \frac{r.\delta\theta}{\delta t} = r\,\frac{d\theta}{dt} = r\omega$$

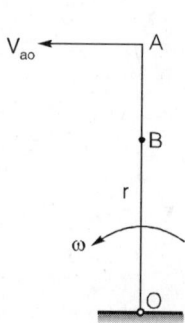

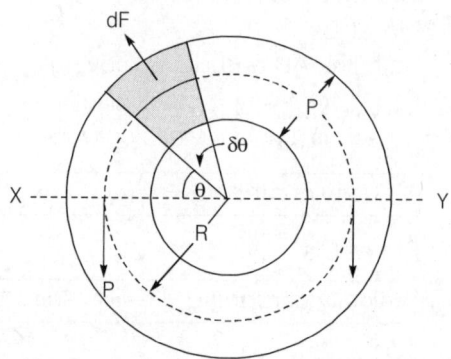

as $\delta t \to o$, AA' will be perpendicular to OA, Thus the velocity of A is ωr and is perpendicular to OA.

Remember: ....

- The velocity of any point relative to any other point on a fixed link is always zero.
- The velocity of an intermediate point on any of links can be found easily by dividing the corresponding velocity vector in the same ratio as the point divides link.
- The angular velocity of a link about one extremity is the same as the angular velocity about the other.'.

2.   Velocity of Rubbing

The velocity of rubbing of the two surface will depend upon the angular velocity of a link, relative to the other.

*   **If pin at A**

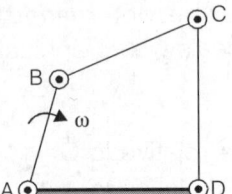

Pin at A joins links AD and AB. AD being fixed the velocity of rubbing will depend only upon angular velocity of AB.

$$\therefore \quad \boxed{\text{Velocity of rubbing} = r_a.\omega_{AB}}$$

Here, r_a = Radius of pin A

*   **Pin at D**

$$\boxed{\text{Velocity of rubbing} = r_d.\omega_{cd}}$$

Here, Radius of pin D

*   **Pin at B**

    Both link AB and BC is moving

$$\omega_{ab} = \omega = \text{clockwise}$$
$$\omega_{bc} = \omega = \text{anticlockwise}$$

$$\therefore \quad \boxed{\text{Velocity of rubbing} = r_b\left(\omega_{ab} + \omega_{bc}\right)}$$

*   **Pin at C**

$$\boxed{\text{Velocity of rubbing} = r_c\left(\omega_{bc} + \omega_{dc}\right)}$$

3.   Theory of Instantaneous Centre

Instantaneous centre of rotation or virtual centre

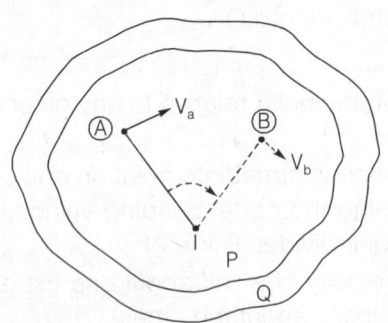

Let a plane body 'P' having non-linear motion relative to another plane body q. At any instant, the linear velocities of the point 'A' and 'B' on the body 'P' are 'V_a' and 'V_b' respectively.

If a line is drawn perpendicular to the direction of V_a at 'A', the body can be imagined to rotate about some point on this line. Similarly for point B. If the intersection of the two lines is at 'I', the body 'P' will be rotating about I at the instant.

This point 'I' is known as instantaneous centre of velocity.

Remember: ....

- If the direction V_a and V_b are parallel to the I-centre of body lies at infinity.

● **Centroid**

As we know, in general the position of instantaneous centre changes throughout the whole motion. The locus of all these instantaneous centre for a particular link is known as "Centroid". It is a *curve*.

● **Axode**

The line passing through instantaneous centre and perpendicular to the plane of motion is known as instantaneous axis. The locus of instantaneous axis for a link during the whole motion is known as "Axode". It is a *surface*.

4. Instantaneous Centre in Different Situation

- If two links are attached with a turning pair, the instantaneous centre of such a pair will be at *pin joint*.

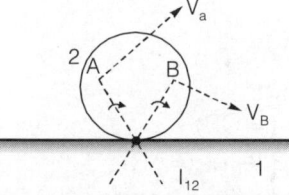

- Rolling of a sphere on a plane.

Instantaneous centre will be at point of content.

- Object sliding on a *plane* surface

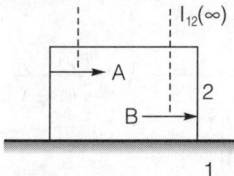

The location of I_{12} will be at infinity but in a direction perpendicular to the *sliding surface*.

4. Object sliding on a *curved* surface

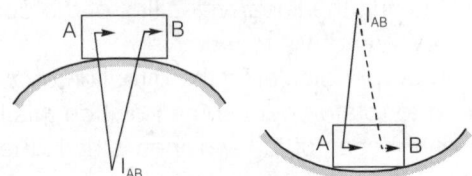

Instantaneous centre will be at the *centre* of curvature of *curved* surface.

5. Number of Instantaneous Centre

$$I = \frac{n(n-1)}{2}$$

Here　　n = Number of Link

6. Crank and Slotted Lever Mechanism

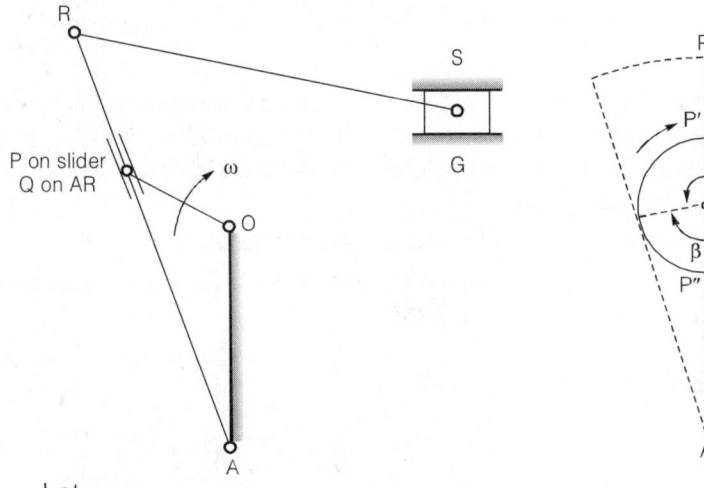

Let,

r = length of crank (= OP)

l = length of slotted lever (=AR)

c = distance between fixed centres (=AO)

ω = angular velocity of crank

Thus during cutting stroke

$$\frac{(V_s)_{max}\ (cutting)}{(V_s)_{max}\ (return)} = \frac{c-r}{c+r}\ ;\quad \frac{Time\ of\ cutting}{Time\ of\ return} = \frac{\theta}{\beta}$$

7. Kennedy's Theorem

For the three bodies having the continuous relative motion there all instaneous centres lies on the same line.

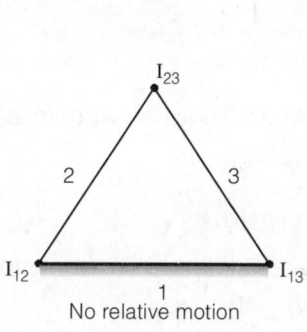

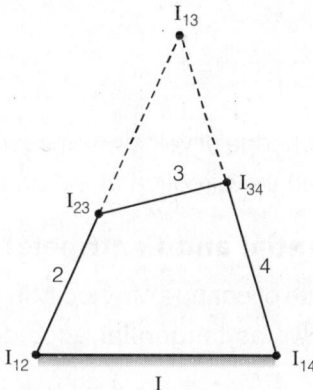

V_{23} = Translational velocity

$$V_{23} = \omega_2(I_{23}\,I_{12}) = \omega_3(I_{23} - I_{13})$$

8. Angular Velocity Ratio Theorem

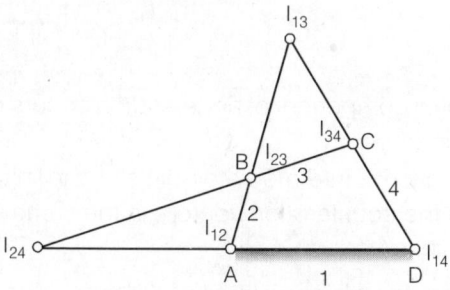

It is used to find angular velocity of a link if angular velocity of another link is known.

The angular velocity ratio of two links relative to third link is *inversely* proportional to the distances of their *common I-centre* from their respective centre of rotation.

$$\omega_2(I_{24}\,I_{12}) = \omega_4(I_{24}I_{14})$$

$$\frac{\omega_4}{\omega_2} = \frac{I_{24}\,I_{12}}{I_{24}\,I_{14}}$$

■■■■

3

Acceleration Analysis

The rate of change of velocity with respect to time is known as *acceleration* and it acts in the direction of the change in velocity.

1. Tangential and Centripetal Acceleration

- The rate of change of velocity in the tangential direction of the motion is known as **tangential acceleration**.

$$a_t = \frac{dv}{dt}$$

- The rate of change of velocity towards the centre of rotation is known as *centripetal* or radial *acceleration*.

$$a_c = \frac{v^2}{r}$$

Remember: ....

- The tangential component of acceleration occurs due to the angular acceleration of link.
- The acceleration of intermediate points on the links can be obtained by dividing the acceleration vectors in the same ratio as the points divide the links.

2. Coriolis Acceleration Component

- Coriolis Acceleration Component

  Let,

  ω = Angular velocity of the link

  α = Angular acceleration of the link

  v = Linear velocity of the slide of the link

  f = Linear acceleration of the slider on the link

  r = Radial distance of point P on the slider.

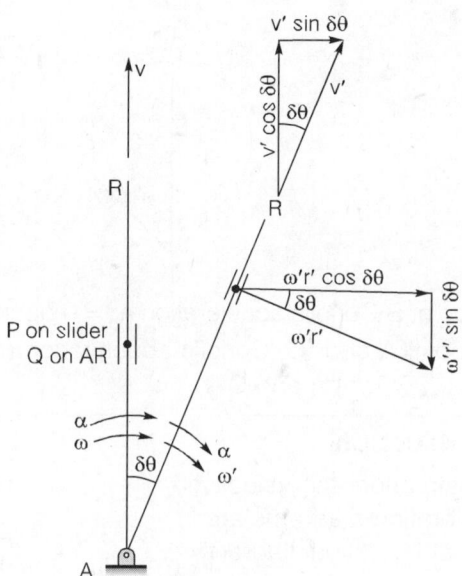

Acceleration of P along AR =

Acceleration of slider – Centrifugal acceleration

$= f - \omega^2 r$

Acceleration of P perpendicular to AR

$= 2\,wV +$ Tangential acceleration

$= 2\,\omega V + r\alpha$

- The component **2 ωv** is known as the *Coriolis acceleration component.*

- The Coriolis component is positive if link AR rotates clockwise and slider moves radially outwards the link rotates counter clockwise and slider moves radially inwards. Otherwise Coriolis component will be negative.

- The direction of the Coriolis acceleration component is obtained by rotating the radial velocity vector 'v' through 90° in the direction of rotation of the link.

- A crank slotted lever mechanism has Coriolis component $2\omega V_{slide}$.

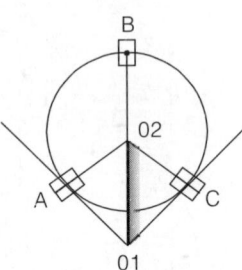

- At point B Coriolis acceleration $a_c = 0$ because $V_{B(slider)} = 0$.
- At point A and C, Coriolis acceleration $(a_c = 0)$ because $w_{(slotted)} = 0$ at A and C.

3. Klein's Construction

In Klein's construction, the velocity and the acceleration diagrams are made on the configuration diagram itself. The line that represents the crank in the configuration diagram also represents the velocity and acceleration of its moving end in the velocity and acceleration diagram respectively.

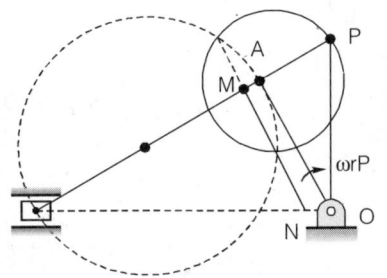

$$\frac{a_{B(Piston)}}{ON} = \frac{a^r_{BA}}{AM} = \frac{a^{tan}_{BA}}{NM} = \frac{a_A}{OA} = \omega^2$$

- **AMNO** is acceleration diagram.

$$\frac{V_{BA}}{AP} = \frac{V_A}{OA} = \frac{V_B}{OP} = \omega_{crank}$$

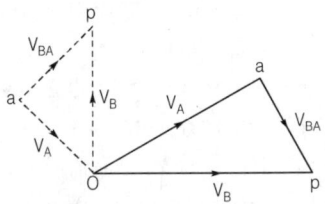

- APO is velocity diagram.

4. Some Other Important Points

- Acceleration images are helpful to find the accelerations of offset points of the links. The acceleration image of a link is obtained in the same manner as a velocity image.
- Acceleration of a point on a link relative to a coincident point on a moving link is the sum of absolute acceleration of the coincident point, acceleration of the point relative to coincident point and the Coriolis acceleration.

A cam is mechanical member used to impart desired motion to a follower by **direct contact**. The cam may be **rotating or reciprocating** whereas the follower may be rotating, reciprocating or **oscillating**.
It is used in automatic machine, IC engine, machine tools, printing control mechanism.

1. Element of Cam

- A *driver* member known as the **cam**
- A *driven* member called the **follower**
- A *frame* which supports the cam and guides the follower.

- A cam and follower combination belong to the category of **higher pair** as it has different contact surface.

Remember

2. Types of Cam

- According to Shape
  - **Wedge and flat cams**
    A wedge cam has a translational motion. The follower can either translate or oscillate.

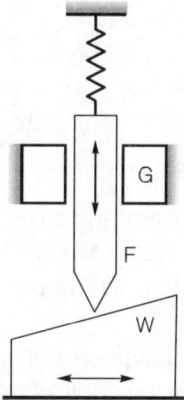

  - **Radial or Disc Cams**
    A cam in which the follower moves radially from the centre of rotation of the cam is known as a radial or disc cam.

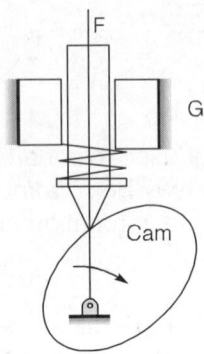

- **Spiral cams**

  A spiral cam is a face cam in which a groove is cut in the form of a spiral. It is used in computer.

- **Cylindrical cams**

  In a cylindrical cam, a cylinder which has a circumferential contour cut in surface, rotates about its axis. It is also known as *barrel* or *drum cams*.

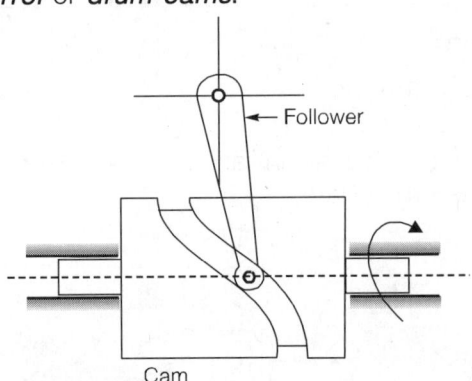

Cam

- **Conjugate cams**

  It is a double disc cam and preferred when the requirements are low wear, low noise, better control of the follower, high speed, *high dynamic loads* etc.

- **Globoidal cams**

  It has two types of surface i.e. convex or concave. It is used when moderate speed and the angle of oscillation of the follower is large.

- **According to follower movement**
  - **Rise-Return-Rise (RRR)**

    In this, there is alternate rise and return of the follower with no period of dwells. The follower has a linear or an angular displacement.

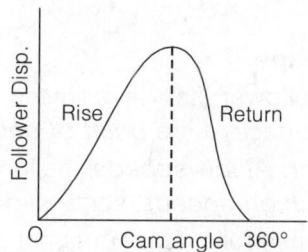

- **Dwell-Rise-Return Dwell (D-R-R-D-S)**
  In this cam, there is rise and return of the follower after a dwell.

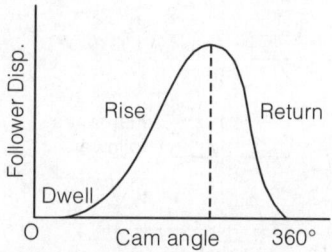

- **Dwell-Rise-Dwell-Return-Dwell (D-R-D-R-D)**
  The dwelling of the cam is followed by rise and dwell and subsequently by return and dwell. In case the return of the follower is by a fall, the motion may be known as **Dwell-Rise-Dwell (DRD).**

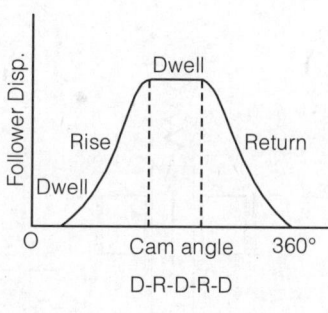

D-R-D-R-D

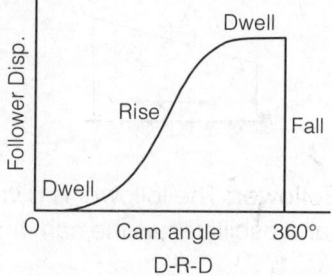

D-R-D

3. Type of Follower

- According to shape
  - **Knife-edge follower:** its use is limited as it produces a *great wear* of the surface at the *point of contact.*
  - **Roller follower:** *At low speeds*, the follower has a *pure* rolling action, but at *high speeds*, some *sliding* also occurs.
- In case of steep rise roller follower is not preferred.
  - **Mushroom follower:** It does not pose the problem of jamming the cam.

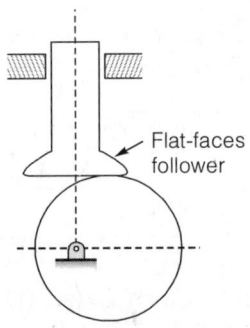

- According to Movement
  - **Reciprocating Follower:** In this type, as the cam rotates the follower reciprocates or translates in the guides.

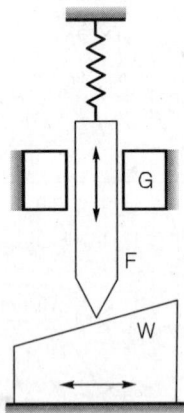

  - **Oscillating Follower:** The follower is pivoted at a suitable point on the frame and oscillates as the cam makes the rotary motion.

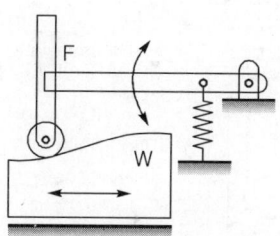

- **According to location of line of movement**
  - **Radial Follower:** The follower is known as a radial follower if the line of movement of the follower passes through the *centre of rotation of the cam*.

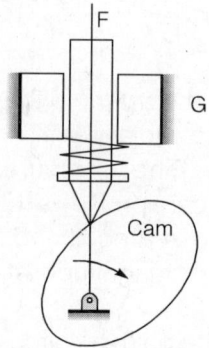

  - **Offset Follower:** If the line of movement of the roller follower *is offset* from the centre of rotation of the cam, the follower is known as an offset follower.

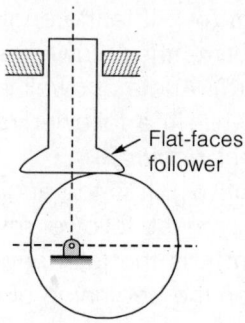

4. Terminology of Cam

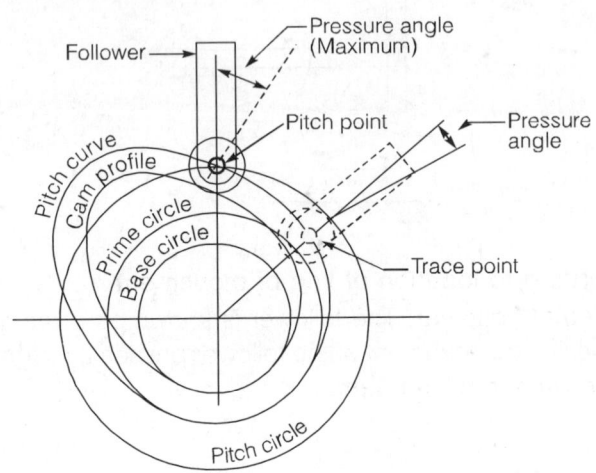

- **Base circle**

  It is the *smallest* circle *tangent* to the cam profile drawn from the *centre of rotation* of a radial cam.

  - **Pressure angle:** The pressure angle, representing the *steepness* of the cam profile, is the angle between the *normal* to the *pitch curve* at a point and the direction of the follower motion. It varies in magnitude at all instants of the follower motion.

  - **Pitch Point:** It is the point on pitch curve at which the pressure angle is *maximum*.

  - **Pitch circle:** It is the circle passing through the pitch point and concentric with the *base circle*.

  - **Prime circle:** The *smallest* circle drawn *tangent* to the *pitch curve* is known as prime circle.

  - **Angle of Ascent (ϕ_a):** It is the angle through which the cam turns during the time the follower rise.

  - **Angle of dwell (δ):** Angle of dwell is the angle through which the cam turns while the follower *remains stationary* at the *highest* or *lowest position*.

  - **Angle of Decent (ϕ_d):** It is the angle through which the cam turns during the time the follower returns to the initial position.

  - **Angle of Action:** It is the total angle moved by cam during the time, between the beginning of rise and the end of return of the follower.

Remember
- The dynamic effects of acceleration (jerks) usually, limit the speed of the cams.
- **Base circle** is smallest circle tangent to **cam profile**.
- **Prime circle** is smallest circle tangent to **pitch curve**.

5. Force Exerted by Cam

The force exerted by a cam on the follower is always normal to the surface of the cam at the point of contact. The vertical component (F cos α) lifts the follower whereas the horizontal component (F sin α) exerts lateral pressure on the bearing.

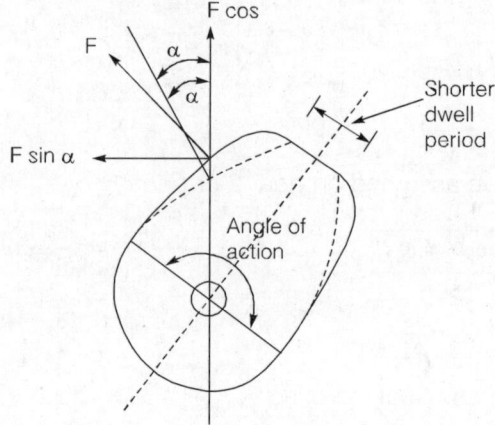

In order to reduce the lateral pressure or F sin α, α has to be decreased which means making the surface more convex and longer. This results in reduced velocity of the follower and more time for the same rise. The minimum value of α cannot be reduced from a certain value.

The increase in the base circle diameter increases the length of the arc of the circle upon which the wedge is to be made. A short wedge for a given rise requires a steep rise or a higher pressure angle, thus increasing the lateral force.

■■■■

Gear

5

1. Introduction to Gear

Gears use *no intermediate link or connector* and transmit the motion by *direct* contact. The two bodies have either a *rolling or a sliding* motion along the *tangent* at the point of contact. No motion is possible along the *common normal* as that will either break the contact or one body will tend to penetrate into the other.

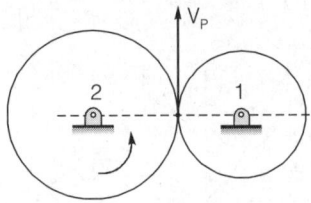

Point P can be assumed on gear 2 or gear 1.

$$\boxed{V_p = \omega_2 r_2 = \omega_1 r_1} \qquad \left[\omega = \frac{2\pi N}{60}\right]$$

$$\frac{\omega_1}{\omega_2} = \frac{r_2}{r_1} = \frac{N_1}{N_2} = \frac{T_2}{T_1}$$

Symbols has usual meaning.

Remember: ....

- Gear is a positive drive because *no slip* occur in its motion.

2. Classification of Gear

- Parallel Shaft
  - Spur Gears: They have *straight* teeth *parallel* to the axes and thus are *not* subjected to axial thrust due to tooth load.

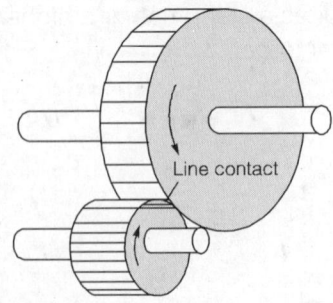

Line contact

Remember

Spur Gear

- At the time of engagement of the two gears, the contact extends across the *entire width on a line parallel* to the axes of rotation. This results in *sudden application* of the load, *high impact stresses* and excessive noise at high speeds.

- **Spur Rack and Pinion:** Spur rack is a special case of spur gear where it is made of in finite diameter so that the *pitch surface* is plane. The spur rack and pinion combination converts *rotary motion* into *translatory motion* or vice-versa. It is used in lathe in which the rack transmits motion to the saddle.

- **Helical gears or Helical spur gears:-** In helical gears, the teeth are *curved*. Two mating gears have the *same helix angle*; but have teeth of opposite lands.

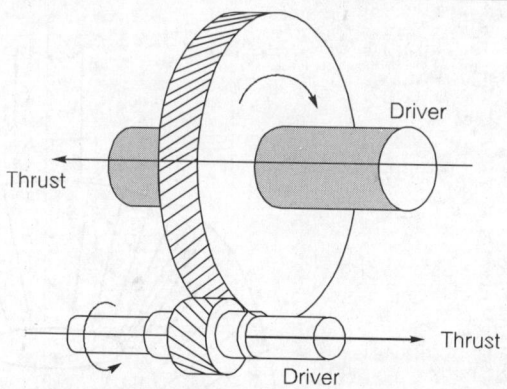

Remember

Helical Gear

- At the beginning of engagement, contact occurs *only at the point* of leading edge of the curved teeth. Thus the load application *is gradual* which results in low impact stresses.

- The helical gears can be used at higher velocities than the spur gears and have greater load - carrying capacity.

- Helical gears have the disadvantage of having *end thrust* as there is a force component along the gear axis.

- **Double-helical and Herring bone Gears:** A double - helical gear is equivalent to *a pair of helical gears* secured together, one having a *right hand helix* and other a left hand helix.

Remember: ....

- No axial thrust is present.
- If the left and the right inclinations of a double - helical gear meet *at a common apex* and there is *no groove* in between the gear is known as *herringbone gear*.
- Intersecting Shaft
  The motion between two intersecting shafts is equivalent to the rolling of *two cones* assuming no slipping.
  - Straight bevel Gears: The teeth are *straight, radial* to the point of inter section of the shaft axes and vary in *cross-section* throughout *their length*.

Remember: ....

- Gears of the *same size* and connecting two shafts at *right angle* to each other are known as *mitre gears*.

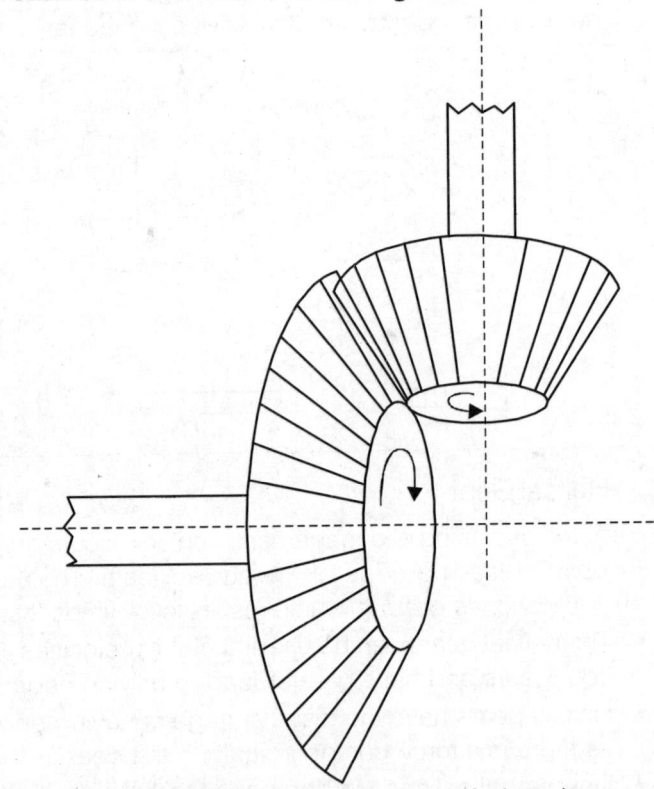

  - Spiral bevel Gears: When the teeth of a bevel gear are *inclined* at angle to the face of the bevel, they are known as spiral bevel or helical bevels.

Remember: ....

- There is *gradual* load application and *low impact* stresses.
- These are used for the drive to the *differential* of an automobile.

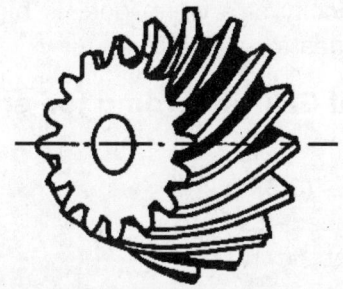

- ■ **Zerol bevel Gears:** Spiral bevel gears with *curved* teeth but with a *zero degree spiral angle* are known as zerol bevel gears.
- **Skew Shafts**
  In case of skew (non-parallel, non-intersecting) shafts, a uniform rotary motion is *not possible* by *pure rolling* contact.

Remember: ....

- If the two hyperboloids rotate on their respective axes, the motion between them would be a combination of rolling and sliding action.
- Angle between two shafts will be qual to the sum of the angles of generation of two hyperboloids.

$$\theta = \psi_1 + \psi_2$$

- ■ **Crossed helical gears:** The use of crossed - helical gears or spiral gears is limited to light loads. These gears are used to drive feed mechanism on machine tools, camshafts and oil pumps in I.C. engine.
- ■ **Worm Gears:** It is a special case of a spiral gear in which the larger wheel usually has a hollow or concave shape.

Remember: ....

- The **smaller** of the two wheels is called the **worm** which also has a large **spiral angle**.
- The sliding velocity of a worm gear is **higher** as compared to other types of gears.

3. Classification of Gear According to Peripheral Velocity of Gear

Low velocity gear – (0-3) m/s
Medium velocity gear – (3-5) m/s
Large velocity gear – (>15) m/s

4. Gear Terminology

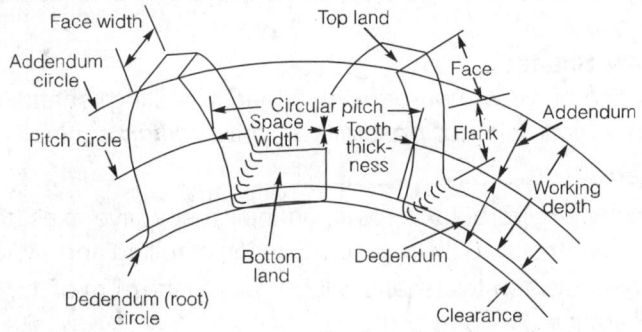

- **Pitch Circle**

  It is an imaginary circle drawn in such a way that a *pure rolling motion* on this circle gives the motion which is exactly similar to the gear motion.

Remember: ....

- These pitch circle always touch each other for the correct power transmission.
- It is not fundamental characteristics of gear.
- **Pitch Point**

  It is a point where the two pitch circle of the mating gear touch each others.
- **Pressure angle (ϕ)**

  It is the angle between common normal at a point of contact at a pitch point.

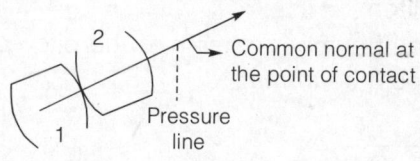

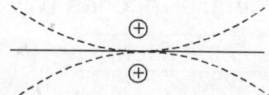

The standard value of pressure angle are 14½ °, 20°, 25°.

- **Module (m)**

  It is defined as the ratio of pitch circle diameter in mm to the number of teeth.

  $$m = \frac{D_{(mm)}}{T}$$

- **Addendum Circle**

  A circle drawn from top of tooth and concentric to pitch circle is known as addendum circle.

  Addendum is Radial distance between pitch circle and addendum circle and it is equal to 1 module.

Remember: ....

- Clearance = 0.157 m
- **Dedendum Circle**

  A circle drawn from bottom of the teeth and concentric with pitch circle.

  Dedendum is Radial distance between pitch circle and dedendum circle and it is equal to *1.157 module*.

- **Circular Pitch (C)**

  It is a distance along a pitch circle from one point on a tooth to the corresponding point on the next tooth.

  $$C = \frac{\pi D}{T}$$

  D = Pitch circle-diameter

  T = Number of Teeth

  $$\boxed{C = \text{space width} + \text{Tooth thickness}}$$

- **Diametral Pitch**

  It is the ratio of number of teeth to the pitch circle diameter but diameter should be *in mm*.

  $$P_d = \frac{T}{D_{mm}}$$

- **Relation between circular pitches (C) diametral pitch (P_d)**

Circular pitch × Diametral pitch = π

- **Tooth Thickness**

  It is the thickness of tooth measured *along pitch circle*.

- **Tooth Space**

  The space between the consecutive teeth measured *along the pitch circle*.

- **Backlash**

  It is difference between *tooth space* and tooth *thickness*, which is generally provided to avoid jamming due to thermal expansion.

- **Face**

  The portion of tooth profile *above* the pitch surface.

- **Flank**

  The portion of tooth profile *below* the pitch surface.

- **Profile**

  The curvature contained by *face and flank*.

- **Path of contact (POC)**

  It is the path travelled by point of contact from the *starting* of engagement to the *end* of engagement.

POC = Path of approach + Path of Recess

- **Arc of Contact (AOC)**

  It is the path traced by a point on the *pitch circle* during starting of engagement to the end of engagement.

- **Gear Ratio (G)**

  $$G = \frac{T}{t}$$

  Here,

       T = Number of teeth of gear.

       t = Number of teeth of pinion (small gear).

- **Velocity Ratio (VR)**

  $$VR = \frac{1}{\text{Gear ratio}}$$

- **Angle of Action**

  Angle turned by gear from the **beginning** of engagement to the **end** of engagement of a pair of teeth.

Angle of Action = Angle of approach + Angle of recess

- **Contact Ratio (C.R.)**

  $$C.R = \frac{\text{Arc of contact}}{\text{Circular pitch}} = \frac{\text{Angle of action}}{\text{Pitch angle}}$$

- Contact ratio **reflects** number of pair of teeth in contact.

5. Law of Gearing

- The law of gearings states the condition which must be fulfilled by the gear tooth profiles to maintain a constant angular velocity ratio between two gears.

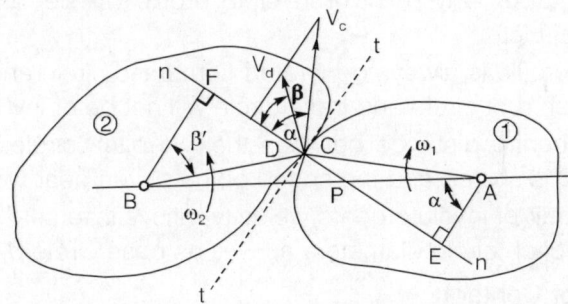

Let ω_1 = angular velocity of gear 1 (clockwise)

        ω_2 = angular velocity of gear 2 (anticlockwise)

$$\frac{\omega_1}{\omega_2} = \frac{PF}{PE} = \frac{BP}{AP}$$

- For **constant angular velocity ratio** of the two gears, the common normal at the point of contact of the two mating teeth must pass through the pitch point.

- **Velocity of sliding**

  If the curved surfaces of the two teeth of the gears are to remain in contact one can have a sliding motion relative to the other along the common tangent.

  = Sum of angular velocities X distance between the pitch point and point of contact.

  $$= (\omega_1 + \omega_2) \, PC$$

6. Type of Profile

- **Involute Profile**

    Involute is a curve generated *by point* on a *tangent* which rolls on a circle *without slipping*. The involute profile on a gear will be generated through a generating circle and this generating circle will be known as *base circle*. It is a fundamental property of a gear its radius will not change in any condition for a gear.

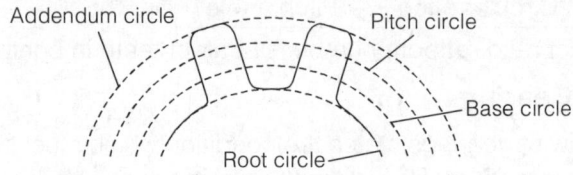

Remember: ....

- A normal on any point of involute profile will be tangent to the base circle.
- Tooth profile is always generated from base circle and the profile between root circle and base circle will not be of involute type.
- If the centre distance between the two pitch circles varies, the point P is shifted and the speed of the driven gear would vary.
- For a pair of involute gears, velocity ratio is inversely proportional to the pitch circle diameters as well as base circle diameters.
- **Path of Contact**

$$= \sqrt{R_a^2 - R^2 \cos^2 \phi} + \sqrt{r_a^2 - r^2 \cos^2 \phi} - (R + r) \sin\phi$$

    Here,

    r = pitch circle radius of pinion
    R = pitch circle radius of wheel
    r_a = addendum circle radius of pinion
    R_a = addendum circle radius of wheel

- **Arc of contact**

$$= \frac{\text{Path of contact}}{\cos \phi}$$

- **Number of pairs of teeth in contact**

$$= \frac{\text{Arc of contact}}{\text{Circular pitch}}$$

- The maximum value of the addendum radius of the wheel to avoid interference.

$$= R\sqrt{1+\frac{r}{R}\left(\frac{r}{R}+2\right)\sin^2\phi}$$

- Maximum value of addendum of the wheel

$$= R\left[\sqrt{1+\frac{r}{R}\left(\frac{r}{R}+2\right)\sin^2\phi}-1\right]$$

- Minimum Number of teeth on the wheel for the given values of the gear ratio the pressure angle and the addendum coefficient (a_w).

$$T = \frac{2a_w}{\sqrt{1+\frac{1}{G}\left(\frac{1}{G}+2\right)\sin^2\phi}-1}$$

Where a_w = Addendum coefficient

Remember: ....

- Point of contact will always be at a line tangent to the base circle.
- Base circle diameter = pitch circle diameter X cos ϕ.
- **Cycloidal Profile Teeth**
  A cycloid is the locus of a point on the circumference of a circle that rolls **without slipping** on the circumference of another circle. In this type, the faces of the teeth are **epicyloids** and flanks the **hypocycloids**.

Type of tooth	Pressure angle (ϕ)	Addendum	Dedendum
Full depth	20°	1 m	1.157 m
	22.5°	1 m	1.35 m
	25°	1 m	1.25 m
Stub	20°	0.8 m	1 m

In cycloidal teeth pressure angle (ϕ) varies while Involute teeth pressure angle (ϕ) is constant.

Remember

7. Terminology / Centre Distance of Helical Gear

- **Helix angle (ψ)**
  It is the angle at which the teeth are inclined to the axis of a gear. It is also known as spiral angle.
- **Normal Circular Pitch (P_n)**
  Normal circular pitch or simply normal pitch is the **shortest distance** measured along the normal to the helix between corresponding points on the adjacent teeth. The normal circular pitch of two mating gears must be same.

$$P_n = P \cos \psi$$

Also, we have, $P = \pi m$ as for spur gear

$$P_n = \pi m_n$$

and $m_n = m \cos\psi$

$$\text{Centre distance} = \frac{m_n}{2}\left[\frac{T_1}{\cos\psi_1} + \frac{T_2}{\cos\psi_2}\right]$$

8. Helical and Spiral Gears

In helical and spiral gears, the teeth are inclined to the axis of a gear. They can be right handed or left handed

Let,

ψ_1 = helix angle for gear 1

ψ_2 = helix angle for gear 2

θ = Angle between shaft

$\theta = \psi_1 + \psi_2$ (for gears of same hand)

$\theta = \psi_1 - \psi_2$ (for gears of opposite hand)

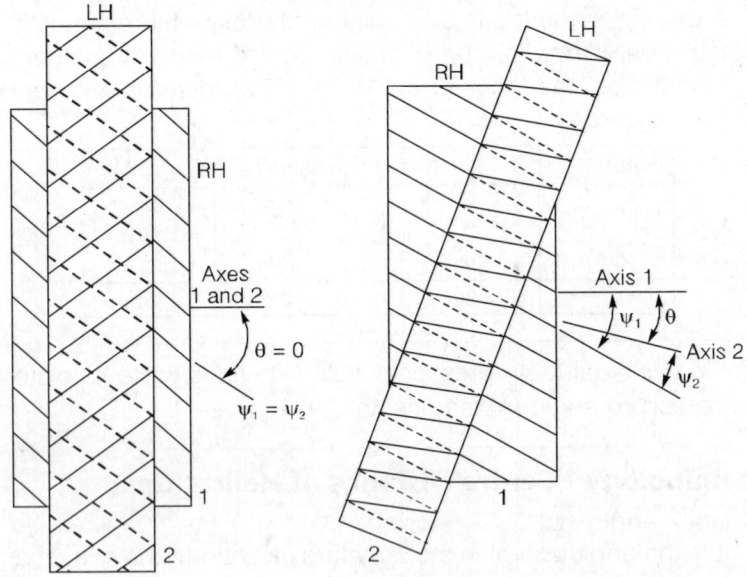

Angle between shafts, $\theta = \psi_1 + \psi_2$ for gears of same hand

$\theta = \psi_1 - \psi_2$, for gears of opposite hand

for $\psi_1 + \psi_2$, $\theta = 0$, a case of helical gears joining parallel shafts.

9. Efficiency of Spiral and Helical Gear

$$\eta = \frac{\cos(\theta + \phi) + \cos(\psi_1 - \psi_2 - \phi)}{\cos(\theta - \phi) + \cos(\psi_1 - \psi_2 - \phi)}$$

$$\eta_{max} = \frac{\cos(\theta + \phi) + 1}{\cos(\theta - \phi) + 1}$$

ϕ = Pressure angle

ψ = Helix angle

θ = Angle between two shaft

10. Terminology of Worm Gear/Velocity Ratio Centre Distance/Efficiency

- **Axial Pitch (p_a)**

  It is the distance between corresponding points on adjacent teeth measured along the direction of the axis.

- **Lead (L):** The distance by which a helix advances along the axis of the gear for one turn around is known as lead.

  In a single helix, the axial pitch is equal to lead. In a double helix, this is one-half the lead, in a triple helix one third of lead, and so on.

- **Lead Angle (λ):** It is the angle at which the teeth are *inclined to the normal* to the axis of rotation, the lead angle is the complement of the helix angle.

  $$\psi + \lambda = 90°.$$

Velocity ratio = $\dfrac{l}{\pi d_2}$ (d = diameter)

Centre distance = $\dfrac{m_2}{2}\left[T_1 \cot\lambda_1 + T_2\right]$

$$\eta = \frac{\tan\lambda_1}{\tan(\lambda_1 + \phi)}$$

$$\eta_{max} = \frac{1 - \sin\phi}{1 + \sin\phi}$$

■■■■

6

Gear Train

A Gear Train is a combination of gears used to transmit motion from one shaft to another. It is required to obtain *large speed reduction* within a small space.

1. Simple Gear Train

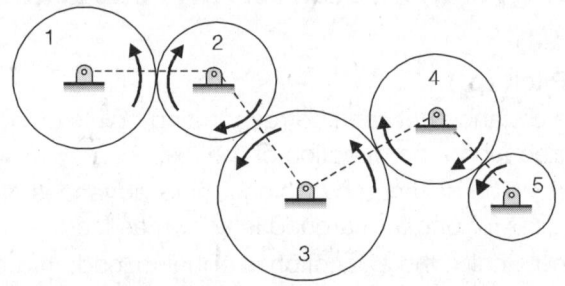

*   A series of gears, capable of receiving and transmitting motion from one gear axes remain fixed relative to the frame and each gear is on separates shaft.

$$\text{Train value} = \frac{\text{number of teeth on driving gear}}{\text{number of teeth on driven gear}}$$

$$= \frac{N_5}{N_1} = \frac{T_1}{T_5}$$

$$\text{Speed ratio} = \frac{1}{\text{train value}}$$

Remember: ....
*   A simple gear train can also have bevel gears.
*   A pair of mated *external* gear always move in opposite direction.
*   All *odd* numbered gears move in *one direction* and all even numbered gears in the opposite direction.
*   *Intermediate* gears have no effect on the *speed ratio* and therefore, they are known as *idlers*.

2. Compound Gear Train

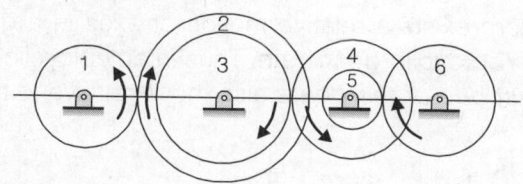

- When a series of gears are connected in such a way that two or more gears rotate about an axis with the same angular velocity; it is known as compound gear train.

$$\text{Train value} = \frac{\text{Product of number of teeth on driving gears}}{\text{Product of number of teeth on driven gears}}$$

$$= \frac{N_6}{N_1} = \frac{T_1 T_3 T_5}{T_2 T_4 T_6}$$

3. Reverted Gear Train

If the axis of the *first* and *last wheel* of a compound gear *coincide* it is called reverted gear train. Such arrangement is used in *clock* and in simple lathe where 'back gear' is used to give slow speed to the chuck.

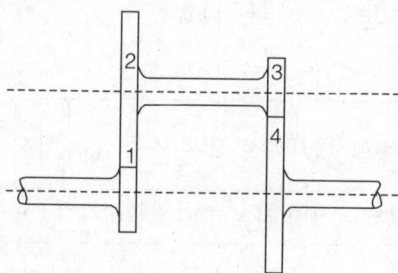

- Train value $= \dfrac{N_4}{N_1} = \dfrac{T_1 T_3}{T_2 T_4}$

If r is the *pitch circle radius* of a gear,

$$r_1 + r_2 = r_3 + r_4$$

$$\frac{m}{2}(T_1 + T_2) = \frac{m'}{2}(T_3 + T_4)$$

Where m is module of gear 1 and 2 while m' is module of gear 3 and 4.

4. Epicyclic Gear Train

- When there exists a relative motion of axes in a gear train, it is called an **epicyclic gear train**. Thus in an epicyclic train, the axis of *at least one of the gears* also move relative to the frame.

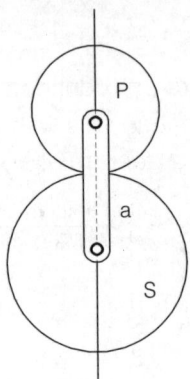

- If arm '**a**' is fixed. Turn '**s**' through x revolutions in the clockwise direction. Assuming clockwise motion of a wheel as positive and counter clockwise motion as negative.

  Revolution made by '**a**' = 0

  Revolution made by '**s**' = x

  Revolution made by '**p**' = $-\left(\dfrac{T_s}{T_p}\right)x$

- When arm is fixed and revolves at x.

$$\text{Revolution made by planet gear(p)} = -\left(\dfrac{T_s}{T_p}\right)x$$

- When arm is revolving at y and s is fixed.

$$\text{Revolution made by planet gear(p)} = y - \left(\dfrac{T_s}{T_p}\right)x$$

Remember: ....

- Large speed reductions are possible with epicyclic gears.
- In general gear trains have two degrees of freedom.

5. Sun and Planet Gear

When an annular wheel is added to the epicyclic gear train, the combination is usually, referred to as sun and planet gear.

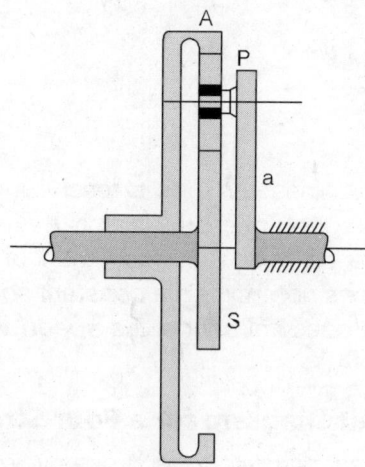

The annular wheel gears with the wheel P which can rotate freely on the arm 'a'. The wheel S and P are generally called the sun and the planet wheels.

If the sun wheel S is fixed, $N_s = 0$

Speed of the arm, $\dfrac{N_a}{N_A} = \dfrac{1}{(T_S / T_A) + 1}$

If the annular wheel A is fixed, $N_A = 0$

$$\frac{N_a}{N_s} = \frac{T_s / T_A}{1 + (T_s / T_A)}$$

6. Differential Gear

When a vehicles takes a turn, the outer wheels must *travel farther* than the *inner wheels*. Since, Both rear wheels are driven by the engine through gearing. Therefore some sort of automatic device is necessary so that the two rear wheels are driven at slightly different speeds. This is accomplished by fitting a *differential gear* on the *power (rear) axle*. Differential gear is a device which adds or subtracts *angular displacements*.

■■■■

Flywheel

7

A *flywheel* used in machines serve as a reservoir, which stores energy during the period when supply of energy is more than the requirement, and release it during the period when the requirement of energy is more than the supply. Flywheel *does not* maintain a *constant speed*, it simply reduce fluctuation of speed. It does not control the *speed variations* caused by the *varying load*.

1. Turning Moment Diagram for a Four Stroke Engine

- A turning moment diagram for a stroke cycle internal combustion engine is shown in fig. We know that in a four stroke internal engine, there is one working stroke after the crank has turned through two revolutions, i.e., 720° (or 4π radians).

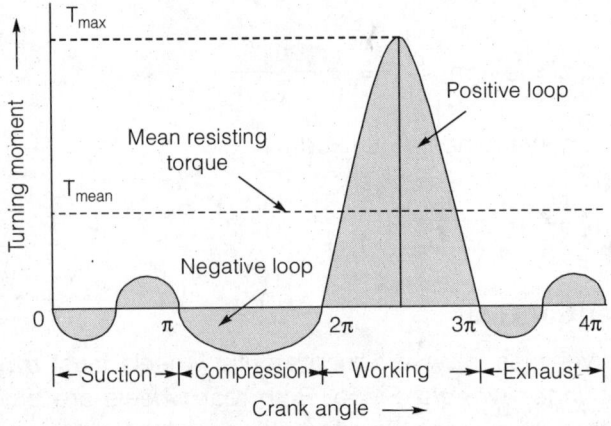

- The pressure inside the engine cylinder is **less than** the **atmospheric pressure** during the suction stroke, therefore a negative loop is formed.
- During the compression stroke, the work is done on the gases, therefore a higher negative loop is obtained.
- During the expansion or working stroke, the fuel burns and the gases expand, therefore a large positive loop is obtained. In this stroke, the work is done by the gases.
- During exhaust stroke, the work is done on the gases, therefore a negative loop is formed.

2. Work Done Per Cycle

$$\boxed{\text{Work done per cycle} = T_{mean} \times \theta}$$

Where, T_{mean} is mean torque,
θ is angle turned is one cycle
$$= 2\pi, \text{ in case of two stroke engine}$$
$$= 4\pi, \text{ in case of 4 stroke engine}$$
- $T_{max} - T_{mean} = I.\alpha_{max.}$

3. Fluctuation of Speed

- The difference between the *maximum* and **minimum** speeds during a cycle is called maximum fluctuation of speed.
- The ratio of maximum fluctuation of speed to the mean speed is called *coefficient of fluctuation of speed* (C_s).

$$\boxed{C_s = \frac{N_{max} - N_{min}}{N_{mean}}}$$

- **Coefficient of Steadiness**
  The reciprocal of the coefficient of fluctuation of speed is known as coefficient of steadiness and is denoted by m.

$$\boxed{m = \frac{1}{C_s} = \frac{N_{mean}}{N_{max} - N_{min}}}$$

4. Maximum Fluctuation of Energy

- The turning moment diagram for multi-cylinder engine is shown in Fig. The horizontal line AG represents the mean line. Let a_1, a_3, a_5 be area above mean **torque line** and a_2, a_4 and a_6 be area below mean **torque line**.

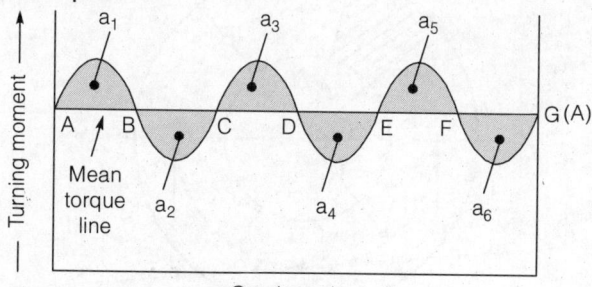

Let the energy is the flywheel at A = E,
∴ Maximum fluctuation of energy

ΔE = Maximum energy – Minimum energy

We can also write

$$\Delta E = E_{max} - E_{min} = \frac{1}{2}I(\omega^2_{max} - \omega^2_{min})$$

$$\Delta E = I\omega^2_{mean}\, C_s$$

Here,

I = Mass moment of inertia of the flywheel about its axis of rotation

ω_{max} = Maximum angular speed during cycle

ω_{min} = Minimum angular speed during cycle

$\omega_{mean} = \dfrac{\omega_{max} + \omega_{min}}{2}$ during cycle

- Energy stored $(E) = I\omega^2$.

- Mass moment of Inertia for ring is MR^2 and for disc $\dfrac{MR^2}{2}$.

Remember

- Selection of flywheel shape is decided so that ω variation should be minimum.

- Coefficient of Fluctuation of Energy (C_E)

$$C_E = \frac{\text{Maximum flucturation of energy}}{\text{Work done per cycle}}$$

5. Dimensions of The Flywheel RIM

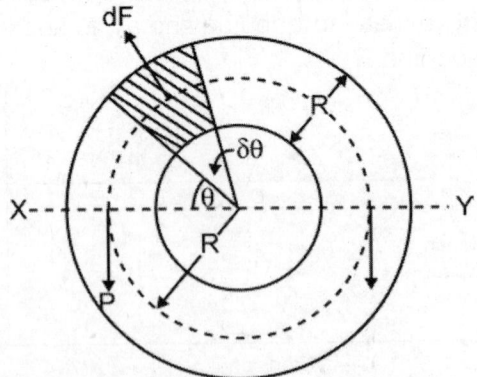

Let   D = Mean diameter of rim in meters,

R = Mean radius of rim in meters,

A = Cross-sectional area of rim in m^2,

ρ = Density of rim material in kg/m^3

N = Speed of the flywheel in r.p.m.,

ω = Angular velocity of the flywheel in rad/s,

v = Linear velocity at the mean radius in m/s

= $\omega . R$

σ = Tensile stress or hoop stress in N/m^2 due to the centrifugal force.

∴ **Total vertical upward force tending of burst the rim across the diameter XY.**

$$= 2\rho.A.R^2.\omega^2$$

This vertical upward force will produce tensile stress of hoop stress and it is resisted by 2P,

$$2P = 2\sigma.A$$

$$2.\rho.A.R^2.\omega^2 = 2\sigma.A$$

or

$$\sigma = \rho.R^2.\omega^2 = \rho.v^2$$

∴

$$\boxed{v = \sqrt{\frac{\sigma}{\rho}}} \; m/s$$

• **Mass of the Rim**

$$\mu = \text{Volume} \times \text{density} = \pi.D.A.\rho$$

∴

$$A = \frac{m}{\pi.D\rho}$$

■■■■

Balancing

8

Balancing is the process of designing or modifying machinery so that the unbalance is reduced to an acceptable level and if possible, eliminate completely.

The most common approach to balancing is by **redistributing the mass** which may be accomplished by **addition or removal of mass** from various machine members.

1. Static Balancing

A system of rotating masses is said to be in static balance if the combined mass centre of the system lies on the **axis of rotation**.

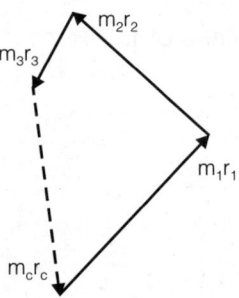

- The rotor is said to be statically balanced. If vector sum of centrifugal force (F) is zero.

$$\Sigma F = 0$$

$$F = m_1 r_1 \omega^2 + m_2 r_2 \omega^2 + m_3 r_3 \omega^3$$

Here, ω = constant angular velocity

m_1, m_2 and m_3 = rotating masses

r_1 r_2 and r_3 = radii of masses.

- If F is not zero, then introduce a counter weight of mass 'm_c' at radius r_c to balance the rotor so that

$$\Sigma mr + m_c r_c = 0$$

and $$\tan \theta_c = \frac{\Sigma mr \sin\theta}{\Sigma mr \cos\theta}$$

and $$m_c r_c = \sqrt{(\Sigma mr \cos\theta)^2 + (\Sigma mr \sin\theta)^2}$$

2. Dynamic Balancing

A system of rotating masses is in *dynamic balance* when there does not exist any *resultant centrifugal* force as well as *resultant couple*.

Force = mr
Couple = mrl

Force = $\Sigma mr\omega^2 = 0$

Couple = $\Sigma mr\omega^2 l =$

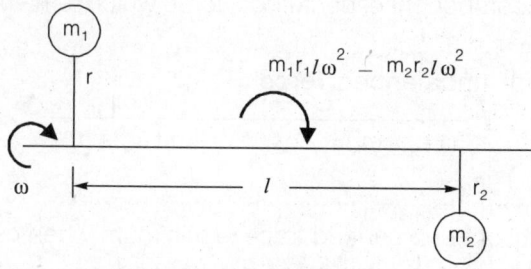

3. Balancing of Reciprocating Mass.

- Acceleration (α) and force (F) of the reciprocating mass of a slider crank mechanism.

$$\alpha = \omega^2 r \left(\cos\theta + \frac{\cos 2\theta}{n} \right)$$

$$\text{Force (F)} = mr\omega^2 \cos\theta + mr\omega^2 \frac{\cos 2\theta}{n}$$

- $mr\omega^2 \cos\theta$ –   primary accelerating force

- $\dfrac{mr\omega^2}{n} \cos 2\theta$ –   secondary accelerating force.

- $n = \dfrac{l}{r} = \dfrac{\text{Length of connecting rod}}{\text{Radius of crank}}$

Remember: ....
- Maximum value of primary force = $mr\omega^2$.

- Maximum value of secondary force = $\dfrac{mr\omega^2}{n}$.

- Secondary force is small as compared to primary force for slow speed.

Primary Balancing

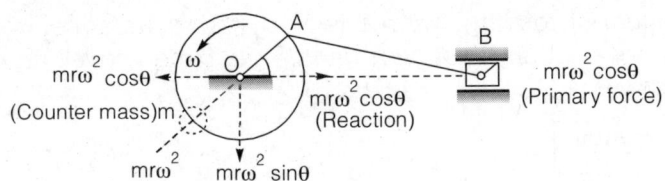

$mr\omega^2 \cos\theta$
(Counter mass)m

$mr\omega^2 \cos\theta$
(Reaction)

$mr\omega^2 \cos\theta$
(Primary force)

$mr\omega^2$ $mr\omega^2 \sin\theta$

- If c is the fraction of the reciprocating mass then,
  Primary force balanced by the mass = $c\, mr\omega^2 \cos\theta$
  Primary force unbalanced by the mass = $(1-c)\, mr\omega^2 \cos\theta$
- Vertical component of centrifugal force which remains unbalanced
  = $cmr\,\omega^2 \sin\theta$.
  Resultant unbalanced force

$$= \sqrt{[(1-c)\, mr\omega^2 \cos\theta]^2 + [cmr\,\omega^2 \sin\theta]^2} \ .$$

Remember: ....
- The resultant unbalanced force is minimum when c = 1/2
- 2/3 of the reciprocating mass is balanced.
- The unbalanced force is zero at the ends of the stroke when $\theta = 0°$ or $180°$ and maximum at the middle when $\theta = 90°$. The magnitude of the unbalanced force remains the same i.e. equal to $mr\omega^2$.
- Secondary Balancing

$$\text{Secondary force} = mr\omega^2\, \frac{\cos 2\theta}{n}$$

$$= mr(2\omega)^2\, \frac{\cos 2\theta}{4n}$$

- For complete balancing of reciprocating mass. Primary force, primary couple, secondary force secondary couple, all must be balanced.

4. Effect of Partial Balancing in Locomotive

- **Hammer Blow**

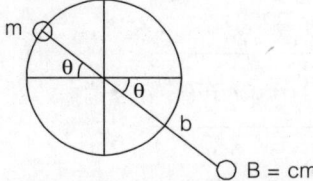

Vertical unbalanced force = $Bb\omega^2 \sin\theta$

$$\boxed{\text{Hammer Blow} = B.b\omega^2 = cmr\omega^2}$$

- It is the maximum vertical unbalanced force caused by the mass provided to balance the reciprocating masses.

Remember: ....

- It varies as square of the speed.
- At high speeds, force of the hammer blow could exceed the static load on the wheels and the wheels can be lifted off the rail when the direction of the hammer blow will be *vertically upwards*.
- **Variation of Tractive Force**
  A variation in the tractive force (effort) of an engine is caused by the unbalanced portion of the primary force which acts along the line of stroke of a locomotive engine.
  - Total unbalanced primary force or the variation in the tractive force = $(1 - c) mr\omega^2 (\cos\theta - \sin\theta)$. This is maximum when $\theta = 135°$ or $315°$

  - $$\boxed{\text{Maximum variation in tractive force} = \pm\sqrt{2}(1 - c)mr\omega^2}$$

- **Swaying Couple**

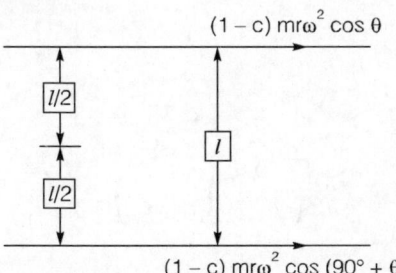

Unbalanced primary forces along the lines of stroke are separated by a distance *l* apart and thus constitute a couple. This tends to make the leading wheels sway from side to side.

- Swaying couple (about engine centre line).

$$= (1 - c)\, mr\omega^2\, (\cos \theta + \sin \theta)\,.\,\frac{l}{2}$$

This is maximum when $\theta = 45°$ or $225°$

Maximum swaying couple $\pm \dfrac{1}{\sqrt{2}}(1 - c)\, mr\omega^2 l$

■ ■ ■ ■

Governors

9

The function of a Governor is to maintain the speed of an engine within specified limits whenever there is a variation of load. The operation of a flywheel is continuous whereas that of a governor is more or less intermittent.

1. Type of Governors

- ### Centrifugal governors
  Its action depends on the change of speed and centrifugal effect produced by the masses, known as governor balls, which rotate at a distance from the axis of rotation.
  The valve is operated by the actual change of engine speed in the case of centrifugal governors.

- ### Inertia Governors
  The positions of the balls are affected by the forces set up by an angular acceleration or deceleration of the spindle, in addition to centrifugal forces on the balls.
  It is by the rate of change of speed in case of inertia governors.

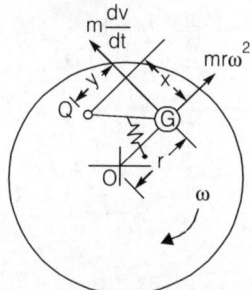

Let, r = Radial distance OG

v = Tangential velocity of G = ωr

ω = Angular velocity of disc

Centrifugal force of the rotating mass,

F = (radially outwards) $mr\omega^2$

If the engine shaft is accelerated due to increase in speed, the ball mass does not get accelerated at the same amount on account of its inertia, the inertia force being equal to

$$F_i = ma = m\frac{dv}{dt}$$

Remember: ....

- Response of inertia governors is *faster than* that of centrifugal governors.
- In inertia governor, both forces i.e., centrifugal force and inertia force are in action.

2. Type of Centrifugal Governor

- Pendulum type — Watt Governor (simplest)
- Loaded type

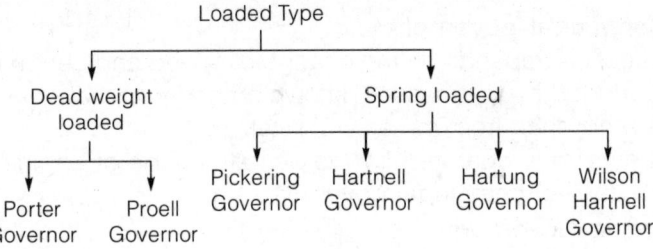

3. Watt Governor

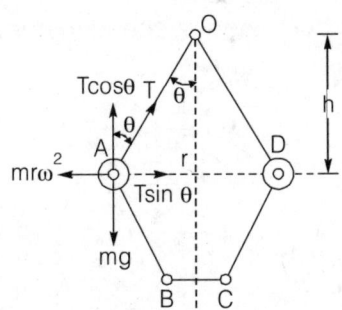

Let,   m = Mass of each ball
       h = Height of each ball
       w = Weight of each ball
       ω = Angular velocity of the balls, arms and the sleeve
       T = Tension in the arm
       r = Radial distance of ball-centre from spindle-axis

- **Assumption**
  - Link is massless.
  - Sleeve is frictionless.

$$h = \frac{895}{N^2} \text{ (metre)}$$

- Variation of height 'Δh' with speed

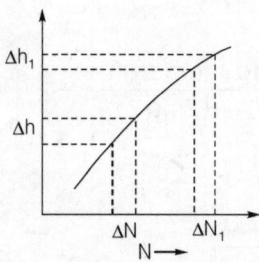

Remember: ....

- Height 'h' is independent of mass of ball.
- At higher speed, *sensitivity* will *decrease*.

4. Porter Governor

If the sleeve of a Watt governor is loaded with a *heavy mass*, it becomes a Porter governor.

Let, M = Mass of the sleeve

m = Mass of each ball

f = Force of friction at the sleeve

h = Height of the governor

r = Distance of the centre of each ball from axis of rotation

θ = Angle between arm and spindle axis

β = Angle between link and spindle axis

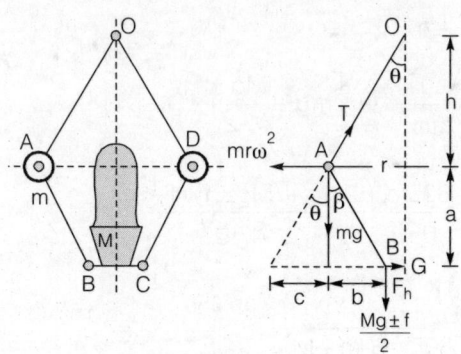

$$\omega^2 = \frac{1}{mh}\left(\frac{2mg+(Mg\pm f)(1+k)}{2}\right)$$

or $\quad N^2 = \dfrac{895}{h}\left(\dfrac{2mg+(Mg\pm f)(1+k)}{2mg}\right)\left[k=\dfrac{\tan\beta}{\tan\theta}\right]$

If $\quad k = 1, \quad f = 0$

$$\boxed{N^2 = \frac{895}{h}\left(\frac{m+M}{m}\right)}$$

5. Proell Governor

A Porter governor is known as a Proell governor if the two balls (masses) are fixed on the upward extensions of the lower links which are in the form of bent links BAE and CDF shown in the below figure.

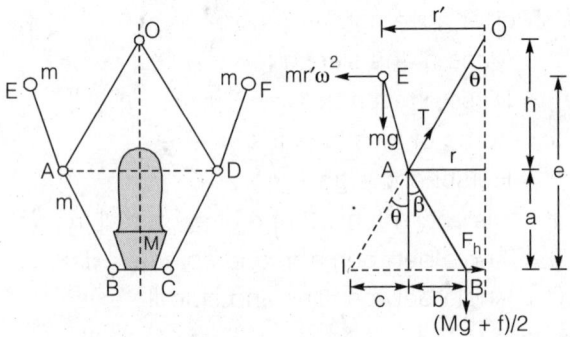

After taking $r' = r$

$$h = \frac{1}{m\omega^2}\cdot\frac{a}{e}\cdot\left[mg+\frac{Mg\pm f}{2}(1+K)\right]$$

$$N^2 = \frac{895}{h}\frac{a}{e}\left(\frac{2mg+(Mg\pm f)(1+k)}{2mg}\right)$$

If $\quad k = 1, \quad f = 0$

$$\boxed{N^2 = \frac{895}{h}\frac{a}{e}\left(\frac{m+M}{m}\right)}$$

6. Hartnell Governor

In this governor, ball are controlled by a *spring*.

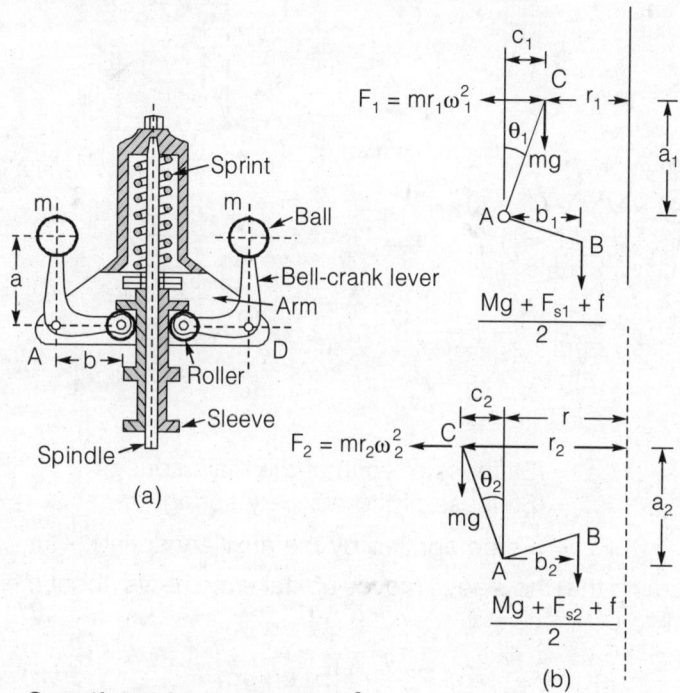

(a)

(b)

Let,   Centrifugal force (F) = $mr\omega^2$

        F_s = Spring force

Taking moments about the fulcrum A,

$$F_1 a_1 = \frac{1}{2}(Mg + F_{s1} + f)b_1 + mgc_1$$

$$F_2 a_2 = \frac{1}{2}(Mg + F_{s2} + f)b_2 + mgc_2$$

Neglect obliquity of the arm in that case,

      $a_1 = a_2 = a$, $b_1 = b_2 = b$, $c_1 = c_2 = 0$

$$F_1 a_1 = \frac{1}{2}(Mg + F_{s1} + f)b$$

$$F_2 a_2 = \frac{1}{2}(Mg + F_{s2} + f)b$$

Let,   K = Stiffness of the spring

$$K = 2\left(\frac{a}{b}\right)^2 \left(\frac{F_2 - F_1}{r_2 - r_1}\right)$$

7. Wilson-Hartnell Governor (Radial-Spring Governor)

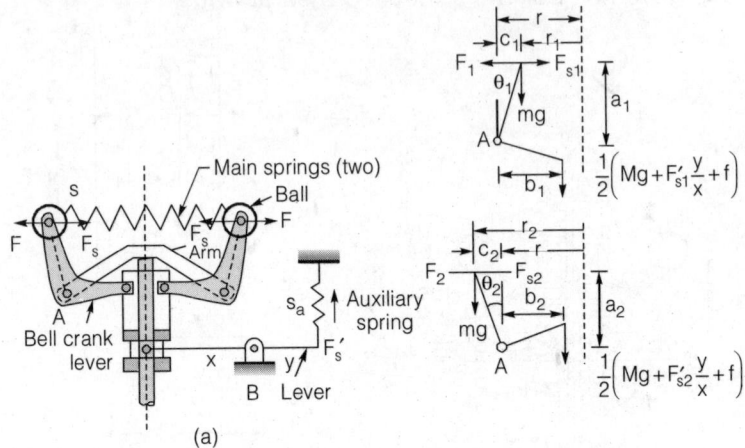

(a)

Let s = Stiffness of each of the main springs

S_a = Stiffness of the auxiliary spring

F_s' = Force applied by the auxiliary spring

Assuming that the sleeve moves up, take moments about the fulcrum A in two positions,

$$F_1 a_1 - F_{s1} a_1 = \frac{1}{2}\left(Mg + F_{s1}'\frac{y}{x} + f\right)b_1 + mgc_1$$

$$F_2 a_2 - F_{s2} a_2 = \frac{1}{2}\left(Mg + F_{s2}'\frac{y}{x} + f\right)b_2 + mgc_2$$

If obliquity effects are neglected,

$$a_1 = a_2 = a, \ b_1 = b_2 = b \text{ and } c_1 = c_2 = 0$$

$$(F_1 - F_{s1})a = \frac{1}{2}\left(Mg + F_{s1}'\frac{y}{x} + f\right)b$$

$$(F_2 - F_{s2})a = \frac{1}{2}\left(Mg + F_{s2}'\frac{y}{x} + f\right)b$$

The main spring consists of two springs. Therefore, the force exerted is given by,

$$F_{s2} - F_{s1} = 2 \times \text{Force exerted by each spring}$$
$$= 2 \times \text{Stiffness of each spring} \times \text{Elongation of each spring}$$
$$= 2 \times s \times 2 \times (r_2 - r_1)$$
$$= 4s\,(r_2 - r_1)$$

and also

$$\frac{F_2 - F_1}{r_2 - r_1} = 4s + \frac{S_a}{2}\left(\frac{b}{a}\frac{y}{x}\right)^2$$

8. Pickering Governor

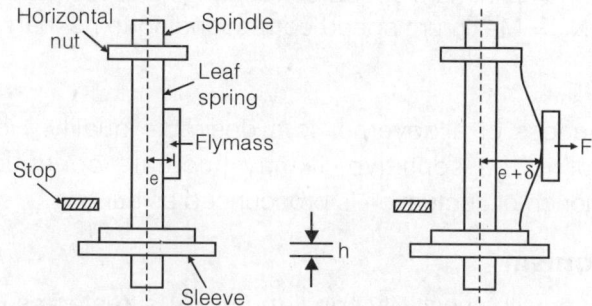

Let   m = Mass fixed to each spring
      e = Distance between spindle axis and centre of mass when the governor is at rest
      ω = Angular speed of the sleeve
      δ = Deflection of the centre of the leaf spring for spindle speed ω

E = Modulus of elasticity of the spring material

I = Moment of inertia of the cross-section of the spring about neutral axis = $\dfrac{bt^3}{12}$, b and t being the width and the thickness of the leaf spring.

Centrifugal force, $F = m(e + \delta)\omega^2$
Assume leaf spring is a beam of uniform cross section fixed at both ends and carrying a load at the centre.

$$\delta = \frac{Fl^3}{192EI} = \frac{m(e+\delta)\omega^2 l^3}{192EI}$$

Remember: ....
- Pickening governor is used in *gramophone*.

9. Sensitiveness of Governor

A governor is said to be sensitive when it readily responds to a small change of speed.

$$\text{Sensitiveness} = \frac{\text{range of speed}}{\text{mean speed}} = \frac{N_2 - N_1}{N}$$

$$\boxed{\text{Sensitiveness} = \frac{2(N_2 - N_1)}{N_1 + N_2}}$$

When N = Mean speed

         N_1 = Minimum speed corresponding to full load conditions

         N_2 = Maximum speed corresponding to no-load conditions

10. Hunting

Sensitiveness of a governor is a desirable quality. However, if a governor is too sensitive, it may fluctuate continuously. This phenomenon of fluctuation is pronounced as hunting.

11. Isochronism

A governor with sensitivity equal *to infinity* is treated as isochronous governor. For *all position of sleeves*, governor has *same* speed.

12. Stability

A governor is said to be stable if it brings the speed of the engine to the required value and there is *not much hunting*. The ball masses, occupy a *definite position* for *each speed* of the engine within the *working range*. The stability and the sensitivity are two opposite characteristics.

13. Effort of Governor

The effort of the governor is the *mean* force acting on the sleeve to raise or lower it for a given *change* of *speed*. At constant speed, the governor is in *equilibrium* and the *resultant* force acting on the sleeve is zero. However, when the speed of the governor increases or decreases, a force is exerted on the sleeve which tends to move it. When the sleeve occupies a new steady position, the resultant force acting on it again becomes zero.

14. Power of Governor

The power of a governor is the work done at the sleeve for a given percentage change of speed.

$$\boxed{\text{Power} = \text{Effort of governor} \times \text{displacement of sleeve}}$$

15. Controlling Force

Controlling force is equal and opposite to the centrifugal force and acts readily inward. It is supplied by

- Gravity of mass of ball in case of watt governor.
- Gravity of mass of ball and dead weight of sleeve in case of porter and proell governor.

- Gravity of ball masses and spring force in **hartnell** and **hartung** governs.
- Controlling force curve for spring loaded governor.

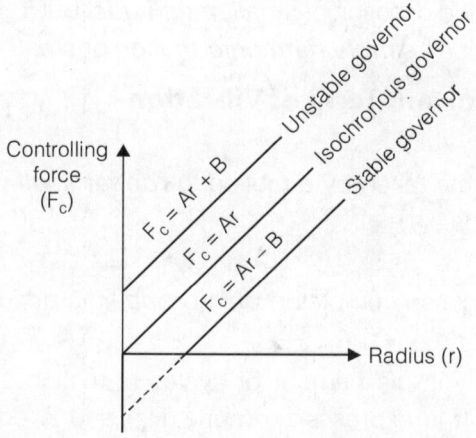

Where, A, B > 0

- Controlling force curve is *parabolic* curve in case of dead weight governor.
- Cons0trolling force curve is *straight line* in case of spring loaded governor.

■■■■

Vibration

10

Vibration is a periodic motion of *small magnitude*. But for sake of simplicity we can assume it as *simple harmonic motion* of *small* amplitude.

1. Some Important Term of Vibration

- **Period**

  It is the time taken by a motion to *repeat itself*, and is measured in seconds.

- **Cycle**

  It is the motion completed during *one* time period.

- **Frequency**

  Frequency is the number of cycles of motion completed in *one second*. It is expressed in hertz (Hz) and is equal to one cycle per second.

- **Resonance**

  When the frequency of the external force is the same as that of the natural frequency of the system, a state of resonance is said to have been reached. Resonance results in large amplitudes of vibrations and this may be dangerous.

2. Type of Vibration

- **Free Vibrations**

  Elastic vibrations in which there are *no friction* and *external* forces after the initial release of the body, are known as free or natural vibrations.

- **Forced Vibrations**

  When a repeated force continuously acts on a system, the vibrations are said to be forced vibration. The frequency of the vibrations is that of the *applied force* and is *independent* of there own *natural frequency* of vibrations.

- **Damped Vibrations**

  When the energy of a vibrating system is *gradually dissipated* by friction and other *resistance the* vibrations are said to be damped vibration.

- **Un-damped Vibrations (Hypothetical)**
  When there is *no friction or resistance* present in system to contract vibration then body execute un-damped vibration.
- **Longitudinal Vibrations**
  If the shaft is elongated and shortened so that the same moves *up and down* resulting in *tensile and compressive stresses* in the shaft, the vibrations are said to be longitudinal.

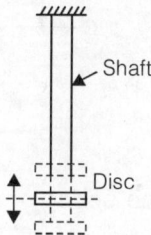

- **Transverse Vibrations**
  When the shaft is *bent* alternately and tensile and compressive stresses due to bending result, the vibrations are said to be transverse vibration

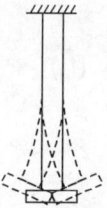

- **Torsional Vibrations**
  When the shaft is *twisted and untwisted* alternately and *torsional shear stresses* are induced, the vibrations are known as torsional vibrations.

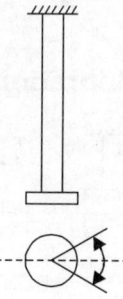

3. Free Longitudinal Vibrations

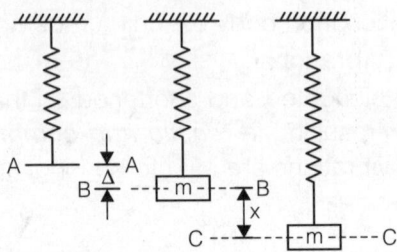

∴ Equation of equilibrium

$m\ddot{x} + sx = 0$

Here,

s = Stiffness of the spring

m = Mass

x = Displacement

- **Natural circular frequency (ω_n)**

$$\omega_n = \sqrt{\frac{s}{m}}$$

- **Natural linear frequency (f_n)**

$$f_n = \frac{1}{2\pi}\sqrt{\frac{s}{m}}$$

Here, mass of spring is neglected.

If we consider mass of spring is 'm_1' then

$$f_n = \frac{1}{2\pi}\sqrt{\frac{s}{m + \dfrac{m_1}{3}}}$$

4. Damped Longitudinal Vibration

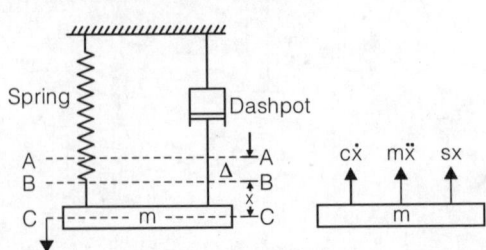

Let   s = Stiffness of the spring

         c = Damping coefficient (damping force per unit velocity)

         ω_n = Frequency of natural undamped vibrations

         x = Displacement of mass from mean position at time t

         $v = \dot{x}$ = velocity of the mass at time t

         $f = \ddot{x}$ = acceleration of the mass at time t

Equation of equilibrium,

$$m\ddot{x} + c\dot{x} + sx = 0$$

● **Damping factor**

$$\boxed{\zeta = \frac{c}{c_c} = \frac{\text{Actual damping coefficient}}{\text{Critical damping coefficient}}}$$

$$\zeta = \sqrt{\frac{(c/2m)^2}{s/m}} = \frac{c}{2\sqrt{sm}}$$

Here $C_c = 2\sqrt{sm}$

Thus when

         $\zeta = 1$, the damping is critical

         $\zeta > 1$, the system is over-damped

         $\zeta < 1$, the system is under-damped

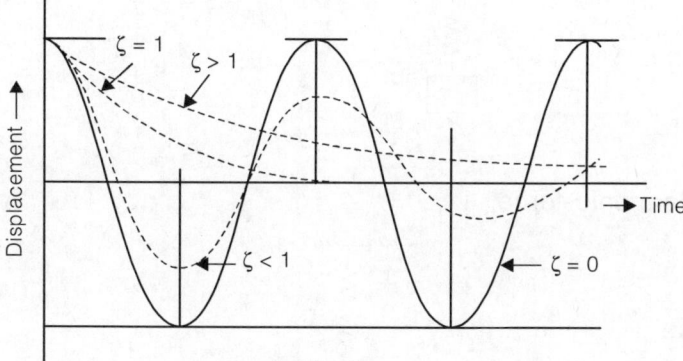

● **Logarithmic Decrement (δ)**

The ratio of two successive oscillations is constant in an underdamped system. Natural logarithm of this ratio is called logarithmic decrement.

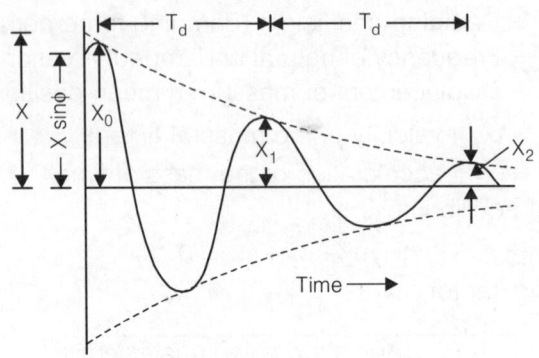

$$\delta = In\left(\frac{X_n}{X_{n+1}}\right) = In\,e^{(\zeta\omega_n T_d)} = \zeta\omega_n T_d$$

$$\boxed{\delta = \frac{2\pi\zeta}{\sqrt{1-\zeta^2}}}$$

5. Forced Vibration

- Step-input force

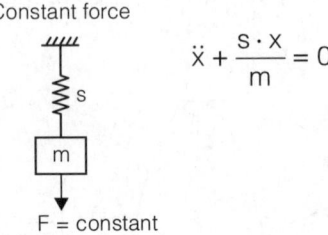

Constant force

$$\ddot{x} + \frac{s \cdot x}{m} = 0$$

F = constant

- Harmonic force

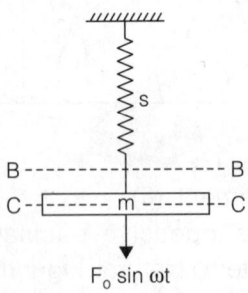

$$m\ddot{x} + sx = F_o \sin \omega t$$

$$x = X\sin(\omega_n t + \phi) + \frac{\dfrac{F_o}{s}}{1-\left(\dfrac{\omega}{\omega_n}\right)^2}\sin\omega t$$

6. Forced-Damped Vibrations

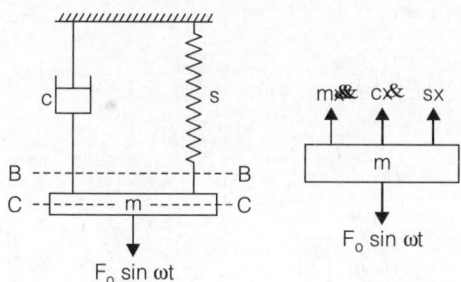

$$m\ddot{x} + c\dot{x} + sx = F_o \sin \omega t$$

Damped Free Response

$$x = Xe^{-\xi\omega_n t} \cdot \sin(\omega_d \cdot t + \phi_1)$$

Steady State Response

$$x = \underbrace{Xe^{-\xi\omega_n t} \cdot \sin(\omega_d \cdot t + \phi_1)}_{\text{Damped Free Response}} + \underbrace{\frac{F_0 \sin(\omega t - \phi)}{\sqrt{(s-m\omega)^2 + (c\omega)^2}}}_{\text{Steady State Response}}$$

- **Magnification Factor**

  The ratio of the amplitude of the steady-state response to the static deflection under the action of force F_o is known as the magnification factor (MF).

$$MF = \frac{1}{\sqrt{\left[1-\left(\dfrac{\omega}{\omega_n}\right)^2\right]^2 + \left[2\xi\dfrac{\omega}{\omega_n}\right]^2}}$$

- **Transmissibility**

  It is defined as the ratio of the force transmitted (to the foundation) to the force applied. It is a measure of the effectiveness of the vibration isolating material.

$$\epsilon = \frac{\sqrt{1+\left(2\xi\dfrac{\omega}{\omega_n}\right)^2}}{\sqrt{\left[1-\left[\dfrac{\omega}{\omega_n}\right]^2\right]^2+\left[2\xi\dfrac{\omega}{\omega_n}\right]^2}}$$

At resonance, $\omega = \omega_n$

$$\epsilon = \frac{\sqrt{1+(2\xi)^2}}{2\xi}$$

No damping $(\xi = 0)$

$$\epsilon = \frac{1}{\pm\left[1-\left(\dfrac{\omega}{\omega_n}\right)^2\right]}$$

Case 1 : $\dfrac{\omega}{\omega_n} < \sqrt{2} \Rightarrow \epsilon > 1$

Case 2: $\dfrac{\omega}{\omega_n} = \sqrt{2} \Rightarrow \epsilon = 1$

Case 3: $\dfrac{\omega}{\omega_n} > \sqrt{2} \Rightarrow \epsilon < 1$

● **TRANSVERSE VIBRATION**

7. Natural Linear Frequency

$$\omega_n = \sqrt{\frac{g}{\Delta}}$$

Δ = Static deflection

8. Whirling of Shaft

y = Displacement of shaft from axis

e = The distance by which centre of mass of shaft is displaced from shaft axis

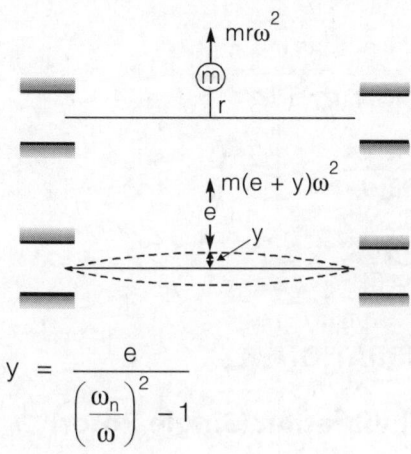

$$y = \frac{e}{\left(\dfrac{\omega_n}{\omega}\right)^2 - 1}$$

Case 1: At resonance, $\omega = \omega_n$

$$\omega_c = \omega_n = \sqrt{\frac{s}{m}} = \sqrt{\frac{g}{\Delta}}$$

Case 2: If $\omega > \omega_n$
         $y < 0$

Shaft will vibrate in opposite direction.

Case 3: If $y = -e$

Shaft will immediately stop vibrating.

9. Dunkerley's Method

Let W_1, W_2, W_3, ... be the concentrated loads on the shaft due to masses m_1, m_2, m_3, ... and δ_1, δ_2, δ_3 ... the static deflections of this shaft under each load when that load acts alone on the shaft. Let the shaft carry a uniformly distributed mass of m per unit length over its whole span and the static deflection at mid-span due to the load of this mass be δ_s. Also, let

      f_n = Frequency of transverse vibration of the whole system

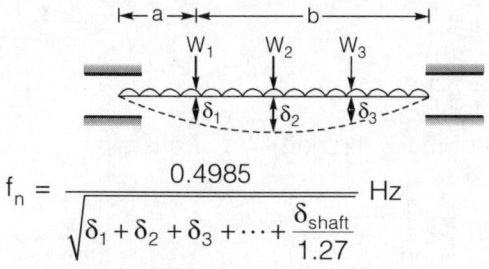

$$f_n = \frac{0.4985}{\sqrt{\delta_1 + \delta_2 + \delta_3 + \cdots + \dfrac{\delta_{shaft}}{1.27}}} \text{ Hz}$$

If mass of shaft is negligible

$$\delta_{shaft} = 0$$

$$\Rightarrow \qquad f_n = \frac{0.4985}{\sqrt{\delta_1 + \delta_2 + \delta_3 + \cdots}}$$

$$\therefore \qquad \delta_1 = \frac{W_1 a^2 b^2}{3EI\, l}$$

$$\delta_{shaft} = \frac{5Wl^4}{384\,EI}$$

- **TORSIONAL VIBRATION**

10. Free Torsional Vibration (Single Rotor)

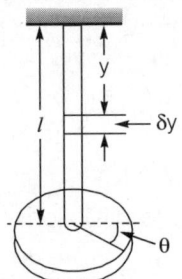

Let θ = Angular displacement of the disc from its equilibrium position at any instant

$$q = \left(\frac{GJ}{l}\right) = \text{Torsional stiffness of the shaft}$$

Here,   G = Modulus of rigidly of the shaft material
        J = Polar moment of inertia of the shaft cross-section

$$\ddot{I}\theta + q\theta = 0$$

$$\omega_n = \sqrt{\frac{q}{I}}$$

$$f_n = \frac{1}{2\pi}\sqrt{\frac{q}{I}}$$

- Inertia effect of mass of shaft

$$f_n = \frac{1}{2\pi}\sqrt{\frac{q}{I + \dfrac{I_1}{3}}}$$

Here, I_1 = Moment of inertia of shaft

11. Free Torsional Vibrations (Two Rotor System)

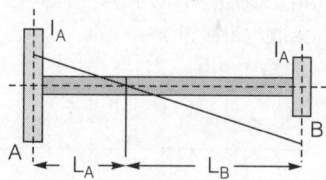

Let

L_a and L_b = Lengths of two portions of the shaft

I_a and I_b = Moment of inertia of rotors A and B respectively

q_a and q_b = Torsional stiffness of lengths I_a and I_b of the shaft respectively

f_{na} and f_{nb} = Natural frequencies of torsional vibrations of rotors A and B respectively.

$$\frac{L_A}{L_B} = \frac{I_B}{I_A} \text{ or } L_A I_A = L_B I_B$$

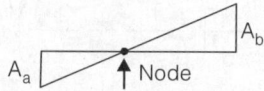

Remember: ....

• At node amplitude will be zero.

12. Free Torsional Vibrations (Three-Rotor System)

• Single Node System

Single node case observed when whether 'A' and 'B' rotate is same direction and 'C' in opposite direction *or* 'B' and 'C' in same direction and 'A' in opposite direction.

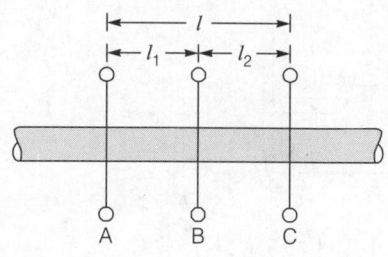

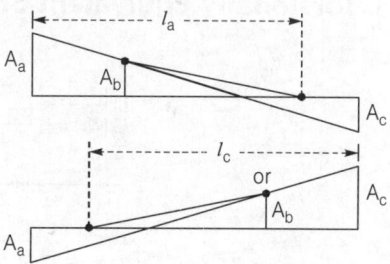

- **Double Node System**

  Double node case occur when 'A' and 'C' are in same direction and 'B' is in opposite direction.

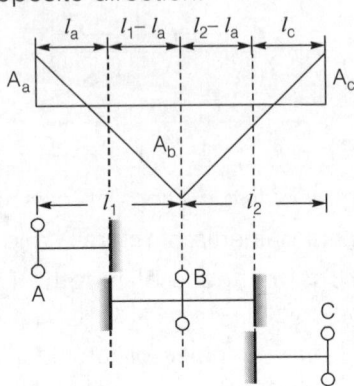

$$f_{na} = \frac{1}{2\pi}\sqrt{\frac{k\theta_A}{I_A}}$$

$$f_{nb} = \frac{1}{2\pi}\sqrt{\frac{GI_p}{I_b}\left(\frac{1}{l_1-l_a}+\frac{1}{l_2+l_c}\right)}$$

$$f_{nc} = \frac{1}{2\pi}\sqrt{\frac{k\theta_c}{I_c}}$$

$$\because \quad f_{na} = f_{nb} = f_{nc}$$

$$l_a\,l_a = l_c\,l_c$$

$$\frac{1}{l_a\,l_a} = \frac{1}{l_b}\left(\frac{1}{l_1-l_a}+\frac{1}{l_2-l_c}\right)$$

13. Torsionally Equivalent Shaft

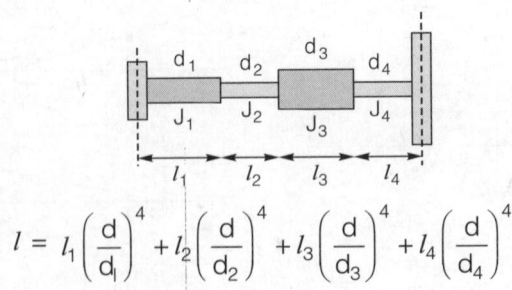

$$l = l_1\left(\frac{d}{d_l}\right)^4 + l_2\left(\frac{d}{d_2}\right)^4 + l_3\left(\frac{d}{d_3}\right)^4 + l_4\left(\frac{d}{d_4}\right)^4$$

■■■■

A Handbook on
Mechanical Engineering

Machine Design

CONTENTS

◀ ☐ ▶

Power Screws

1. Power Screw

It is mechanical device used for converting *rotary motion* into *linear motion* and *transmitting power*.

- Part of Power Screw
  - Screw
  - Nut
  - A part to hold either the screw and nut in its place
- Application of Power Screw
  - To raise the load.
  - To obtain accurate motion in machining operation i.e. lead-screw of lathe.
  - To clamp a work piece e.g. vice.
  - To load a specimen e.g. universal testing machine.

- In lathe lead screw: Screw rotates in bearing and nut moves in axial direction while, In screw jack : nut is stationary and screw(spindle) moves in axial direction.

- Advantage of Power Screw
  - *Large load* carrying capacity
  - Overall dimension is *small* result in *compact* construction.
  - It provides *large mechanical advantage*. A load of 15 kN can be raised by applying an effort as small as 400 N.
  - It can give highly accurate linear motion.
- Disadvantage of Power Screw
  - Poor efficiency as low as 40%.
  - High friction in thread cause rapid wear.

- square thread-nut is usually made of soft material.
- In trapezoidal threads, split-type of nut is used to compensate for the wear.

- Application and Efficiency
  - In power transmission high efficiency is required so lead screw, press is used.
  - For self locking purpose, Screw Jack, Clamp, Vice are used.

• Forms of Threads

V-threads are *not suitable* for *power screw* due to high friction. Screws with *smaller angle* of thread, such as trapezoid thread, are preferred for power transmission.

2. Type of Thread

• Square Thread (Screw Jack, Presses, Clamping device)

(i) Efficiency of square thread is more.

(ii) No radial pressure or side thrust on the nut. Life of nut is more.

(iii) Difficult to manufacture. Usually turned on lathe with a single-point cutting tool.

(iv) Square *thread* has, less *thickness* at the core diameter thus reduces load carrying capacity.

(v) Wear is compensated by making nut of soft material.

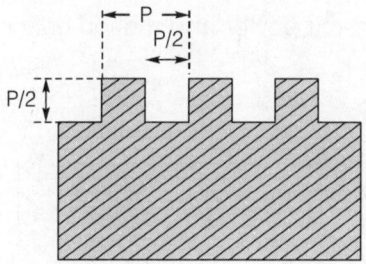

• Trapezoidal ISO Metric Thread

(i) Less efficiency

(ii) Side thrust is present

(iii) Easy to manufacture. Thread milling machine is used which has multi point cutting tool so result in less cost.

(iv) More thread thickness at core diameter more strength.

(v) The axial wear on the surface can be compensated by means of split-type of nut.

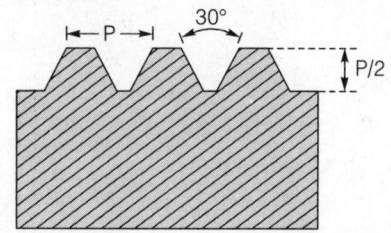

Lead Screw, Power Transmission device

3. Special Type of Thread

- **Acme Thread (Power transmission screw, Lead screw)**

  It is identical thread with respect to trapezoidal thread in all respect but thread angle is **29°** instead of 30°. Its advantage and disadvantage are same as trapezoidal threads.

- **Buttress Thread**

  It is used in *vices*. It combines the advantage of *square* and *trapezoidal* threads.

  It is used where a *heavy axial* force acts along the screw axis in *one direction only*.

- **Advantages**

  (i) It has higher efficiency compared with trapezoidal thread.

  (ii) Thread milling machine is used to manufacture it.

  (iii) *Stronger* than *square* and *trapezoidal* thread both.

- **Disadvantages**

  (i) It can transmit power and motion only in *one direction*.

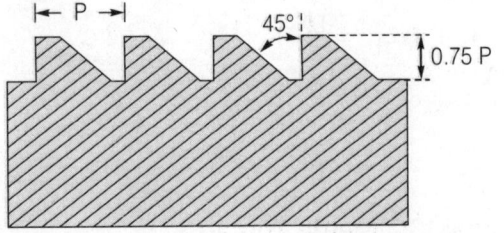

 ● Buttress threads are ideally suited for connecting *turbular component* (barrel housing in anti-aircraft gun).

Remember ● Acme thread transfers power in both direction while buttress in one direction only.

4. Designation of Screw

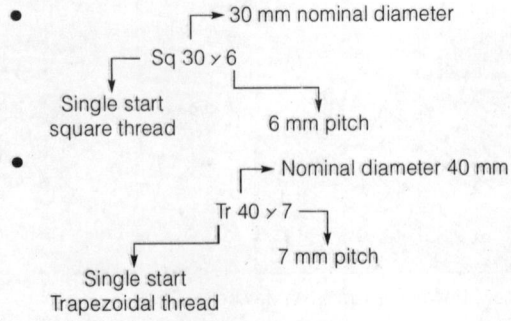

Remember: ....

- In double start thread, nominal diameter is followed by lead.

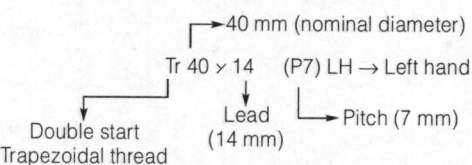

$$\therefore \text{ Number of start} = \frac{\text{Lead}}{\text{Pitch}} = \frac{14}{7} = 2$$

5. Multiple Threaded Screws

It is used where *higher travelling speed* is required. The travelling speed is twice that of single start screw.

- Efficiency of multi-start screw is higher than efficiency of single-start screw. This is due to increase in helix angle (α).
- Mechanical advantage of multi-start screw is less as compare to mechanical advantage of single-start screw. Thus effort required to raise a particular load is more in multiple threaded screw.
- It is likely that the self-locking property may be lost in a multi-threaded screws.
- Multiple threaded screws are used in high speed actuators and sluice valve.

- In multiple screw thread α increases thus condition for self locking $\phi > \alpha$ may get lost due to increase in α.

6. Terminology of Power Screw

- **Pitch (P)**

  It is defined as the distance measured parallel to the axis of screw from a point on one thread to the corresponding point on the adjacent thread.

- **Lead (*l*)**

  It is defined as the distance measured parallel to the axis of the screw which the nut will advance in *one revolution* of the screw. For *single-threaded* screw, the lead is same as the pitch.

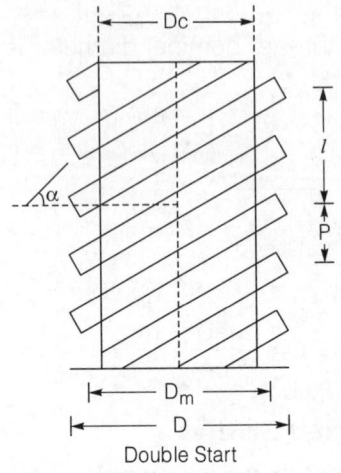

Double Start

Height of thread = P/2

- ## Nominal Diameter (D)
  It is the largest diameter of screw. It is also called major diameter.
- ## Core Diameter (D_c)
  It is the smallest diameter of screw thread. It is also called minor diameter.

$$D_c = D - P$$

- ## Mean Diameter (D_m)

$$D_m = \frac{D + D_c}{2} = D - 0.5P$$

- ## Helix Angle (α) or Lead Angle
  It is defined as the angle made by the helix of thread with a *plane perpendicular to* the axis of the screw.

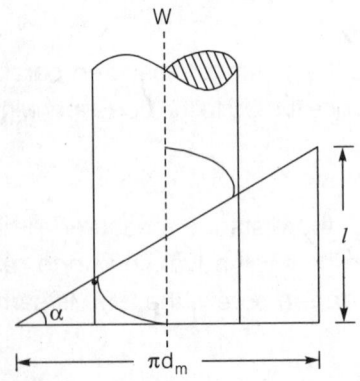

$$\tan \alpha = \frac{l}{\pi d_m}$$

7. Torque Requirement for Lifting and Lowering

Diagram for lifting   Diagram for lowering

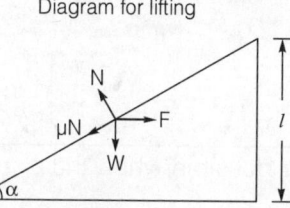 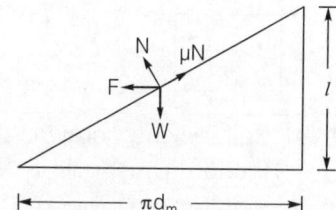

Here,   W = Load

         N = Normal reaction

         μN = Frictional force

         F = Effort

         μ = Friction coefficient

         ϕ = Friction angle

- **Effort Required to Lower the Load**

$$F = \frac{W(\mu \cos \alpha - \sin \alpha)}{(\cos \alpha + \mu \sin \alpha)} = \frac{W(\mu - \tan \alpha)}{(1 + \mu \tan \alpha)}$$

$$= W \tan(\phi - \alpha) \qquad\qquad [\because \mu = \tan \phi]$$

- **Effort Required to Raise the Load**

$$F = \frac{W(\mu \cos \alpha + \sin \alpha)}{(\cos \alpha - \mu \sin \alpha)} = \frac{W(\mu + \tan \alpha)}{(1 - \mu \tan \alpha)}$$

$$= W \tan(\phi + \alpha)$$

- **Torque Required to Lower the Load.**

$$T = W \cdot \frac{d_m}{2} \tan(\phi - \alpha)$$

- **Torque Required to Raise the Load.**

$$T = W \cdot \frac{d_m}{2} \tan(\phi + \alpha)$$

8. Self Locking Screw

Torque required to lower the load

$$T = W \cdot \frac{d_m}{2} \cdot \tan(\phi - \alpha)$$

For $\phi < \alpha$, torque required will be negative. This condition is known as overhauling or back driving. This is important in "Yankee Screwdriver".

For $\phi \geq \alpha$, Positive torque is required to lower the load. This condition is known as "self-locking". For self-locking

$$\phi > \alpha \; ; \text{ i.e.}$$

$$\boxed{\mu > \frac{l}{\pi d_m}} \qquad \left[\tan \alpha = \frac{l}{\pi d_m} \right]$$

Remember

- Self-locking screw is not possible when the coefficient of friction (μ) is low.
- Self-locking property is lost when lead is large. Thus single-threaded screw is better than multiple-threaded screw.

9. Efficiency of Square Threaded Screw

$$\therefore \qquad \eta = \frac{\text{work output}}{\text{work input}} = \frac{Wl}{F \times D_m} = \frac{W}{F} \tan \alpha$$

but we know that,

$$F = W \tan (\phi + \alpha)$$

$$\therefore \qquad \boxed{\eta = \frac{\tan \alpha}{\tan(\phi + \alpha)}}$$

Remember: ....

- Efficiency depends upon helix angle (α), friction angle (ϕ).
- Helix angle (α) depends upon lead (l) and mean diameter (d_m).
- Above relation consider frictional loss *only at the contacting surface* between screw and the nut. It does not take collar frictional loss.
- **Graph for Load Lifting**

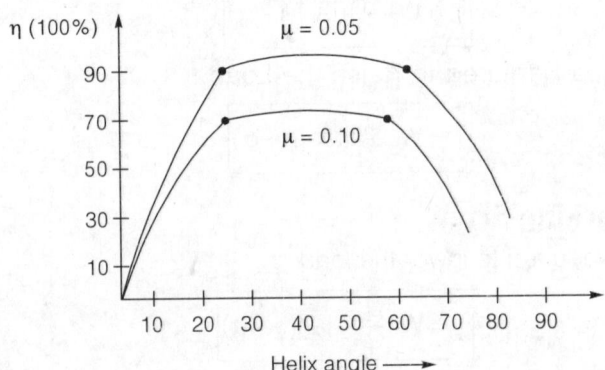

- Efficiency increases rapidly upto 20° helix angle (α).
- Efficiency is maximum for $\alpha = 40°\text{-}45°$.
- Efficiency decreases rapidly when helix angle (α) goes beyond 60°.
- Efficiency decreases when coefficient of friction increases.

- **Condition for Maximum Efficiency**

for $\eta_{maximum}$, $\boxed{\alpha = 45° - \dfrac{\phi}{2}}$

$\therefore \eta_{maximum} = \dfrac{1 - \sin\phi}{1 + \sin\phi}$

Remember

- **Efficiency of Self-Locking Screw**

$$\boxed{\eta \leq \dfrac{1}{2} - \dfrac{\tan^2\phi}{2}}$$

- Maximum efficiency depends only on friction angle (ϕ) i.e., coefficient of friction.
- Efficiency (h) must be less than or equal to 50%.

---

10. Efficiency of Trapezoidal & ACME Threads

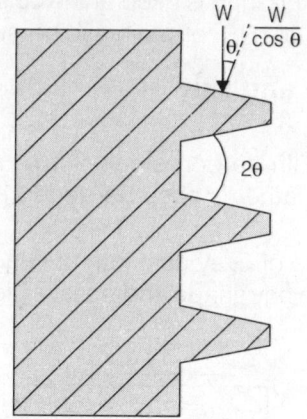

For Acme Thread $2\theta = 29°$

Here,

Axial force = W

Normal force $= W/\cos\theta = W \sec\theta$

Friction force will depend upon normal force thus, the effect of thread angle is to increase the friction force by 'sec θ'. To account for this effect the coefficient of friction is taken as $\mu \sec\theta$ instead of μ in case of trapezoidal threads.

- Force Required to Lift the Load

$$F = \frac{W(\mu \sec \theta + \tan \alpha)}{1 - \mu \sec \theta \cdot \tan \alpha}$$

- Force Required to Lower the Load

$$F = \frac{W(\mu \sec \theta - \tan \alpha)}{1 + \mu \sec \theta \cdot \tan \alpha}$$

- Efficiency

$$\eta = \frac{\tan \alpha (1 - \mu \sec \theta \cdot \tan \alpha)}{\mu \sec \theta + \tan \alpha}$$

11. Coefficient of Friction

Coefficient of friction will

- **Depends upon:**
  Workmanship in cutting the thread and lubricant
- **Practically Independent upon:**
  Load, Rubbing velocity, Material

Remember: ....

- Average coefficient of friction for mineral oil = 0.15.
- When thrust ball-bearing is used at the collar surface. Its coefficient of friction is about 1/10, of plain sliding surface.

12. Design of Screw and Nut

- **Screw**
  It is made of plain carbon steel. Screw are case hardened e.g. lead screw of a lathe is case hardened by *Nitriding Process*.
- **Frame**
  It is usually made of Grey cast Iron of Grade FG200, because cast iron can be given any shape and posses high compressive strength.

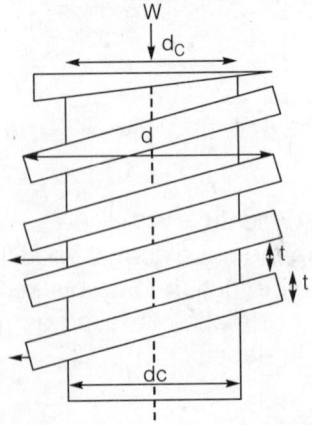

Axial Froce = W

Torsional moment (torque) = T

- **Compressive stress**

$$\sigma_c = \frac{W}{\frac{\pi}{4}d_c^2}$$

- **Torsional shear stress**

$$\tau = \frac{16T}{\pi d_c^3}$$

- **Principal shear stress**

$$\tau_{maximum} = \sqrt{\left(\frac{\sigma_c}{2}\right)^2 + \tau^2}$$

The *screw* will tend to shear off the thread at core diameter

$$\therefore \quad \tau_s = \frac{W}{\pi d_c \cdot t \cdot Z}$$

Here,

t = Thread thickness at the core diameter

Z = Number of thread engage within nut

Similarly for Nut

$$\tau_N = \frac{W}{\pi dt \cdot Z}$$

- **Bearing Pressure (S_b)**

Bearing area between Nut and screw for one thread.

$$= \frac{\pi}{4}\left(d^2 - d_c^2\right)$$

$$S_b = \frac{W}{\frac{\pi}{4}(d^2 - d_c^2)Z}$$

Remember

- Bearing pressure depends upon material and rubbing velocity.

■■■■

Rolling Contact Bearing

2

CLASSIFICATION OF BEARING

- Depending upon direction of force
  - Radial Bearing

    A radial bearing support the load, which is perpendicular to the axis of the shaft.

    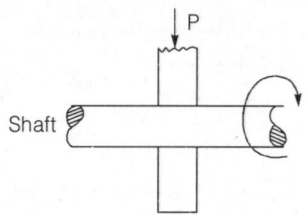

  - Thrust Bearing

    It support the load which act along the axis of the shaft.

    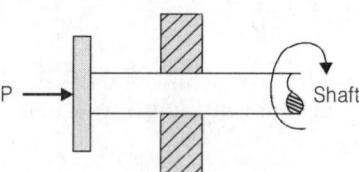

- Depending upon type of friction
  - Sliding contact bearing e.g. plain bearing, journal bearing, sleeve bearing.

    In this surface of the shaft slides over the surface of bush. To reduce friction, these two surface are separated by *film* of lubricating oil.

    Bush is made of white metal or Bronze.
  - Rolling contact bearing (*Anti-friction* bearing).

    In this *sliding friction* is replaced by *rolling friction*. e.g. bearing used in machine tool spindles, automobile axles, gear box, small size electric motor.

- A rolling contact bearing is called as antifriction bearing as it offers very low friction $f_{i.e.}$ $(0.005 \leq f_{R.C.} \leq 0.003)$.

Remember

1. Type of Rolling Contact Bearing

For *starting* condition and *at moderate speeds* the frictional loss in rolling contact bearing are *lower than* that of equivalent *hydrodynamic journal bearing*.

(i) Deep Groove Ball Bearing

In this the radius of ball is slightly less than the radii of curvature of the Groove in the races.

It gives a *point contact*. Due to point contact there will be less friction loss result in less temperature rise and less noise, so maximum permissible speed is high.

- Advantage
  - Due to large ball size the load carrying capacity will be high.
  - It can takes load in axial as well as radial direction.

Disadvantage :
  - Self alignment is not possible.
  - Poor rigidity due to point contact.

(ii) Cylindrical Roller Bearing

It is used when maximum load carrying capacity is required in a given space.

Advantage:
  - It gives a line contact so more rigid thus radial load carrying capacity is high.
  - Frictional loss is less in *high speed* application.

Disadvantage :
  - In general, it can not take thrust load.
  - Self alignment is not possible.

(iii) Angular Contact Bearing

In this, the grooves in inner and outer races are so shaped that the line of reaction at the contact between balls and races makes an angle with the axis of bearing.

- Advantage
  - It can take *both* radial and axial load.
  - Load carrying capacity is *more* than deep groove ball bearing.

- Disadvantage
  - It must be mounted without axial play.
  - Two bearing are required to take thrust in both direction.
  - This is assembled with specific magnitude of pre-load.

(iv) Self-aligning bearing

In self-aligning bearing, the external surface of the bearing bush is made spherical. The centre of this spherical surface is at the centre of the bearing. Therefore, the bush is free to roll in its seat and align itself with the journal.

Arrangement is made to provide lubricant between the spherical surfaces of the bush and its seat in order to reduce the friction. It is used in Agricultural machinery, railway axle-boxes.

Remember
- Self aligning bearing is used to compensate misalignment and can take radial and axial load both.
- This has less load carrying capacity compare with spherical roller bearing.

(v) Taper Roller Bearing

They are arranged in such a way that the axes of individual rolling elements intersects in a common apex point on the axis of the bearing. This is essential requirement of *pure rolling motion* between conical surfaces.

When it is subjected to pure radial load it induces a thrust component and vice-versa.

- Advantage
  - It can take heavy radial and thrust load
  - It is more rigid
  - It can be easily dissembled/assembled.
- Disadvantage
  - Two bearing is required to balance axial force.
  - Pre-load is required.
  - It can't tolerate mis-alignment.
- Application
  Car, truck, differential, propeller shaft, railroad axle-box, large size bearing in rolling mills.

(vi) Thrust ball bearing

A thrust ball bearing consists of a row of balls running between two rings-the shaft ring and the housing ring. Ball and inner/outer races are made of high carbon chromium steel and roller made of case hardened steel. It is used in gear boxes.

- Advantage
  - Large number of ball — High thrust load carrying capacity.
- Disadvantage
  - It carry thrust in only one direction.

- It can not carry radial load.
- It is not self-aligning.
- It performance is good only for low and medium speed.
- It require application of *continuous* pressure applied by spring to hold the rings together.

  Use : Worm gear boxes, core hooks

2. Selection of Bearing

Application	Type of Bearing
Low/medium radial load	Ball bearing
Heavy load	Roller bearing
Mis-alignment	Self-aligning bearing, spherical roller bearing
Medium thrust	Thrust ball bearing
Heavy thrust	Cylindrical thrust bearing
Load in either direction	Double acting thrust bearing
Radial and Thrust load both	Taper roller bearing, Deep groove bearing, angular contact bearing, spherical-roller bearing
High speed	Deep groove ball bearing, angular contact bearing
Rigidity	Double row cylindrical roller bearing, taper roller bearing.
Less Noise	Deep groove ball bearing

3. Static Load Carrying Capacity

- **Static Load**

  It is defined as the load acting on the bearing when the shaft is *stationary*.

- **Static load carrying capacity**

  It is defined as the static load which corresponds to a total permanent deformation of *ball and races* at the *most* heavily stressed point of contact equal to **0.0001** *of the ball diameter*.

4. Dynamic Load Carrying Capacity

It is defined as the radial load in radial bearing (Thrust load in thrust bearing) that can be carried for a *minimum life of one million revolution*. It is based assumption that inner race is rotating and outer race is stationary.

Remember
- Basic load rating of a ball bearing is the radial load at which 90% of the group of apparently identical bearing run for 1 million revolutions before failure occurs.

5. Equivalent / Actual Bearing Load

The equivalent dynamic load is defined as the *constant* radial load in radial bearing (Thrust load in thrust bearing), which if applied to the bearing would give *same life* as that which the bearing will attain in *actual condition*.

$$F_{eq.} = XVF_r + YF_a$$

$F_{eq.}$ = Equivalent dynamic load (N)

F_r = Radial load

F_a = Axial load

X = Radial factor

Y = Thrust factor

V = Race-rotation factor

= 1 when inner race rotates

= 1.2 when outer race rotates.

Remember: ....

- In case of pure radial load $F_{eq.}$ will be equal to radial load and in case of pure axial load $F_{eq.}$ will be equal to axial load.

6. Bearing Life Under Variable Load

$$L_{10(MR)} = \left(\frac{C}{Pe}\right)^n$$ million revolution

Here,   C = Dynamic load capacity

Pe = Equivalent dynamic load

$\dfrac{C}{Pe}$ = Loading rate

n = 3 for ball bearing

n = $\dfrac{10}{3}$ for roller bearing

$$L_{10(mR)} = \frac{L_{h_{10}} \times 60 \times N_{rpm}}{10^6}$$

L_{h10} = Bearing life in hour

N_{rpm} = Rev/n per minute

$L_{10(mR)}$ = Bearing life in million revolution.

Remember
- The value of n = 3 for ball bearing and n = 10/3 for roller bearing.

7. Designation of Ball-Bearing

$$P \quad Q \quad R \quad S$$

Here,

    P = Type of bearing

    Q = Indicate series

        1 — Extra light 2 — Light

        3 — Medium   4 — Heavy

   R S = When multiply by 5, it gives shaft diameter in mm.

8. Bearing Failure

- Abrasive Wear
  Cause:
  - Dust, Rust, Spatter.

  Remedies:
  - Oil seal, increase surface hardness, uses of high viscosity oil.

- Corrosive wear
  Cause:

  Water, moisture, corrosive element in lubricant.

  Remedies:
  - Complete enclosure, Proper additive.

- Pitting wear

  Pitting is the main cause of the failure of *anti friction* bearing. It occur when load on the bearing part exceeds the surface *endurance strength of material*.

  Pitting depends upon number of hertz contact stress and number of stress cycle.

  Surface endurance increases by increase in surface hardness.

- Scoring

  Excessive surface pressure, high surface speed and inadequate supply of lubricant result in breakdown of the lubricant film. It is also called stick-slip phenomenon.

9. Needle Bearing

It has cylindrical roller having very small diameter and relatively *long length*. It is also called quill bearing.

$$\frac{l}{d} > 4$$

This can be used *with or without inner* and outer race. This is suitable where limited *radial space* is available. It is used in oscillatory motion, piston-pin bearing, rocks arms, universal joint.

- **Advantage**
  - It is used to replace sleeve bearing because outer diameter is less.
  - It has large load carrying capacity compared to their size.
  - Large load carrying capacity particularly *at low* peripheral speed.

Remember

- It is also suitable for continuous rotation where load is *variable or intermittent*.
- $\mu = 0.0011$ – Cylindrical roller bearing
  $\mu = 0.0045$ – Needle bearing.
- Needle bearing is used to take light radial load with space limitation.
- Scoring is breakdown of lubricant film due to excessive pressure while pitting occurs when load exceeds endurance strength.

■■■■

Sliding Contact Bearing

3

1. Thick Film Lubrication

In this two surface of bearing in relative motion are *completely* separated by a *film of fluid*. Thus *surface friction* has no role, it has two types:

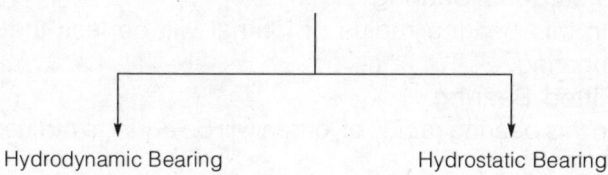

Hydrodynamic Bearing   Hydrostatic Bearing

2. Hydrodynamic Bearing

In hydrodynamic bearing *load-supporting fluid film* is created by *shape and relative motion* of sliding surface. In this *not neccesary* to supply the lubricant under pressure. The only requirement is sufficient and continuous supply. Its main application is in engine, centrifugal pump.

Remember: ....

- Bearing which operate without lubricant is called zero film bearing.
- Journal Bearing
  It is a *sliding contact bearing* working on hydrodynamic lubrication, support the load in radial direction. It has two type

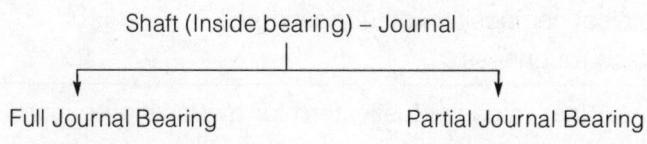

Shaft (Inside bearing) – Journal

Full Journal Bearing

- Angle of contact between journal and bearing is 360° thus take radial load in any direction.

Partial Journal Bearing

- Angle of contact between journal and bearing is less than 180°, can take load in only one direction.

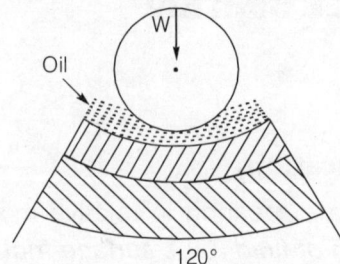

- Some Another Type of Bearing Related to Journal Bearing
    - Clearance Bearing
      In this bearing radius of journal will be less than radius of bearing.
    - Fitted Bearing
      In this bearing radius of jouran will be equal to radius of bearing.

3. Hydro-Static bearing

In this load supporting fluid film, separating two surface is created by external source like pump, supplying sufficient fluid under pressure. Thus it is also called externally pressurized bearing, it is used in centrifuges, ball mills and vertical turbo generator.

- Advantage
    - High load carrying capacity
    - Starting friction is absent
    - There are no rubbing action

4. Thin Film Lubrication

Lubricant film is *relatively thin* and there is *partial* metal to metal contact. In this lubrication under excessive load, less oil supply, low speed, misalignment, boundary lubrication will occur. Performance of bearing under boundary lubrication will depend upon.

(i) Chemical composition of lubricating oil.

(ii) Surface roughness.

Remember

- Wear rate is virtually zero for hydrosafic bearing.
- Under excessive load and low speed. Hydro-dynamic bearing also operates under boundary lubrication.

---

5. Measurement of Viscosity

Saybolt Universal Second (SUS)

$$Z_K = \left[0.22t - \frac{180}{t} \right]$$

Here, Z_k = Kinematic viscosity

        t = Viscosity in SUS

6. MCKEE'S Investigation

It is done for hydro-dynamic bearing.

- **Transition Graph From Thin Film to Thick Film**

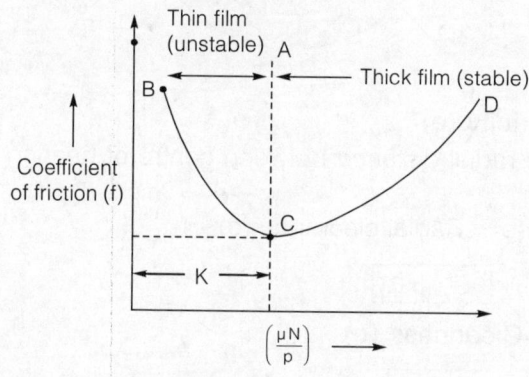

Bearing characteristics number = $\dfrac{\mu N}{p}$, it is dimensionless.

Here,   p0 = Unit bearing pressure

      μ = Absolute viscosity

      N = Speed (RPM)

At point C viscosity is minimum.

- The value of bearing characteristics number corresponding to minimum value of **coefficient of friction(f)** is called **bearing modulus (K)**.
- Bearing should not be operated near critical value of 'k' because slight drop in the speed, or increase in load result in boundary lubrication.

Ideally it should be **5 K**. But for impact fluctuating load it is **15 K**.

6. Raimondi and Boyd Method

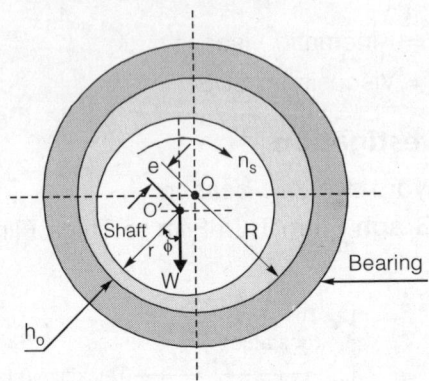

- **Eccentricity (e)**

  It is the radial distance between centre of bearing and centre of journal.

  ecentricity(e) = Radial clearance (c) − h_0

  $$e = c - h_0$$

- **Radial Cleanness (c)**

  $$c = R - r$$

- **Minimum Film Thickness (h_0)**

  $$h_0 = R - (r + e)$$

- **Eccentricity Ratio (ε)**

  $$\varepsilon = 1 - \frac{h_0}{c} = \frac{e}{c}$$

  Here, $\dfrac{h_0}{c}$ = Minimum thickness variable.

- **Sommerfeld Number**

  $$S = \left(\frac{r}{c}\right)^2 \cdot \frac{\mu \cdot n_s}{P}$$

  Here, μ = Viscosity (Pa − S)

        n_s = Journal speed in r.p.s.

        P = Unit bearing pressure in pascal (Pa)

        r = Radius of journal

        c = Radial clearance

- Petroff's equation

$$f = 2\pi^2\left(\frac{\mu n s}{P}\right)\left(\frac{2r}{C}\right)$$

- Attitude Angle (ϕ)

  It locate the position of minimum film thickness with respect to direction of load.

- Unit Bearing Pressure (p)

  The load per unit of projected area of the bearing in running condition.

- Flow Variable (FV)

$$FV = \frac{Q}{Nrcl}$$

Here,

           N = Speed (RPM)

           Q = Total Flow of lubricant

           r = Shaft radius

           l = Length of bearing

           c = clearance

- Co-efficient of Flow Variable (CFV) :

$$CFV = \left(\frac{r}{c}\right)\mu.$$

- Temperature Rise

$$\Delta t = \frac{8.3p\,(CFV)}{FV}$$

         p = Unit bearing pressure

7. Bearing Design (Selection of Parameter)

- Length to Diameter Ratio $\left(\dfrac{l}{d}\right)$

  as $\left(\dfrac{l}{d}\right)$ increases, film pressure increases. Thus long bearing has more load carrying capacity. But shorter bearing has greater side flow of lubricant. Which improves heat dissipation. Long bearing are more susceptible to metal to metal contact.

- In general $0.5 < \left(\dfrac{l}{d}\right) < 2$

- For long bearing $\left(\dfrac{l}{d}\right) > 1$

- For short bearing $\left(\dfrac{l}{d}\right) < 1$

- For square bearing $\left(\dfrac{l}{d}\right) = 1$

- **Start-up Load**
  Unit bearing pressure for starting conditions should not exceed 2N/mm^2.

- **Radial Clearance**
  Radial clearance should be small to provide necessary velocity gradient.

  ideally $c = \dfrac{r}{1000}$

- **Minimum Oil Film Thickness**

  $h_{0_{(min.)}} = \dfrac{r}{5000}$

- Maximum oil film temperature $< 120°$, to prevent oxidation of oil.

8. Bearing Material

It should have
- High compression strength
- High *endurance* strength to avoid failure due to *pitting*.
- Ability to yield and adopt its shape to that of journal (called conformability).
- The bearing material should be soft to allow dirt particles to get embedded in the lining and avoid further trouble. This property is called *embeddability*.
- *Comformability* is ability of bearing material to accommodate shaft deflection and bearing in accuracy by plastic deformation without excessive wear and heating.
- *Bondability* is ability of bearing material to accommodate bonding of series of layer of bearing material to steel shell and the strength of bond is called bondability.

Remember

- Babbitt is most popular bearing material. Due to its silvery appearance it is called 'white metal'. There are two varieties of babbitt, lead-based and tin-based.
- Tin babbitt poses excellent comformobility and embeddability.

9. Comparison of Rolling and Sliding Contact Bearing

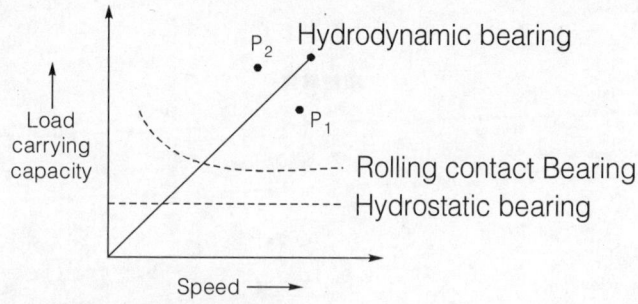

- **Hydrodynamic Bearing**
  - Higher starting friction (torque)
  - Load carrying capacity vary directly with speed.
  - It is suitable for shock, fatigue.
  - It is used for high load, high speed application.
- **Hydro-static Bearing**
  - Load carrying capacity is independent of speed.
- **Rolling Contact Bearing**
  - Poor damping capacity.
  - Required low starting torque.
  - Under full running capacity, friction loss is *more* as compare to *hydrodynamic bearing*.
  - Suitable for *frequent start*.
  - It require more radial space [but hydrodynamic require more axial space]
  - It is preferred for precise location of journal axis.

10. Heat Generated in a Bearing

- Power Loss Due to Friction is Heat Generated

$$P_{loss} = \mu W V$$

Here, W = Radial load

        V = Rubbing velocity

- Heat Dissipation Equation

$$H_d = \frac{(\Delta t + 18)^2 \times A}{K} \, watt$$

Δt = Bearing temperature

$A = l \times d$

$K = 0.484$ For unventilated bearing

     $= 0.273$ For well-ventilated bearing

■■■■

Design Against Static Load

4

A static load is defined as a force, which is gradually applied to a mechanical component and which does not change its magnitude or direction with respect to time.

1. Factor of Safety

While designing a component, it is necessary to provide sufficient *reserve strength* in case of an accident. This is achieved by taking a suitable factor of safety (fos).

The factor of safety is defined as

$$FOS = \frac{\text{failure stress}}{\text{allowable stress}}$$

or

$$FOS = \frac{\text{failure load}}{\text{working load}}$$

Magnitude of factor of safety depends upon type of load, degree of accuracy in force analysis, material of component, reliability of component, service condition etc.

2. Cotter Joint

A cotter joint is used to connect two rod either subjected to *tensile force* or *compressive axial force* with no relative motion between them.

- Application of Cotter Joint
  - Joint between the piston rod and the cross head of a steam engine.
  - Joint between the slide spindle and the fork of the valve mechanism.
  - Joint between the piston rod and the tail or pump rod.
  - Foundation bolt.
- Design of Cotter Joint

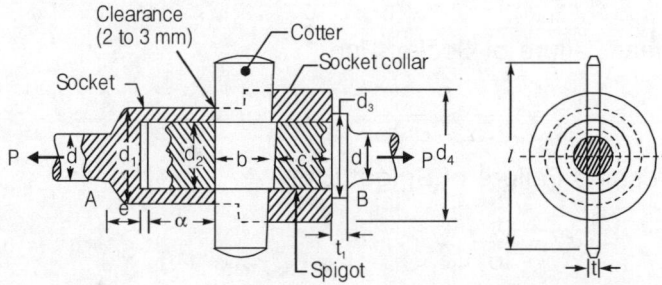

Here, P = Tensile force acting on rods

d = Diameter of each rod

d_1 = Outside diameter of socket

d_2 = Diameter of spigot or inside diameter of socket

d_3 = Diameter of spigot-collar

d_4 = Diameter of socket-collar

a = Distance from end of slot to the end of spigot on rod-B

b = Mean width of cotter

c = Axial distance from slot to end of socket collar

t = Thickness of cotter

l = Length of cotter

t_1 = Thickness of spigot collar

- **Tensile Failure of Rods**

$$\sigma_t = \frac{P}{\left[\frac{\pi}{4}d^2\right]}$$

- **Tensile Failure of Spigot**

$$\sigma_t = \frac{P}{\left[\frac{\pi}{4}d_2^2 - d_2 t\right]}$$

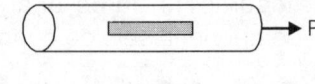

Remember: ....

- Thickness of cotter is usually determined by t = 0.31 d relation.
- **Shear Failure of Cotter**

$$\tau = \frac{P}{2bt}$$

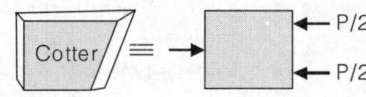

Here, τ = Permissible shear stress

- **Shear Failure of Spigot End**

$$\tau = \frac{P}{2ad_2}$$

- **Shear Failure of Socket End**

$$\tau = \frac{P}{2(d_4 - d_2)c}$$

- **Crushing Failure of Spigot End**

$$\sigma_c = \frac{P}{td_2}$$

- **Crushing Failure of Socket End**

$$\sigma_c = \frac{P}{(d_4 - d_2)t}$$

- **Bending Failure of Cotter**

$$\sigma_b = \frac{\dfrac{P}{2}\left[\dfrac{d_2}{4} + \dfrac{d_4 - d_2}{6}\right]\dfrac{b}{2}}{\left(\dfrac{tb^3}{12}\right)}$$

Cotter is provided with slight taper because:

(a) Cotter get tightened by wedge action.

Remember   (b) Taper provides easy removed of cotter and dismantle of the joint.

3. Knuckle Joint

Knuckle joint is used to connect two rods whose axes either *coincide* or *intersect* and *lie in one plane*. The knuckle joint is used to transmit axial *tensile* force. The construction of this joint permits *limited* angular movement between rods, about the axis of the pin.

- Knuckle joints is used to transmit axial tensile force and permits only limited angular movement between rods about the axis.

Remember

- **Application**
  - Joints between the tie bars in roof trusses.
  - Joints between the links of a suspension bridge.
  - Joints in valve mechanism of a reciprocating engine.
  - Fulcrum for the levers.
  - Joints between the links of a bicycle chain.
- **Stress Analysis**

  For the purpose of stress analysis of a knuckle joint, the following assumptions are made:
  - The rods are subjected to *axial tensile* force.
  - The effect of *stress concentration* due to holes is *neglected*.
  - The force is uniformly distributed in various parts.

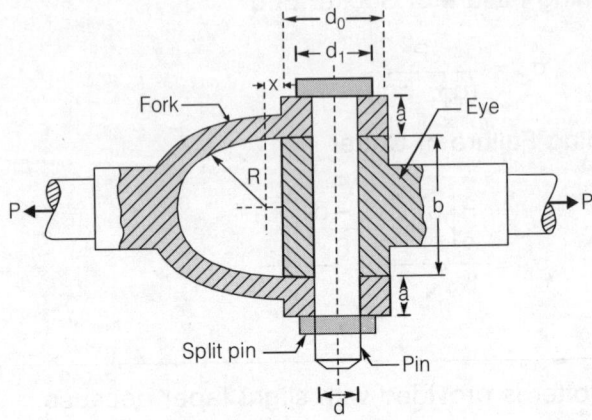

Here,   D = Diameter of each rod
D_1 = Enlarged diameter of each rod
d = Diameter of knuckle pin
d_0 = Outside diameter of eye of fork
a = Thickness of each eye of fork
b = Diameter of eye end of rod-B
d_1 = Diameter of pin head
x = distance of the centre of fork radius R
from the eye

- **Tensile Failure of Rod**

$$\sigma_t = \frac{P}{\frac{\pi}{4}D^2}$$

σ_t = Permissible tensile stress

- **Shear Failure of Pin**

$$\tau = \frac{P}{\frac{\pi}{2}d^2}$$

- **Crushing Failure of Pin in Eye**

$$\sigma_c = \frac{P}{l \times d}$$

- **Crushing Failure of Pin in Fork**

$$\sigma_c = \frac{P}{2ad}$$

- **Bending Failure of Pin**

$$\sigma_b = \frac{32}{\pi d^3} \times \frac{P}{2} \left[\frac{b}{4} + \frac{a}{3} \right]$$

- **Tensile Failure of Eye**

$$\sigma_t = \frac{P}{b(d_0 - d)}$$

- **Shear Failure of Eye**

$$\tau = \frac{P}{b(d_0 - d)}$$

- **Tensile Failure of Fork**

$$\sigma_t = \frac{P}{2a(d_0 - d)}$$

- **Shear Failure of Fork**

$$\tau = \frac{P}{2a(d_0 - d)}$$

Remember: ....

- Standard proportion for dimension 'a' and 'b' are a = 0.75 D and b = 1.25 D.

4. Design Against Fluctuating Load

- **Soderberg and Goodman Lines**
  When a component is subjected to fluctuating stresses as shown.

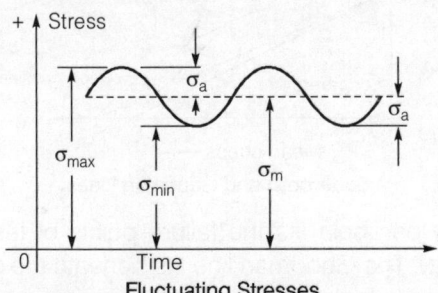

Fluctuating Stresses

There is mean stress (σ_m) as well as stress amplitude (σ_a). It has been observed that the mean stress component has an effect on fatigue failure when it is present in combination with an alternating component. The fatigue diagram for this general case in shown. In this stress amplitude is plotted on the ordinate. The magnitudes of (σ_m) and (σ_a) depend upon the magnitudes of maximum and minimum force acting on the component. When stress amplitude (σ_a) is zero, the load is purely static and the criterion of failure is S_{ut} or S_{yr}. These limits are plotted on the abscissa. When the mean stress (σ_m) is zero, the stress is completely reversing and the criterion of failure is the endurance limit S_e that is plotted on the ordinate. When the component is subjected to both components of stress, viz., (σ_m) and (σ_a), the actual failure occurs at different scattered points shown in the figure. There exists a border, which divides safe region from unsafe region for various combinations of (σ_m) and (σ_a). Different criterions are proposed to construct the borderline dividing safe zone and failure zone. They include Gerber line, Soderberg line and Goodman line.

- **Gerber Line**

  A parabolic curve joining S_e on the ordinate to S_{ut} on the abscissa is called the Gerber line.

- **Soderberg Line**

  A straight line joining S_e on the ordinate to S_{yt} on the abscissa is called the Soderberg line.

- **Goodman Line**

  A straight line joining S_e on the ordinate to S_{ut} on the abscissa is called the Goodman line.

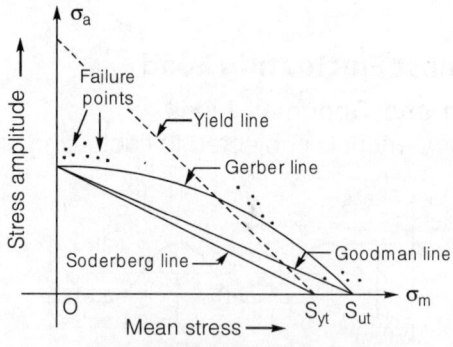

Soderberg and Goodman Lines

The Gerber parabola fits the failure points of test data in the best possible way. The Goodman line fits beneath the scatter of this data. Both Gerber parabola and Goodman line intersect (S_e) on the ordinate to (S_{ut}) on the abscissa. However, the Goodman line is more safe from design considerations because it is completely inside the Gerber parabola and inside the failure points. The Soderberg line is a more to consider even yielding in this case. A yield line is constructed connecting (S_{yt}) on both axes. It is called the limit on 'first cycle' of stress. This is because if a part yields, it has failed, regardless of its safety in fatigue.

We will apply following form for the equation of a straight line,

$$\frac{x}{a} + \frac{y}{b} = 1$$

where a and b are the intercepts of the line on the X and Y axes respectively.

Applying the above formula, the equation of the Soderberg line is given by,

$$\boxed{\frac{\sigma_m}{S_{yt}} + \frac{\sigma_a}{S_e} = 1}$$ Soderberg equation

Similarly, the equation of the Goodman line is given by,

$$\boxed{\frac{\sigma_m}{S_{ut}} + \frac{\sigma_a}{S_e} = 1}$$ Goodman equation

where σ_m = Permissible mean stress.

             σ_a = Permissible amplitude stress.

■■■■

Welding can be defined as a process of joining metallic parts by heating to a suitable temperature with or without the application of pressure.

1. Advantage and Disadvantage of Welded joint

- Advantage
  - The welded structure are usually *lighter* than riveted structures.
  - The welded joints provide *maximum efficiency* (may be 100%) which is not possible in case of riveted joints.
  - Alterations and addition can be easily made in existing structure.
  - In welding connections, the *tension members* are not weakened as in case of riveted joints.
  - A welded joint has a great strength. Often a welded joint has the strength of the parent metal itself.
  - The welding provides very rigid joints.
  - The process of welding takes less time than the riveting.
- Disadvantage
  - Since there is an uneven heating and cooling during fabrication, therefore the members may get *distorted* or *additional stresses may develop*.
  - It requires a highly skilled labour and supervision.
  - Since no provision is kept for expansion and contraction in the frame, therefore there is a *possibility of cracks* developing in it.
  - The inspection of welding work is more difficult than riveting work.

2. Butt Joints

Butt joint is defined as a joint between two component lying *approximately* in the *same plane*.

- Strength of Butt Welds

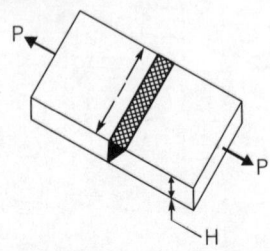

$$\sigma_t = \frac{P}{hl}$$

Where, σ_t = Tensile stress in the weld (N/mm^2)

P = Tensile force on the plates (N)

h = Throat of the butt weld (mm)

l = Length of the weld (mm)

Equating the throat of the weld 'h' to the plate thickness 't'.

$$\sigma_t = \frac{P}{tl}$$

Remember: ....

- For **unfired** pressure vessels, reduction in strength of a butt welded joint by a factor called efficiency of the joint (η) is necessary

$$P = \sigma_t\, l\, t\, \eta$$

3. Fillet Joints

A fillet joint, also called a **lap joint**, is a joint between two **overlapping** plates or components. A fillet weld consists of an approximately triangular cross-section joining two surfaces **at right angles** to each other.

- **Type of Fillet Joint**

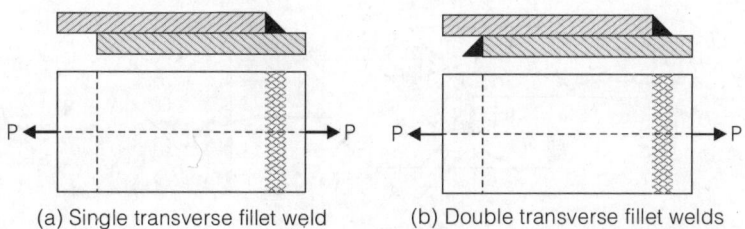

(a) Single transverse fillet weld   (b) Double transverse fillet welds

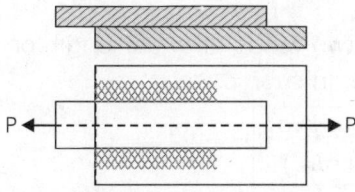

(c) Double parallel fillet welds

4. Strength of Parallel Fillet Weld

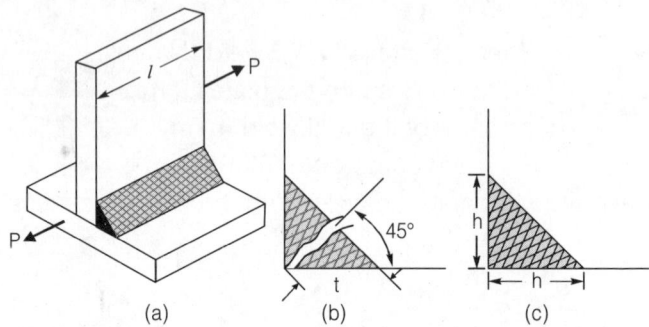

(a)   (b)   (c)

Where,

P = Tensile force on plates (N)

h = Leg of the weld (mm)

l = Length of the weld (mm)

τ = Permissible shear stress for the weld (N/mm²)

Usually, there are *two welds* of equal length on two sides of the vertical plate. In that case,

or

$$P = 2(0.707\, h l \tau)$$
$$P = 1.414\, h l \tau)$$

5. Strength of Transverse Fillet Weld

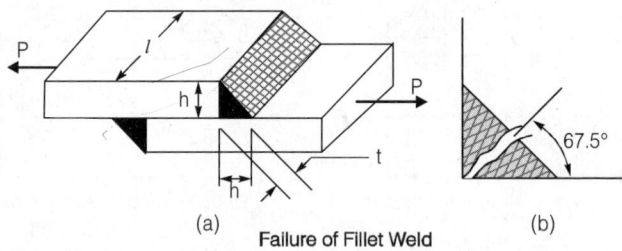

(a)   (b)

Failure of Fillet Weld

Usually, there are two welds of equal length on two side of the plate as shown in figure. In such cases,

or

$$P = 2(0.707\, h l \sigma_t)$$
$$P = 1.414\, h l \sigma_t)$$

Here,

σ_t = Permissible tensile stress for the weld (N/mm²)

Remember: ....

- Theoretically, it can be proved that for transverse fillet weld, the inclination of the plane, where maximum shear stress is induced, is 67.5° to the leg dimension.

6. Axially Loaded Unsymmetrical Welded Joint

Sometimes unsymmetrical section such as angles, channels, T-sections etc. welded on the flange edges are loaded axially. In such cases the lengths of weld should be proportioned in such a way that the sum of resisting moment of the welds about the gravity axis is zero.

Let, l_a = length of weld at top

l_b = length of weld at the bottom

l = total length of weld = $l_a + l_b$

P = axial load

a = distance of top weld from gravity axis

b = distance of bottom weld from gravity axis.

f = Resistance offered by the weld per unit length.

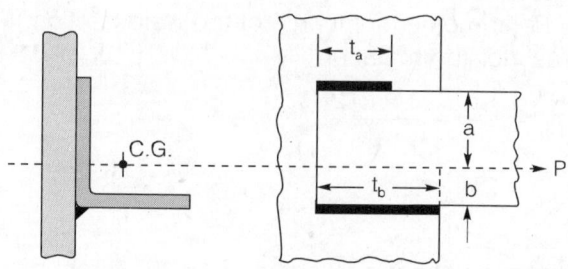

Moment of top weld about gravity axis

$$= l_a \times f \times a$$

Moment of bottom weld about gravity axis

$$= l_b \times f \times b$$

Since sum of the moments of the weld about the gravity axis must be zero.

$l_a \times f \times a - l_b \times f \times b = 0$

but $l = l_a + l_b$

$$\therefore \quad l_a = \frac{l.b}{a+b} \quad \text{and} \quad l_b = \frac{l.a}{a+b}$$

7. Circular Fillet Weld Subjected to Torsion

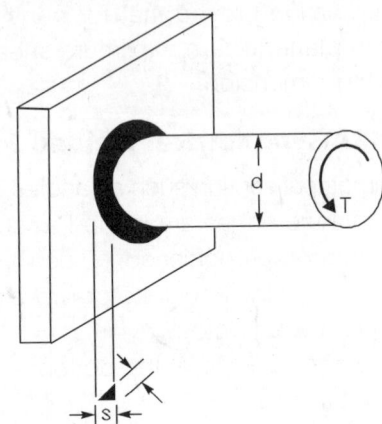

Let, d = diameter of rod

     r = Radius of rod

     T = Torque acting on the rod

     S = size of weld

     t = throat thickness

     J = Polar moment of inertia of the weld section.

Shear stress for the material

$$\tau_s = \frac{T.r}{J} = \frac{T \times d/2}{J}$$

$$\boxed{\tau_s = \frac{2T}{\pi t d^2}}$$

- The maximum shear stress occurs on the throat of weld which is inclined at **45°** to the horizontal plane.

$$\boxed{(\tau_s)_{max} = \frac{2T}{\pi \times 0.707 \times S \times d^2} = \frac{2.83}{\pi S d^2}}$$

8. Circular Fillet Weld Subjected to Bending Moment

Let,   M = Bending Moment acting on the rod

           Z = section modulus of the weld section

$$= \frac{\pi t d^2}{4}$$

Bending stress,

$$\boxed{\sigma_b = \frac{M}{Z} = \frac{4M}{\pi t d^2}}$$

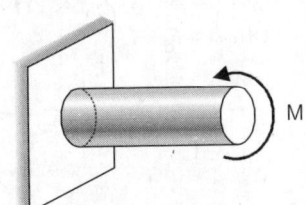

- The maximum bending stress occurs on the throat of weld which is inclined at 45° to the horizontal plane

$$(\sigma_b)_{max} = \frac{4M}{\pi \times 0.707 S.d^2} = \frac{5.66M}{\pi S d^2}$$

9. Long Fillet Weld Subject to Torsion

Let,   T = torque acting on the vertical plate

l = length of weld

J = polar moment of inertia of weld section

$$= \frac{2 \times t l^3}{12} = \frac{t l^3}{6}$$

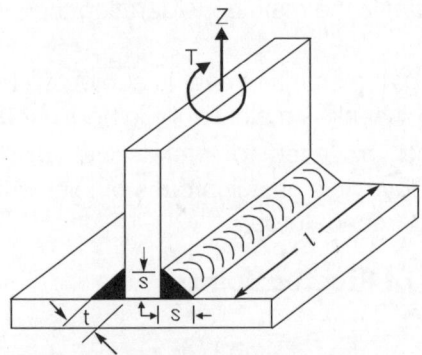

This rotation is resisted by shearing stresses developed between two fillet welds and the horizontal plate. It is assumed that these horizontal shearing stresses vary from zero at the z-axis and maximum at the ends of the plate. This variation of shearing stress is analogous to the variation of normal stress over the depth (l) of a beam subjected to pure bending.

$$\text{Shear stress} = \frac{T \times l/2}{t l^3/6} = \frac{3T}{t.l^2}$$

Maximum shear stress occurs at throat

$$\tau_{max} = \frac{4.242T}{S.l^2}$$

◆ RIVETED JOINT

A rivet consist of a cylindrical shank with a head at one end. This head is formed on shank by an upsetting process. In rivet terminology, the closing head is called point. The **cylindrical portion** of the rivet is called **shank** or body and **lower portion** of shank is known as **tail**. The rivets are used for **permanent fastening**. The riveted joints are widely used for joining **light metals**.

- The function of rivets in a joint is to make a connection that has strength and *tightness*.

10. Type of Riveting

- When a *cold rivet* is used, the process is known as *cold riveting* and when a *hot rivet* is used the process is known as *hot riveting*.
- The cold riveting process is used for *structural joint* while hot riveting is used to make *leak proof joints*.

Remember: ....

- The material of the rivets must be *tough* and *ductile*. They are usually made of steel, brass, aluminium, but when *strength and fluid tight joint* is the main consideration then the *steel rivets* are used.
- In hot riveting, the shank of rivet is subjected to *tensile stress*. In cold riveting, shank is mainly subjected to *shear stress*.
- Riveted joints are used for metals with *poor weldability* like aluminium alloys. It is useful in case of joint which is subjected to vibration and impact force.

11. Terminology of Riveted Joints

- Pitch (p)

  The pitch of the rivet is defined as the distance between the centre of one rivet to the centre of the adjacent rivet in the same row. Usually

  $$p = 3d$$

  Where   d = Shank diameter the rivet

- Margin (m)

  The margin is the distance between the edge of the plate to the centre line of rivets in the nearest row. Usually,

  $$m = 1.5\,d$$

- Transverse Pitch (p_t)

  Transverse pitch, also called back pitch or row pitch, is the distance between two consecutive rows of rivets in the same plate. Usually,

  $$p_t = 0.8\,p \qquad \text{(for chain riveting)}$$
  $$= 0.6\,p \qquad \text{(for zig-zag riveting)}$$

- Diagonal Pitch (p_d)

  Diagonal pitch is the distance between the centre of one rivet to the centre of the adjacent rivet located in the adjacent row.

12. Strength Equation for Rivet

The strength of riveted joint is defined as the force that the joint can withstand without causing failure.

In analysis of riveted joints, mainly three types of failure are considered. They are as follows:

    (i)   shear failure of the rivet

    (ii) tensile failure of the plate between rivets; and

    (iii) crushing failure of the plate.

- **Shear Strength of Rivet**

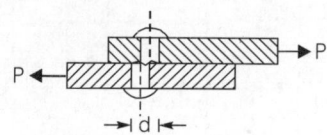

$$P_s = \frac{\pi}{4} d^2 \tau n$$

Here,   d = Shank diameter of rivet

            τ = Permissible shear stress for rivet

            n = Number of rivet per pitch length

$$P_s = 2\left[\frac{\pi}{4} d^2 \tau n\right] \qquad \text{[For \textit{double} shear]}$$

- **Tensile Strength of Plate Between Rivets**

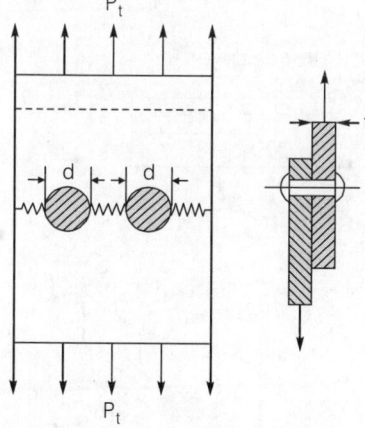

$$P_t = (p - d)\, t\, \sigma_t$$

Here,

    P_t = Tensile resistance of plate per pitch length (N)

p = Pitch of rivets (mm)

t = Thickness of plate (mm)

σ_t = Permissible tensile stress of plate material (N/mm²)

- **Crushing Strength of Plate**

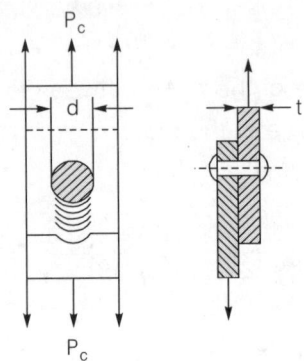

$$P_c = dt\,\sigma_c\,n$$

Where, P_c = Crushing resistance of plate per pitch length (N)

σ_c = Permissible compressive stress of plate material (N/mm²)

13. Efficiency of Joint

The efficiency of the riveted joint is defined as the ratio of the strength of riveted joint to the strength of unriveted solid plate.

The strength of solid plate of width, equal to the pitch p and thickness t, subjected to tensile stress σ_t is given by,

$$P = pt\sigma_t$$

Therefore, the efficiency is given by,

$$\eta = \frac{\text{Lowest of } P_s, P_t, \text{and } P_c}{P}$$

14. Caulking and Fullering

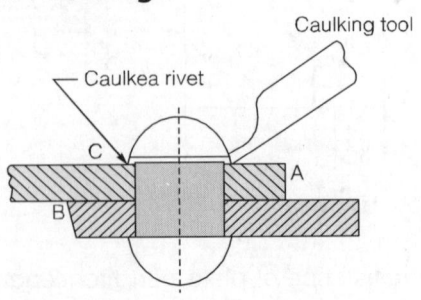

Caulking tool

Caulkea rivet

- In order to make the joints *leak proof or fluid tight in pressure vessels* like steam boilers, air receivers and tanks etc. a process known as caulking is employed. In this process a narrow blunt tool called **caulking** tool, about 5 mm thick and 38 mm in breadth is used. The edge of the tool is ground to an angle of 80°. The tools burrs down the plate in forming a metal to metal Joint.

- A more satisfactory way of making the joints *staunch* is known as **fullering** which has largely super sided caulking. In this case a fullering tool with a thickness at the end equal to that of the plate is used in such a way that the greatest pressure due to the blows occur near the joint, giving a clean finish with less risk of damaging the plate.

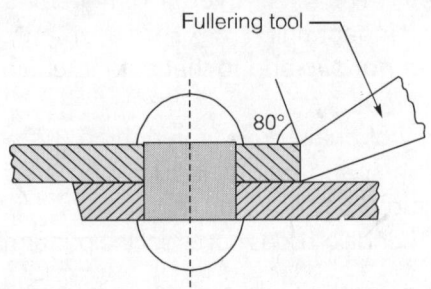

Fullering tool ⎯

80°

15. Design of Boiler Joint

- The load on Joints is equally shared by all the rivets.
- The tensile stress is equally distributed over the section of metal between the rivets.
- The shearing stress in all rivets is uniform.
- The crushing stress is uniform.
- There is no bending stress in the rivets.
- The holes into which the rivets are driven do not weaken the member.
- The friction between the surfaces of the plate is neglected.

- Thickness of boiler shell, $t = \dfrac{P.D}{2 f_t . \eta_l} + 1$ mm as corrosion allowance.

  where, η_l = Efficiency of the longitudinal joint.

- Diameter of rivets- using *unwin's empirical formula* $d = 6\sqrt{t}$ (t > 8 mm).

■■■■

Chain Drive

A chain is a series of links connected by pin joint. A chain drive consists of an endless chain wrapped around two sprockets. The chain drive is intermediate between *belt drive* and *gear* drive.

1. Advantage of Chain Drives Compared with Belt and Gear Drives

- Chain drives can be used over a *wider range* of centre distance.
- Chain drives have small overall dimensions than belt drives, resulting in compact unit.
- A chain *does not slip* and to that extent, chain drive is a positive drive.
- The efficiency of chain drive is very high (upto 98%).
- It permits high speed ratio of 8-10 in *one step*.
- It transmits more power than belts.
- Atmospheric condition does not affect the performance of chain drive.

2. Disadvantage of Chain Drive Over Belt or Rope Drive

- It operate without full lubricant film between the joints thus result in more wear at joint.
- It is not suitable for *non-parallel* shaft.
- It require *precise* alignment of shaft.

3. Terms Used in Chain Drive

- Pitch of Chain

  It is the distance between the hinge centre of link and the corresponding hinge centre of the adjacent link.
- Pitch Circle Diameter of Chain Sprocket

  It is the diameter of the circle on which the hinge centres of the chain lie, when the chain is wrapped round a sprocket.

4. Relation Between Pitch and Pitch Circle Diameter

$$D = p \cosec\left(\frac{180°}{T}\right)$$

Here, D = Diameter of the pitch circle

T = Number of teeth on the sprocket

p = Pitch of chain

5. Velocity Ratio of Chain Drive

$$V.R = \frac{N_1}{N_2} = \frac{T_2}{T_1}$$

Where

      N_1 = Speed of rotation of smaller sprocket in rpm

      N_2 = Speed of rotation of larger sprocket in rpm

      T_1 = Number of teeth on the smaller sprocket

      T_2 = Number of teeth on the larger sprocket.

6. Length of Chain and Centre Distance

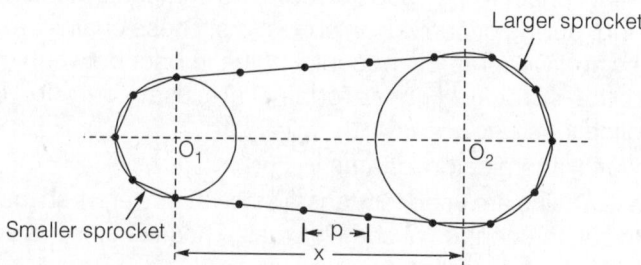

- Length of Chain (L)

      L = K.P

Here,

      K = Number of chain links

      p = Pitch of chain

- Number of chain link (K)

$$K = \frac{T_1 + T_2}{2} + \frac{2x}{p} + \left[\frac{T_2 - T_1}{2\pi}\right]^2 \frac{p}{x}$$

- Centre distance (x)

$$x = \frac{p}{4}\left[K - \frac{T_1 + T_2}{2} + \sqrt{\left(K - \frac{T_1 + T_2}{2}\right)^2 - 8\left(\frac{T_2 - T_1}{2\pi}\right)^2}\right]$$

7. Classification of Chain

- ### Load Lifting Chain

  Load lifting chains are used for suspending, raising or lowering loads in materials handling equipment. e.g 'link' chain.

- **Advantage**
  - They have good flexibility in all direction.
  - Link chains can operate with small diameter pulleys and drums.
  - They are simple to design and easy to manufacture.
  - They produce low noise and are practically noiseless at low speeds of less than 0.1 m/s.
- **Disadvantage**
  - Link chain is heavy in weight.
  - It is susceptible to jerks and overloads.
  - The failure of link chain is sudden.
  - Link chains operate at low speed.
- **Hauling Chains**

  Hauling chains are used for carrying materials continuously by sliding, pulling or carrying in conveyors. These chains are relatively noisy and wear rapidly because of the impact between the blocks and the sprocket. These chains are used only for conveyor applications.
- **Power transmission chains**

  These chains are used for transmission of power when the distance between the centres of shaft is small. These chains have provision for efficient lubrication.

8. Power Transmitted By Chains

- **On the Basis of Breaking Load**

$$P = \frac{W_B \times v}{n \times K_s} \text{ (in watts)}$$

Here,

W_b = Breaking load in Newton
v = Velocity of chain in m/s
n = Factor of safety
K_s = Service factor

- **On the Basis of Bearing Stress**

$$P = \frac{\sigma_b \times A \times v}{K_s}$$

Here, σ_b = Allowable bearing stress (N/mm^2)
A = Projected bearing area in mm^2
v = Velocity of chain in m/s
K_s = Service factor

■■■■

The clutch is a mechanical device, which is used to connect to or disconnect the source of power from the remaining parts of the power transmission system at the will of the operator.

1. Type of Clutches

- **Positive Contact Clutches**

  In these clutches, power transmission is achieved by *means of interlocking* of jaws or teeth. Their main advantage is *positive engagement* and once coupled, they can transmit *large torque with no slip*. e.g. square jaw clutch, spiral jaw clutch, toothed clutch.

- **Friction Clutches**

  In these clutches, power transmission is achieved by *means of friction* between contacting surface. e.g. single and multi-plate clutch, cone clutches, centrifugal clutches.

- **Electromagnetic Clutches**

  In these clutches, power transmission is achieved by *means of the magnetic field*. These clutches have many advantages, such as rapid response time, ease of control, and smooth starts and stops.

2. Friction Clutches

The force of friction is used to start the driven shaft from rest and gradually brings it up to the proper speed without excessive slipping of the friction surfaces.

The main requirement in friction clutch is that

- The contact surfaces should develop frictional force that may pick up and hold the load with reasonably low pressure between the contact surfaces.
- The heat of friction should be rapidly dissipated and tendency to grab should be at minimum.
- The surfaces should be backed by a material, stiff enough to ensure a reasonably uniform distribution of pressure.

Remember

- Friction clutch is very useful in transmission of power of shafts and machines which must be started and stopped frequently.

3. Properties of Material Used for Friction Clutches

- It should have a *high* and *uniform* coefficient of friction.
- It should not be affected by moisture and oil.
- It should have the ability to withstand high temperature caused by slippage.
- It should have high heat conductivity.
- It should have high resistance to wear and scoring.

4. Advantage of Friction Clutch

- The engagement is smooth.
- Slip occurs *only during engaging* operation and once the clutch is engaged, there is no slip between the contacting surfaces. Therefore, power loss and consequent heat generation do not create problems, unless the operation requires frequent starts and stops.
- In certain cases, the friction clutch serves as a safety device. It slips when the torque transmitted through it exceeds a safe value. This prevents the breakage of parts in the transmission chain.

Common type of friction clutch are:

(a) Disc(Plate) clutch

(b) Cone clutches

(c) Centrifugal clutch

5. Torque Carrying Capacity

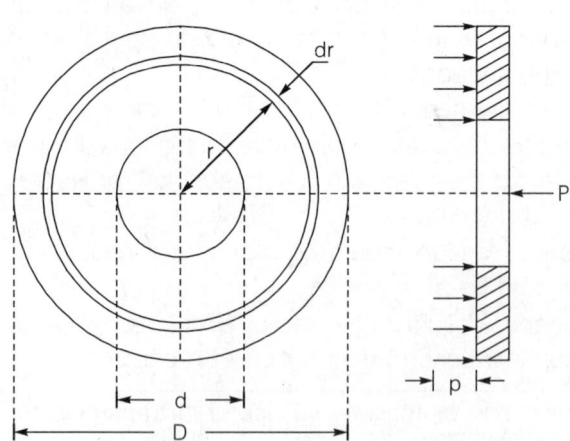

Friction force on ring tangentially:

$$F_r = \mu \delta w = (\mu)P \times 2\pi r dr$$

where, δw is normal force on ring area $2\pi r dr$

$$M_t = 2\pi\mu \int_{d/2}^{D/2} pr^2 \, dr$$

Here,   D = Outer diameter of friction disk (mm)
         d = Inner diameter of friction disk (mm)
         p = Intensity of pressure at radius r (N/mm^2)
         P = Total operating force (N)
         M_t = Torque transmitted by the clutch (N-mm)

Two theories are used to obtain the torque capacity of the clutch

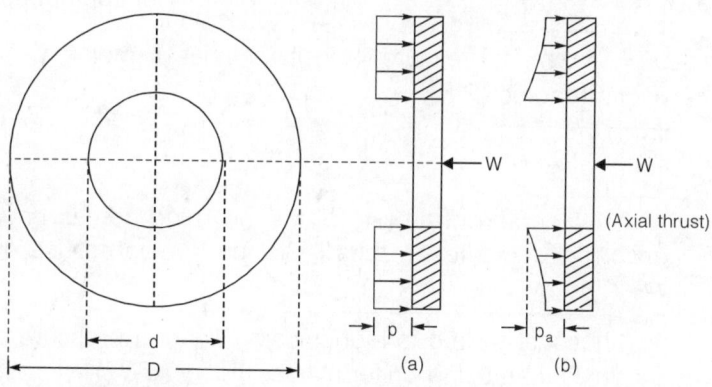

(Axial thrust)

(a)   (b)

- **Uniform Pressure Theory (Diagram a)**
  - **Assumption:** Pressure remain constant over entire surface area of the friction disk.
  - This theory is applicable for new clutches pressure (p) = constant.

  - $$M_t = \frac{\mu W}{3}\left[\frac{D^3 - d^3}{D^2 - d^2}\right]$$ (UPT) [For single pair of contact surface]

  $$M_t = \mu W R_f \quad \text{where,} \quad R_f = \frac{1}{3}\frac{(D^3 - d^3)}{D^2 - d^2} = \frac{2}{3}\left(\frac{R_0^3 - R_i^3}{R_0^2 - R_i^2}\right)$$ (UPT)

  - Friction radius for new clutches is slightly greater than that of worn-out clutches.

- **Uniform Wear Theory (Diagram b)**
  - **Assumption:** The wear is uniformly distributed over the entire surface area of the friction disk.
  - This theory is used for ***worn-out*** clutches.

    p.r = constant

$$M_t = \mu W \left(\frac{D+d}{4} \right) \quad \text{(UWT)}$$

$$M_t = \mu W R_f \quad \text{where,} \quad R_f = \frac{R_0 + R_i}{2} \quad \text{(UWT)}$$

  - $$M_t = \frac{\mu P}{4}(D+d) \qquad \text{[For single pair of contact surface]}$$

    p_a = Maximum pressure intensity at inner diameter
  - Friction radius (R_f)

$$R_f = \frac{1}{4}(D+d)$$

  - A major portion of the life of friction lining comes under the uniform wear criterion thus it is more logical to use uniform wear theory.

Remember

- Frictional torque is higher using uniform pressure theory (LPT) than using uniform wear theory (UWT).

6. Method Used to Increase Torque Transmitting Capacity

- Use friction material with a higher coefficient of friction (μ).
- Increase the plate pressure (p)
- Increase the mean radius of the friction disk (r_m).

7. Multi Disk Clutches

Number of pairs of contacting surface

$$z = z_1 + z_2 - 1$$

Here, z_1 = Number of disks on driving shaft

z_2 = Number of disks on driven shaft

- **Torque transmitting capacity**
  - For the Uniform-Pressure Criterion

$$M_t = \frac{\mu P z (D^3 - d^3)}{3(D^2 - d^3)}$$

- For the Uniform-Wear Criterion

$$M_t = \frac{\mu P z}{4}(D + d)$$

8. Difference Between Single-Disk Clutches and Multi-Disk Clutches

Single - disk clutches	Multi - disk clutches
(i) Torque transmitting capacity is less.	(i) Torque transmitting capacity is more
(ii) Single-disk clutches are dry because heat generation is less	(ii) Multi-disk clutches are wet clutches because heat generationdue to friction is more
(iii) It is used where large radial space is avaibale e.g. Trucks and cars.	(iii) It is used where compact construction is desirable e.g. scooter motor cycle.

9. Cone Clutches

In cone clutches, the conical surface results in considerable friction force even with a small engaging force due to the wedge action. The recommended semi-cone angle (α) is 12.5°.

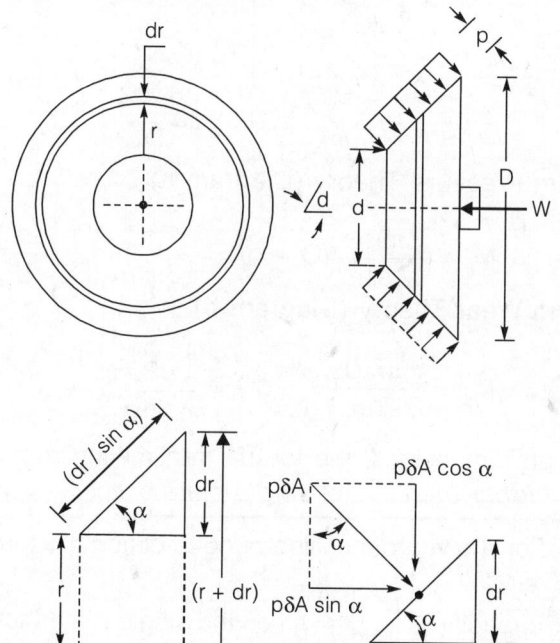

Remember: ....

- α is semi cone angle/face angle of cone/ angle of friction surface with axis of the clutch and its general value range is 12.5° to 15°.

- **Torque Transmitted**

$$M_t = \frac{2\pi\mu}{\sin\alpha} \int_{d/2}^{D/2} p \cdot r^2 \cdot dr$$

- **Total Operating Force**

$$P = 2\pi \int_{d/2}^{D/2} p \cdot r \cdot dr$$

The theories are used to obtain the torque capacity of the cone clutch.

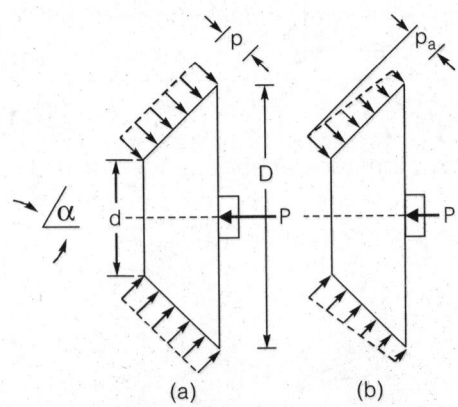

(a)   (b)

- **Uniform Pressure Theory (Diagram a)**

  ∴ $\quad M_t = \dfrac{\pi\mu p}{12\sin\alpha}(D^3 - d^3) = \dfrac{\mu P}{3\sin\alpha}\dfrac{(D^3 - d^3)}{(D^2 - d^2)}$

- **Uniform Wear Theory (Diagram b)**

  ∴ $\quad M_t = \dfrac{\pi\mu p_a d}{8\sin\alpha}(D^2 - d^2) = \dfrac{\mu P}{4\sin\alpha}(D + d)$

For a given dimensions, the torque transmitting capacity of *cone clutch* is *higher than* that of *single plate* clutch.

Remember

- For a given dimension of cone clutch the torque carrying capacity is $\left(\dfrac{1}{\sin\alpha}\right)$ times the single plate clutch.

10. Centrifugal Clutch

When it is required to engage the load after the driving member has *attained a particular speed*, a centrifugal clutch is used.

These clutches are particularly useful with internal combustion engines, which cannot start under load.

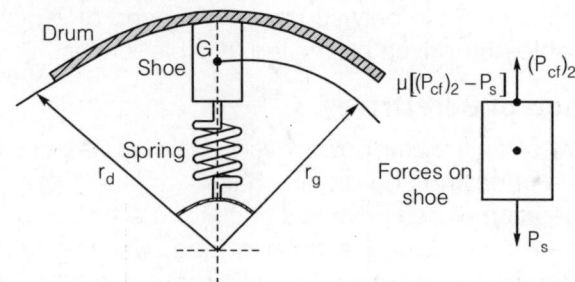

Where,

The centrifugal forces corresponding to speed ω_1 and ω_2 are given by,

$$(P_{cf})_1 = \frac{m\omega_1^2 r_g}{1000} \quad \text{and} \quad (P_{cf})_2 = \frac{m\omega_2^2 r_g}{1000}$$

Where, r_d = Radius of the drum (mm)

r_g = Radius of the centre of gravity of the shoe in engaged position (mm)

m = Mass of each shoe (kg)

P_{cf} = Centrifugal force (N)

P_s = Spring force (N)

z = Number of shoes

ω_2 = Running speed (rad/s)

ω_1 = Speed which engagement starts (rad/s)

The centrifugal clutch works on the principle of centrifugal force.

- Net force on drum = $\dfrac{mr_g\left(\omega_2^2 - \omega_1^2\right)}{1000}$

- Friction force = $\dfrac{\mu mr_g\left(\omega_2^2 - \omega_1^2\right)}{1000}$

- Friction torque = $\dfrac{\mu mr_g r_d\left(\omega_2^2 - \omega_1^2\right)}{1000}$

■■■■

8

Belt Drive

Belt drive is flexible drive. In flexible drives, there is an intermediate link such as belt, rope or chain between the driving and driven shafts. Since this link is flexible, the drives are called 'flexible' drives.

1. Advantage of Belt Drive

- Belt drives can transmit power over considerable distance between the axes of driving and driven shafts.
- The operation of belt drive is smooth and silent.
- They can transmit only a definite load, which if exceeded, will cause the belt to slip over the pulley, thus protecting the parts of the drive against overload.
- They have the ability to absorb the shocks and damp vibration.

2. Disadvantage of Belt Drive

- Belt drives have large dimensions and occupy more space.
- The velocity ratio is not constant due to belt slip.
- They impose heavy loads on shafts and bearings.
- There is considerable loss of power resulting in low efficiency.
- Belt drives have comparatively short service life.

3. Difference Between Flat Belt and V-Belt

Flat Belt	V-Belt
1. They do not require precise alignment of shaft and pulleys.	1. They require precise alignment of shaft and pulleys.
2. It can be used in dusty and abrasive atmosphere.	2. It cannot be used in dusty and abrasive atmosphere and requires casing.
3. Design is simple and inexpensive.	3. Design is relatively difficult and expensive.
4. It can be used for long centre distances, even upto 15 metre.	4. It can not be used for long centre distance.
5. The efficiency of flat belt drive is more than V-belt drive.	5. The efficiency of V-belt drive is comparatively lower.
6. Power transmission capacity is low.	6. Power transmission capacity is high.
7. Its velocity ratio is lower.	7. Its velocity ratio is high.
8. High speed reduction is not possible.	8. High speed reduction is possible.
9. In this, slip may occur.	9. This drive is approximately positive because the slip is negligible due to wedge action.

4. Desirable Properties of Belt Material

- The belt material should have high coefficient of friction with the pulleys.

- The belt material should have high tensile strength to withstand belt tensions.
- The belt material should have high wear resistance.
- The belt material should have high flexibility and low rigidity in bending in order to avoid bending stresses while passing over the pulley.

5. Difference Between Open Belt Drive and Cross Belt Drive

Open Belt Drive	Cross-belt Drive
1. In this both driving and driven pulleys rotate in the same direction.	1. In this, both driving and driven pulleys rotate in opposite direction.
2. Power transmission capacity is less as compare to cross-beld drive.	2. Power transmission capacity is more.
3. When centre distance is more belt whips.	3. Crossed belt drive do not have these limitations.

6. Length of Open Belt Drive/Cross Belt Drive

- Open Belt Drive

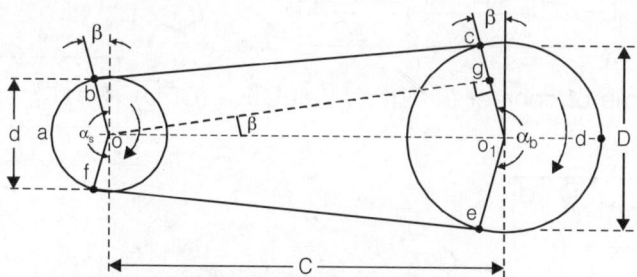

Here, α_s = Wrap angle for small pulley (degree)
        α_b = Wrap angle for big pulley (degree)
        D = Diameter of big pulley (degree)
        d = Diameter of big pulley(degree)
        C = Centre distance (degree)

∴
$$L = 2C + \frac{\pi(D+d)}{2} + \frac{(D-d)^2}{4C}$$

Angle of contact for open belt drive $\boxed{\theta = (180° - 2\beta)\dfrac{\pi}{180}}$ radian

$$\sin\beta = \frac{D-d}{2C}$$

- For all calculation we consider smaller pulley is considered.

  Angle of contact for Open Belt Drive (OBD) $\boxed{\theta = (180° - 2\beta)\dfrac{\pi}{180}}$ radian.

- **Cross-Belt Drive**

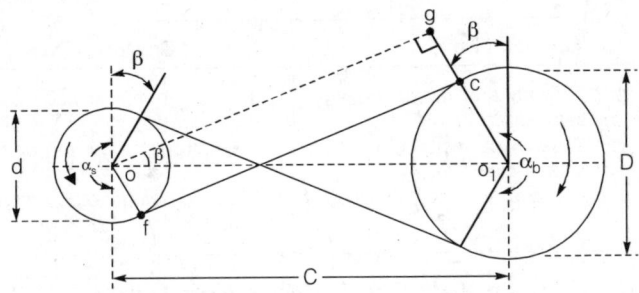

$$\boxed{L = 2C + \frac{\pi(D+d)}{2} + \frac{(D+d)^2}{4C}}$$

Angle of contact for Cross Belt Drive (CBD) $\boxed{\theta = (180° + 2\beta)\dfrac{\pi}{180}}$ radian.

$$\boxed{\sin\beta = \frac{D+d}{2C}}$$

- In open belt drive angle of lap for smaller pulley is smaller than. Larger pulley. ($\alpha_s < \alpha_b$)
- In cross belt drive angle of lap for smaller pulley is equal to larger pulley. ($\alpha_b = \alpha_s$).

Remember

7. **Belt Tension**

- **For Open Belt**

$$\frac{P_1 - mv^2}{P_2 - mv^2} = e^{f\alpha}$$

- **For Cross Belt**

$$\left[\frac{P_1 - mv^2}{P_2 - mv^2}\right] = e^{f\alpha/\sin(\theta/2)}$$

Here,

P_1 = Belt tension in the tight side

P_2 = Belt tension in the loose side (N)

m = Mass of the one meter length of belt (kg/m)

v = Belt velocity (m/s)

f = Coefficient of friction

α = Angle of wrap for belt (radians)

- Power Transmitted = $(P_1 - P_2)$ Velocity
  - Condition for maximum power

  (i) $V = \sqrt{\dfrac{P_i}{3m}}$,   ($P_i$ = Initial tension)

  ■ $V = \sqrt{\dfrac{P_{max}}{3m}}$

  - The maximum permissible tension in the belt should be three times the tension due to centrifugal force ($P_{max} = 3P_c$).
  - Alternatively, the tension in the tight side of the belt should be twice the tension due to centrifugal force ($P_1 = 2P_c$).
  - In case when maximum stress is considered total maximum tension $\boxed{T_t = \sigma \cdot b \cdot t}$.

  - $\boxed{T_t = P_1 + P_c}$

8. Reason of Power Loss in Belt Drive

- Power loss due to belt creep on the pulley.
- Power loss due to internal friction between the particles of the belt in alternate bending and unbending over the pulley.
- Power loss due aerodynamic resistance to the motion of pulleys and belt.
- Power loss due to friction in bearings of pulleys.

■■■■

SHAFT

It is a rotating machine element, circular in cross-section, which supports transmission elements like gears, pulleys and sprockets and transmits power. Shafts are given specific names in typical applications, although all applications involve transmission of power, motion and torque.

1. Types of Shaft

- **Axle**

  Axle is used for a shaft that supports rotating elements like wheels, hoisting drums, which is fitted to the housing by means of bearings. *In general*, an axle is subjected to bending moment due to transverse loads like bearing reactions and does not transmit any useful torque, e.g., rear axle of a railway wagon. *Occasionally*, the axle also transmits torque, e.g., automobile real axle.

- **Spindle**

  Spindle is short rotating shaft. Spindles are used in all machine tools such as the small drive shaft of a lathe or the spindle of a drilling machine.

- **Countershaft**

  It is a secondary shaft, which is driven by the main shaft and from which the power is supplied to a machine component.

- **Line Shaft**

  A line shaft consists of a number of shafts, which are connected in axial direction by means of couplings. A number of pulleys are mounted on the line shaft and power is transmitted to individual machines by different belts.

Remember: ....

- Ordinary transmission shafts are made of medium carbon steels, commercial shafts are made of low carbon steels.
- Cold-drawing produces a stronger shaft than hot-rolling.

2. Shaft Design on Strength Basis

Transmission shafts are subjected to axial tensile force, bending moment or torsional moment or their combinations.

- When the shaft is subjected to axial tensile force (P), the tensile stress is given by,
  - For solid shaft

$$\sigma_t = \frac{4P}{\pi d^2}$$

(d = diameter of shaft)

  - For hollow shaft

$$\sigma_t = \frac{4P}{\pi d_o^2 (1 - C^2)}$$

$$\left(C = \frac{d_i}{d_o} \right)$$

Here, d_i = Inside diameter

         d_o = Outside diameter

- When the shaft is subjected to pure bending moment (M_b), the bending stresses is given by
  - For solid shaft

$$\sigma_b = \frac{32 M_b}{\pi d^3}$$

  - For hollow shaft

$$\sigma_b = \frac{32 M_b}{\pi d_o^3 (1 - C^4)}$$

- When the shaft is subjected to pure torsional moment (M_t), the torsional shear stress is given by
  - For solid shaft

$$\tau = \frac{16 M_t}{\pi d^3}$$

  - For hollow shaft

$$\tau = \frac{16 M_t}{\pi d_o^3 (1 - C^4)}$$

- When the shaft is subjected to a combination of bending and torsional moments without any axial force.
- **According to maximum principal stress theory**
  - For solid shaft

$$\sigma_1 = \frac{16}{\pi d^3} \left[M_b + \sqrt{(M_b)^2 + (M_t)^2} \right]$$

- Value of equivalent bending moment

$$M_e = \frac{1}{2}\left[Mb + \sqrt{M_b^2 + b_t^2} \right]$$

- For hollow shaft

$$\sigma_1 = \frac{16}{\pi d_o^3 (1 - C^4)}\left[M_b + \sqrt{(M_b)^2 + (M_t)^2} \right]$$

- **According to maximum shear stress theory**
  - For solid shaft

$$\tau_{max} = \frac{16}{\pi d^3} \sqrt{(M_b)^2 + (M_t)^2}$$

- Value of equivalent twisting moment

$$T_e = \sqrt{M_b^2 + b_t^2}$$

- For hollow shaft

$$\tau_{max} = \frac{16}{\pi d_o^3 (1 - C^4)} \sqrt{\left[(M_b)^2 + (M_t)^2 \right]}$$

Remember

- The maximum *shear stress* theory is applicable to *ductile* materials.

3. Shaft Design of Torsional Rigidity Basis

A transmission shaft is said to be rigid on the basis of torsional rigidity, if it does not twist too much under the action of an external torque. In certain applications, like *machine tool* spindles, it is necessary to design the shaft on the basis of torsional rigidity.

- **The Angle of Twist θ_r (radian),**

$$\theta_r = \frac{M_t l}{JG}$$

- For circular shaft

$$J = \frac{\pi d^4}{32}$$

- For hollow circular shaft

$$J = \frac{\pi(d_0^4 - d_i^4)}{32} = \frac{\pi d_0^4 (1 - C^4)}{32}$$

Here,

l = Length of shaft subjected to twisting moment

M_t = Torsional moment

G = Modulus of rigidity

4. Advantage of Hollow Shaft

- The stiffness of the hollow shaft is more than that of solid shaft with same weight.
- The strength of hollow shaft is more than that of solid shaft with same weight.
- The natural frequency of hollow shaft is higher than that of solid shaft with same weight.

KEYS

A key can be defined as a machine element which is used to connect the transmission shaft to rotating machine elements like pulleys, gears, sprockets or flywheels. The keyway results in stress concentration in the shaft.

5. Function of Keys

- The primary function of the key is to transmit the torque from the shaft to the hub.
- The second function of the key is to prevent relative rotational motion between the shaft and the joined machine element like gear or pulley.

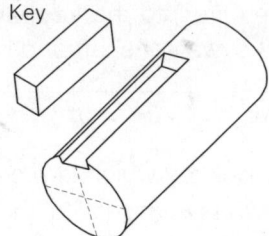

Key

6. Saddle Keys

A saddle key is a key which **fits** in the keyway of the hub **only**. In this case, there is **no keyway** on the shaft.

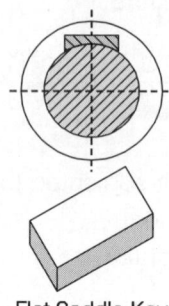

Flat Saddle Key

In this *friction* between the shaft, key and hub prevents relative motion between the shaft and the hub. The power is transmitted by means of *friction*. It is suitable for *light* duty *or low* power transmission.

7. Sunk Keys

A sunk key is a key in which *half* the thickness of the key fits into the keyway on the shaft and the *remaining half* in the keyway on the hub. In sunk key, power is transmitted due to *shear resistance of the key.*

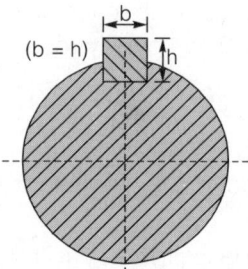

It is a positive drive.

- A saddle key fits into the keyway of hub while sunk key fits into the keyway of shaft and hub both.

Remember

8. Feather Key

A feather key is a parallel key which is *fixed* either to the shaft or to the hub and which *permits* relative axial movement between them. It is a type of sunk key with *uniform width* and *height*. It is used where the parts mounted on the shaft are required to slide along the shaft such as clutches or gear shifting devices.

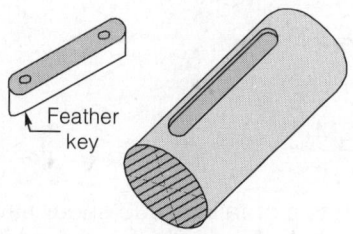

9. Woodruff Key

A Woodruff key is a sunk key in the form of an almost *semicircular* disk of *uniform thickness*.

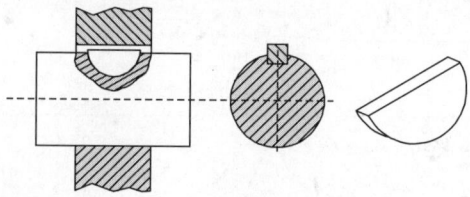

- **Advantage**
  - It can be used on *tapered* shaft.
  - The extra depth of key in the shaft prevents its *tendency* to slip over the shaft.

10. Design of Square and Flat Keys

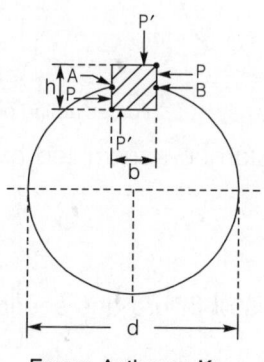

Forces Acting on Key

Here,

      h = Height

      b = Width

      P = Applied force (assume tangential to the shaft)

M_t = Torque transmitted
d = Diameter of a shaft
l = Length of key

$$P = \frac{2M_t}{d}$$

- Design on the basis of failure due shear stress

$$\tau = \frac{2M_t}{dbl}$$

- Design on the basis of failure due compressive stress

$$\sigma_c = \frac{4M_t}{dhl}$$

Remember: ....
- For square key σ_c = 2t.

11. Design of Kennedy Key

The Kennedy key consists of two square keys.

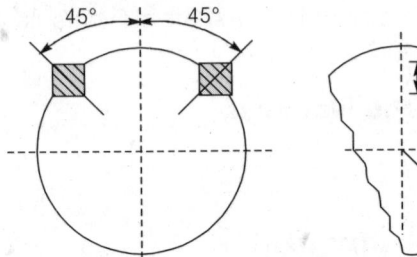

Kennedy Key   Forces acting on Kennedy key

Since there two keys torque transmitted by each key is one half of the total torque.

$$P = \frac{M_t}{d}$$

- Design on the basis of failure due shear stress

$$\tau = \frac{M_t}{\sqrt{2}\,dbl}$$

- Design on the basis of failure due compressive stress

$$\sigma_c = \frac{\sqrt{2}M_t}{dbl}$$

COUPLING

A coupling can be defined as a mechanical device that *permanently* joins two rotating shafts to each other.

12. Muff Coupling / Design Procedure For Muff Coupling

It consists of a sleeve or a hollow cylinder, which is fitted over the ends of input and output shafts by means of a sunk key.

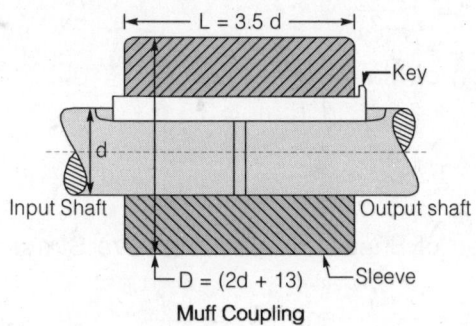

Muff Coupling

$$D = (2d + 13) \text{ mm}$$
and $\quad L = 3.5 \, d$

Here, D = Outer diameter of the sleeve
L = Axial length of the sleeve
d = Diameter of shaft

- **Calculation of Diameter**

$$\tau = \frac{16M_t}{\pi d^3}$$

- For $\quad l = \dfrac{L}{2}$,

$$\tau = \frac{2M_t}{dbl} \quad \text{and} \quad \sigma_c = \frac{4M_t}{dhl}$$

13. Clamp Coupling

It is a rigid type of coupling. In this coupling, the sleeve is made of two halves, which are split along a plane passing through the axes of shafts.

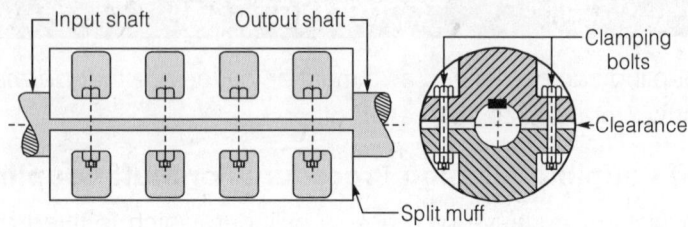

- **Calculation of Diameter**

$$\tau = \frac{16M_t}{\pi d^3}$$

- **Calculation of Main Dimension**

$$D = 2.5\,d \quad \text{and} \quad L = 3.5\,d$$

$$l = \frac{L}{2}$$

- **Calculation of Shear and Compressive Stress**

$$\tau = \frac{2M_t}{db l} \quad \text{and} \quad \sigma_c = \frac{4M_t}{dh l}$$

■■■■

A Handbook on
Mechanical Engineering

9

Strength of Materials

CONTENTS

◀ ☐ ▶

Properties of Metals, Stress and Strain

■ IMPORTANT MECHANICAL PROPERTIES

● Elasticity

It is the property by virtue of which a material deformed under the load is *enabled* to return to its original dimension when the load is removed.

Remember: ....

- If body regains *completely* its original shape it is called *perfectly elastic body*
- *Elastic limit* marks the *partial* break down of elasticity beyond which removal of load result in a degree of *permanent deformation*.
- Steel, Aluminium, Copper, Concrete may be considered to be perfectly elastic *within certain limit*.

● Plasticity

The characteristics of the material by which it undergoes *inelastic strain* beyond those at the *elastic limit* is known as plasticity.

Remember: ....

- This property is particularly useful in operation of *pressing* and *forging*.
- When large deformation occurs in a *ductile* material loaded in *plastic* region, the material is said to undergo *plastic flow*.

● Ductility

It is the property which permits a material to be drawn out *longitudinally* to a reduced section, under the action of *tensile force*.

Remember: ....

- A ductile material must posses a high degree of plasticity and strength.
- Ductile material must have *low* degree of elasticity.
- This is useful in *wire drawing*.

● Brittleness

It is lack of ductility. Brittleness implies that it can *not* be drawn out by tension to smaller section

Remember: ....

- In brittle material failure take place under load *without* significant deformation.
- Ordinary *Glass* is nearly *ideal* brittle material.
- Cast iron, *concrete* and ceramic material are brittle material.

Malleability

It is the property of a material which permits the material to be *extended* in *all direction* without rupture.

Remember: ....

- A malleable material posses a *high degree* of plasticity, but *not* necessarily *great strength*.

Toughness

It is the property of material which enables it to absorb energy *without fracture*.

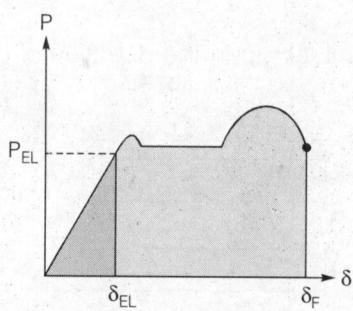

Remember: ....

- It is desirable in material which is subject to *cyclic* or *shock loading*.
- It is represented by area under *stress-strain* curve for material upto fracture.
- *Bend test* used for common comparative test for toughness.

Hardness

It is the ability of a material to resist *indentation* or *surface abrasion*.

Remember

- Brinell hardness test is used to check hardness.

- Brinell hardness number = $\dfrac{P}{\dfrac{\pi D}{2}\left[D - \sqrt{D^2 - d^2}\right]}$

Here, P = Standard load
D = Diameter of steel ball (mm)
d = Diameter of indent (mm)

- ## Strength

  This property enables material to resist fracture under load.

  - This is most important property from *design* point of view.
  - Load required to cause fracture, divided by area of test specimen, is termed as *ultimate strength*.

STRESS AND STRAIN

1. ## Stress (N/m²)

   It is the resistance offered by the body to deformation

   - *Nominal* stress (*Engineering* stress) $= \dfrac{\text{Load}}{\text{Original Area}}$

   - Actual/*True* stress $= \dfrac{\text{Load}}{\text{Changed (Actual) Area}}$

2. ## Strain

   Deformation per unit length in the direction of deformation is known as strain.

   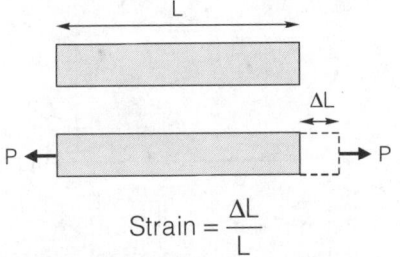

   $$\text{Strain} = \frac{\Delta L}{L}$$

   - It is a *dimensionless* quantity.

3. ## Engineering curve for mild steel for tension under static-loading.

   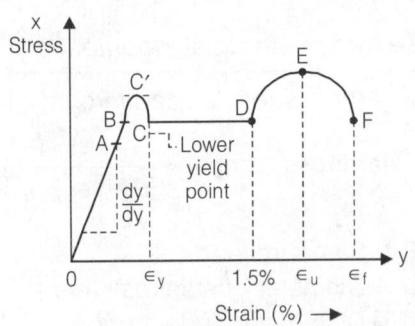

OA — Straight line (proportional region, *Hookes law is valid*)
OB — Elastic region
BC — Elasto plastic region
CD — Perfectly plastic region
DE — Strain hardening
EF — Necking region
A — Limit of proportionality
B — Elastic limit
C′ — Upper yield point
C — Lower yield point
D — Strain hardening starts
E — Ultimate stress
F — Fracture point

- **Limit of Proportionality**
  It is the stress at which the stress-strain curve *ceases* to be a straight line.

Remember: ....

- *Hooke's law* is valid upto proportional limit.

- **Elastic Limit**
  It is the point in the stress-strain curve upto which the materials remains elastic.

- Upto this point there is *no permanent* deformation after revmoal of load.

- **Plastic Range**
  It is the region of the stress-strain curve between the elastic limit and point of rupture.

- **Yield Point**
  This point is just beyond the elastic limit, at which the specimen undergoes an appreciable increase in length *without* further increase in the load.

- **Rupture Strength**
  It is the stress corresponding to the failure point 'F' of the stress-strain curve.

- **Proof Stress**
  It is the stress necessary to cause a *permanent extension* equal to defined percentage of gauge length.

- Slope of OA= Modulus of elasticity
  (*Young's Modulus*).
  It is constant of proportionality which is defined as the intensity of stress that causes unit strain.
- Plastic strain is 10 to 15 times elastic strain.
- Fracture strain (ϵ_f) depends on *percentage carbon* in steel.
- When carbon percentage increases then fracture strain decreases and yield stress increases.

4. Type of Tension failure in Metal

A. Ductile metal (*Shear* failure)

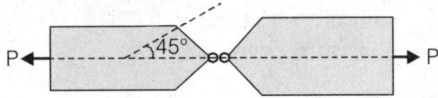

- Failure plane = *45°*
- Cup-cone fracture

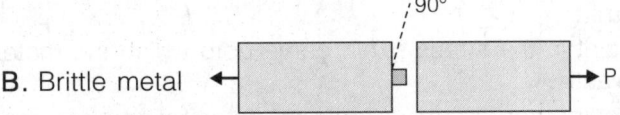

B. Brittle metal

- Failure plane at *90°*
- Failure is due to *principal* stress.

5. Type of failure in compression.

A. Ductile material

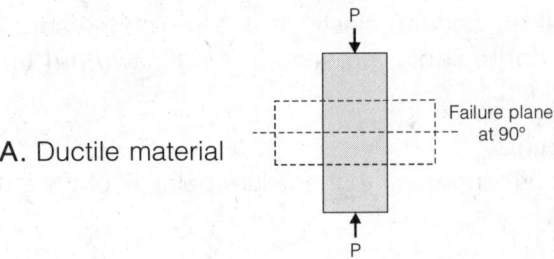

Failure plane at 90°

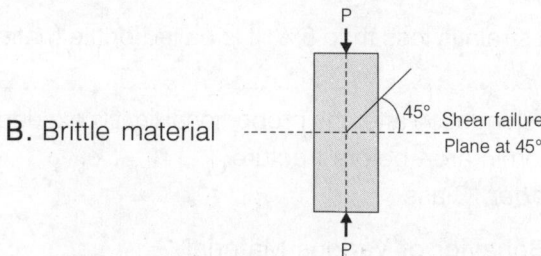

B. Brittle material

6. **Stress-strain graph for Various type of steel/material**

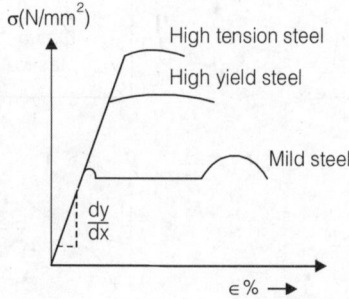

All the grade of steel have same young modulus but different yield stress.

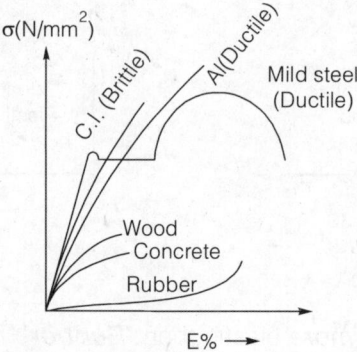

7. **Ductile material**

If strain, post elastic strain is greater than 5%. It is called ductile material.

Remember: ....

- Undergo large permanent strains before failure
- Large reduction in area before fracture
- e.g. *lead*, mild steel, copper

Brittle Material

If strain, post elastic strain is less than 5%. It is called brittle material.

Remember: ....

- Fail with only little elongation after the proportional limit is exceeded.
- Very less reduction in area before fracture
- e.g. Bronze, *Rubber*, Glass

Behavior of Various Material

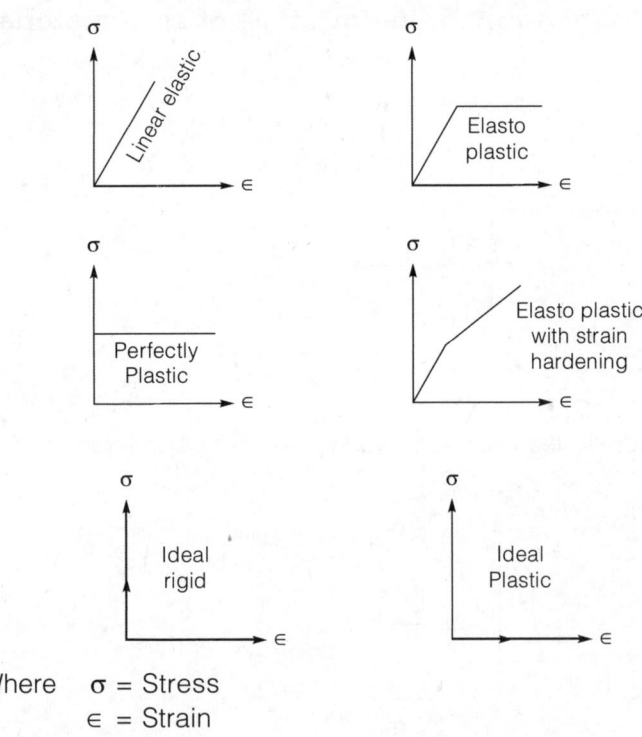

Where   σ = Stress
         ∈ = Strain

Remember: ....

- *'Mild steel'* is *more* elastic than *'Rubber'*.

8. Hooke's Law

When a material behaves elastically and exhibits a linear relationship between stress and strain, it is called linearly elastic. In this stress (σ) is directly proportional to strain (∈).

$$\sigma = E \cdot \in$$

Here,   σ = Stress
         p = load applied

Δ = Change in length

L = Original length

E = Young modulus of elasticity

Remember

- $E_{castiron} \approx \dfrac{1}{2} E_{steel}.$

- $E_{Aluminium} \approx \dfrac{1}{3} E_{steel}$

Where 'E' is young modulus.

9. Axial elongation (Δ) of bar prismatic bar due to external load

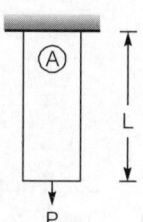

$$\Delta = \frac{PL}{AE}$$

Here, P = Load applied

L = Length of bar

A = Area of bar

E = Young modulus

$$\Delta = \frac{P}{\dfrac{EA}{L}} = \frac{P}{K}$$

K = Axial stiffness of bar

AE = Axial rigidity

EI/L = Flexural stiffness

10. Deflection of bar (Δ) due to self-weight

A. *Prismatic bar*

$$\Delta = \frac{PL}{2AE} = \frac{\gamma L^2}{2E}$$

Here, P = Self weight

B. *Conical bar*

$$\Delta = \frac{\gamma L^2}{6E} = \frac{1}{3} \times \text{Deflection of prismatic bar}$$

γ = Specific weight
L = Length of bar
E = Young's modulus

11. Deflection (Δ) of tapered bar

A. Circular tapering bar

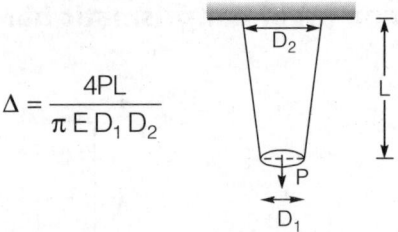

$$\Delta = \frac{4PL}{\pi E D_1 D_2}$$

Here,   P = Load applied
D = Diameter
L = Length of bar

B. Rectangular tapering bar

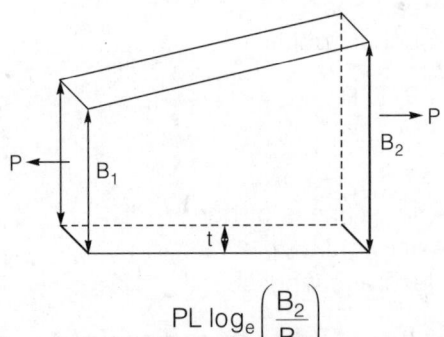

$$\Delta = \frac{PL \log_e\left(\dfrac{B_2}{B_1}\right)}{E \cdot t(B_2 - B_1)}$$

t = thickness
P = Load applied
E = Young modulus

12. Equivalent young modulus of parallel composite bar

$$E_{equivalent} = \frac{A_1 E_1 + A_2 E_2}{A_1 + A_2}$$

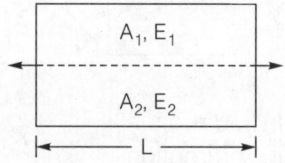

Here,   P = Load
A$_1$ = Area of first bar
A$_2$ = Area of second bar
E$_1$ = Young modulus of first bar
E$_2$ = Young modulus of second bar
L = Length of bar

ELASTIC CONSTANT

Elastic constant are those factor which determine the deformation produced by a given stress system acting on material.

- Modulus of elasticity (E) = $\dfrac{\text{Longitudinal stress}}{\text{Longitudinal strain}}$

- Modulus of rigidity (G) = $\dfrac{\text{Shear stress}}{\text{Shear strain}}$

- Bulk modulus (K) = $\dfrac{\text{Direct stress}}{\text{Volumetric strain}}$

13. Poisson Ratio (μ)

$$\mu = \frac{-\text{Lateral strain}}{\text{Longitudinal Strain}}$$

Remember: ....

- μ = 0 to 0.5(under uni-axial loading)
- μ = 0 for cork
- μ = 0.5 For perfectly plastic body *(Rubber)*

14. Volumetric strain under tri-axial loading

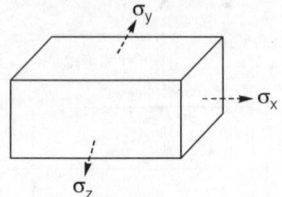

Here, σ_x = Stress in x-direction
σ_y = Stress in y-direction
σ_z = Stress in z-direction
$\in_V$ = Volumetric strain

$$\in_V = \in_x + \in_y + \in_z$$

$$= \frac{\sigma_x + \sigma_y + \sigma_z}{E}(1 - 2\mu)$$

•   Under hydrostatic loading
$$\sigma_x = \sigma_y = \sigma_z = \sigma$$

∴ $$\in_V = \frac{3\sigma}{E}(1 - 2\mu)$$

15. Volumetric strain of cyclindrical bar

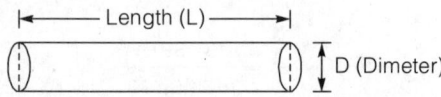

$$\in_V = \text{Longitudinal Strain} + (2 \times \text{Diametral strain})$$

16. Volumetric strain of sphere

$$\in_V = 3 \times \text{Diametric strain}$$

17. Relation between E, G, K, μ

• $E = 3K(1 - 2\mu)$
• $E = 2G(1 + \mu)$

• $E = \dfrac{9KG}{3K + G}$

• $\mu = \dfrac{3K - 2G}{6K + 2G}$

Here,   E = Young's modulus, G = shear modulus
K = Bulk modulus, μ = Poisson ratio

Remember

- For steel $\mu = 1/3$ then $\boxed{E = K = \dfrac{8}{3}G}$

- Theoretical limit of Poisson rato $\boxed{-1 < \mu < \dfrac{1}{2}}$

- By using elastic constant relation

  $E = 2G(1 + \mu) > 0$ **Always** then $\boxed{\mu > -1}$

  $E = 3K(1 - 2\mu) > 0$ **Always** than $\boxed{\mu > \dfrac{1}{2}}$

18.

Material	Number of Independent elastic constant
Homogeneous & Isotropic	2
Orthotropic (Wood)	9
Anisotropic	21

19. Strain Energy

It is the ability of material to absorb energy when it is strained

$$= \frac{1}{2}P \times \delta = \frac{1}{2}T \times \theta$$

Here

       P = Applied load

       δ = Elongation due to applied load

       T = Applied torque

       θ = Angle of twist due to applied torque

- **Resilience**

  Ability of a material to absorb energy in the *elastic region* when it is strained.

  $$= \text{Area under P-}\delta \text{ curve} = \frac{1}{2}P \times \delta$$

- **Proof Resilience**

  Maximum energy absorbing capacity of a material in the *elastic region* is called proof resilience.

$$= \text{Area under P-}\delta \text{ curve} = \frac{1}{2}P_{EL} \times \delta_{EL}$$

Here

$$P_{EL} = \text{Load at elastic limit}$$
$$\delta_{EL} = \text{Elongation upto elastic limit}$$

$$\text{Modulus of Resilience} = \frac{\text{Proof Resilience}}{\text{Volume}} = \frac{\sigma_{EL}^2}{2E}$$

Here

$$\sigma_{EL} = \text{Strain at elastic limit}$$
$$E = \text{Modulus of elasticity}$$

20. Thermal stress and strain

$$\sigma_{Th \cdot stress} = E\alpha T$$

$$\Delta = L\,\alpha\,T$$

$$\text{Strain} = \frac{L\,\alpha\,T}{L} = \alpha T$$

Here,

$$\sigma = \text{Thermal stress}$$
$$\alpha = \text{Coefficient of thermal expansion}$$
$$T = \text{Temperature change}$$
$$\Delta = \text{Change in length}$$

$$\alpha_{steel} = \alpha_{concrete} = 12 \times 10^{-6}/°C$$

$$\alpha_{Aluminium} > \alpha_{Brass} > \alpha_{Copper} > \alpha_{Steel}$$

Remember

• When bar is *free* to expand, there will be *no thermal* stress due to change in temperature.

■■■■

Shear Force and Bending Moment

2

1. Types of Beam

- ### SIMPLY SUPPORTED BEAM
  If the ends of a beam are made to rest freely on supports beam it is called a simply (freely) supported beam.

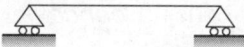

- ### FIXED BEAM
  If a beam is fixed at both ends it is called fixed beam its another name is encastre or built-in beam.

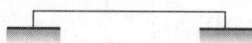

- ### CANTILEVER BEAM
  If a beam is fixed at one end while other end is free, it is called cantilever beam.

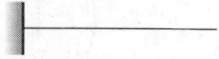

- ### CONTINUOUS BEAM
  If more than two supports are provided to beam, it is called continuous beam.

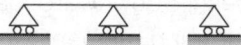

2. Shear Force

It is an *internal force* acting *tangentially* to the section which is *normal* to *longitudinal* direction.

Remember: ....

- It may be horizontal or vertical.
- It is that internal force at a section which is needed to maintain *free body equilibrium either* to the left or right of section
- Shear force at any section is *algebraic* sum of all transverse forces *either* from left or right of that section
- ### SIGN CONVENTION
  Shear force having an upward direction to the left hand side of *section* or downward direction to the right hand side of section will be taken positive and vice-versa.

3. Bending Moment

Bending moment at any section is the internal reaction due to all the *transverse* force *either* from left side or right side of that section.

Remember: ....

- It is equal to *algebraic* sum of moments at that section either from left or from sight side of that section.
- Bending moment is different from twisting moment.

SIGN CONVENTION OF BENDING MOMENT

- A bending moment causing *concavity upward* will be taken as positive and called *sagging* bending moment.
- A bending moment causing *convexity upward* will be taken as *negative* and will be called a *hogging* bending moment.

4. Relationship Between Bending Moment (M) and Shear Force (S) and Load (W)

- Rate of change of shear force is equal to load

$$\boxed{\dfrac{ds}{dx} = -w}$$

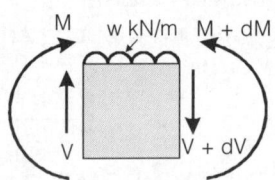

- **Valid only** when there is **no concentric load** between ends.
  Here,   w = Load per unit length
- Negative slope represent downward loading.
- Rate of change of bending moment *along the length* of beam is equal to shear force.

$$\boxed{\dfrac{dM}{dx} = S_x}$$

- **Valiid only** when there is **no couple** between ends considered.

Remember

- At hinge, bending moment will be zero.
- Bending moment is maximum when shear force is zero.
- If degree of loading curve = n then
  degree of shear force = n + 1
  degree of bending moment = n + 2
- Point of contra-flexture/inflection is that point where bending moment *changes its sign*.

■■■■

Torsion of Shaft

3

1. Torsion

Torsion means twisting of a structural member when it is loaded by *couple* that produces rotation about *longitudinal* axis.

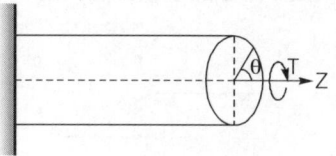

Remember: ....

- Torsion causes rotation of all the fibre about *longitudinal/polar* axis.
- Force required for torsion is *normal* to *longitudinal* axis having certain *eccentricity* from centroid.
- It there are no *shear force* and no *bending moment* is present in structural member then it will be a case of *pure* torsion.
- For pure torsion, shaft is *prismatic.*
- If torque applied in non-circular section then *warping* will occur.
- A plane section *before* twisting remains plane *after* twisting.

2. Equation of Torsion

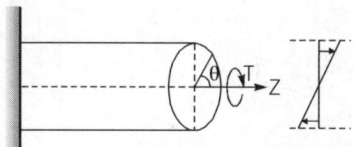

Shear stress distribution

$$\frac{\tau}{r} = \frac{T}{J} = \frac{G \cdot \theta}{L}$$

Here, τ = Shear stress
   r = Distance from centre of shaft
   D = Diameter of shaft
   T = Torque
   J = Polar moment of inertia
   G = Shear modulus
   θ = Angle of twist
   L = Length of shaft

Remember

- $\phi = \dfrac{\tau}{G} = \dfrac{r \cdot \theta}{L}$ Shear strain for any given plane of given section from the fixed end.
- Shear stress should be maximum at extreme fibre of shaft.
- Angle of twist is maximum at the *free end* of shaft.

3. Moment of Inertia About Polar Axis

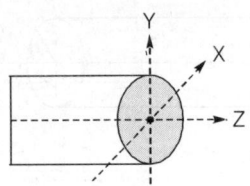

(i) For Solid Circular Section

$$J = I_z = I_x + I_y = \frac{\pi}{64} d^4 + \frac{\pi}{64} \cdot d^4 = \frac{\pi}{32} d^4$$

- Polar section modulus/torsional strength (Z_p)

$$\frac{\tau}{r} = \frac{T}{J}$$

$$\tau_{(max)} = \frac{T}{J/r_{max}}$$

$$\tau_{max} = \frac{T}{Z_p}$$

$$Z_p = \frac{J}{r_{max}} = \frac{\pi}{16} D^3$$

$$\tau_{max} = \frac{16T}{\pi d^3}$$

(ii) For Hollow Circular Section

$$I_{z\,(Hollow)} = \frac{\pi}{32} (D^4 - d^4)$$

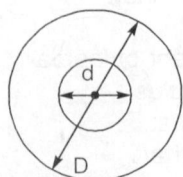

         D = Outside diameter

         d = Inside diameter

- Polar section modulus/torsional strength (Z_p)

$$J = \frac{\pi}{32} D^4 (1 - K^4)$$

$$z_p = \frac{\pi}{16} \cdot D^3 (1 - K^4) \quad K = \frac{d}{D}$$

4. Compound Shaft

- Series Connection

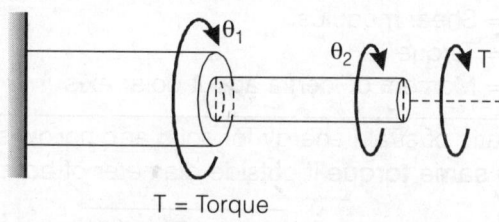

T = Torque

     θ_1 = Angular deformation of 1st shaft

     θ_2 = Angular deformation of 2nd shaft

   θ_{total} = Total angular deformation of free end of shaft from the fixed end

        = $\theta_1 + \theta_2$

Remember: ....

- Torque will be *same* on both shaft.
- Total angular deformation of free end of shaft from the fixed end will be equal to *sum* of angular deformation of first shaft and angular deformation second shaft.

- Parallel Connection

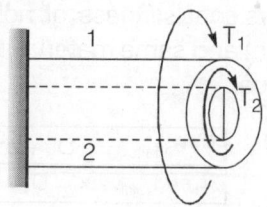

Here, θ_1 = Deflection of first shaft

           θ_2 = Deflection of second shaft

T_1 = Torque on first shaft

T_2 = Torque on second shaft

$\theta_1 = \theta_2$

Remember: ....

- Total torque will be equal to *sum* of torque acting on first shaft and torque acting on second shaft.
- Angular deflection will be *same* for both shaft.

5. Strain energy (U) stored in shaft due to torsion

$$U = \frac{1}{2}T \cdot \theta = \frac{1}{2}\frac{T^2L}{G \cdot J} = \frac{\tau_{max}^2}{2G} \times \text{Volume of shaft}$$

Here,   G = Shear modulus

T = Torque

J = Moment of inertia about polar axis

Remember

- Ratio of strain energy for solid and hollow shaft subjected to **same torque** if outside diameter of both shaft is equal.

$$\boxed{\frac{U_{Hollow}}{U_{solid}} = \frac{D^2 + d^2}{D^2}\frac{1}{2}}$$

d = Inside diameter of hollow shaft

- Ratio of torque in case of hollow and solid shaft subjected to *same maximum shear stress*.

$$\boxed{\frac{T_{Hollow}}{T_{Solid}} = \frac{D^4 - d^4}{D^4}}$$

Commonly used terms in torsion

- $\dfrac{T}{\theta} \rightarrow$ Torsional stiffness and $G.I_p \rightarrow$ Torsional rigidigy.

Remember

- Ratio of torsional stiffness of hollow to the solid shaft of same weight and same material if outside diameter of both shaft is equal.

$$\boxed{\frac{T/\theta_{(Hollow)}}{T/\theta_{(solid)}} = \frac{D^2 + d^2}{D^2}}$$

6. Effect of pure bending on shaft

$$\sigma = \frac{32\,M}{\pi D^3}$$

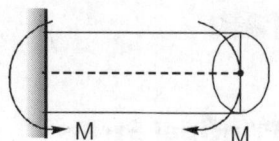

σ = **Principal** stress

D = Diameter of shaft

M = Bending moment

7. Effect of Pure Torsion on Shaft

$$\tau_{max} = \frac{16T}{\pi D^3}$$

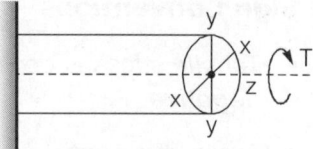

τ_{max} = Maximum **shear** stress

T = Torque

D = Diameter of shaft

8. Combined effect of bending and torsion

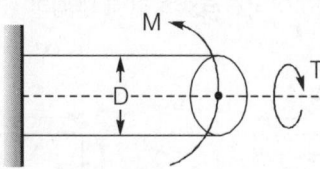

Principal Stress $= \dfrac{16}{\pi D^3}\left[M \pm \sqrt{M^2 + T^2}\right]$

Maximum **shear** stress $= \dfrac{16}{\pi D^3}\sqrt{M^2 + T^2}$

Remember

- Equivalent **bending moment**

$$= \frac{1}{2}\left[M + \sqrt{M^2 + T^2}\right]$$

- Equivalent torque $= \sqrt{T^2 + M^2}$

■■■■

Principal Stress/Principal Strain

1. Principal Stress

Principal stress are maximum or minimum *normal* stress which may be developed on a loaded body.

Remember: ....

- The plane of principal stress carry *zero shear stress*.

2. Sign Conventions

- *Tensile* stress is considered *positive* and *compressive* stress is *negative*.

- Angle 'θ' is considered *positive* if it is in *anti-clockwise direction*.

- *Shear* stress acting on a positive face of an element is considered positive if it is acts in positive direction of one of the coordinate axes and negative if acts in the negative direction of the axes. Similarly on a negative face of an element is positive if it act in negative direction of the axes and negative if it act in the positive direction.

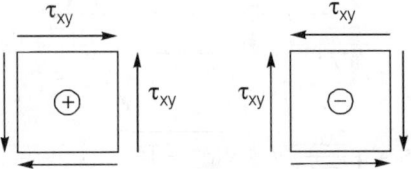

Positive shear stress   Negative shear stress

Remember: ....

- Normally the reference plane taken are major principal plane or vertical plane.

3. Analytical Method of Analysis

(i) If σ_1 and σ_2 are given *principal* stress as shown in figure, then normal and shear stress on plane a-a which is inclined at angle 'θ' from major principal plane ($\sigma_1 > \sigma_2$)

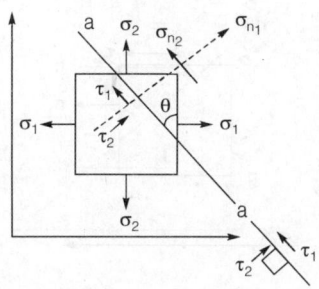

$$\therefore \quad \sigma_{n_1} = \sigma_1 \cos^2\theta + \sigma_2 \sin^2\theta$$

$$\sigma_{n_1} = \frac{\sigma_1 + \sigma_2}{2} + \left(\frac{\sigma_1 - \sigma_2}{2}\right) \cdot \cos 2\theta$$

$$\sigma_{n_2} = \frac{\sigma_1 + \sigma_2}{2} - \left(\frac{\sigma_1 - \sigma_2}{2}\right) \cdot \cos 2\theta$$

$$\tau_1 = \tau_2 = -\left(\frac{\sigma_1 - \sigma_2}{2}\right) \sin 2\theta$$

Remember

- $\sigma_{n_1} + \sigma_{n_2} = \sigma_1 + \sigma_2 = $ constant
- If $\theta = 45°$ or $135°$ then,

$$\tau_1 = \tau_2 = \tau_{max} = -\left(\frac{\sigma_1 - \sigma_2}{2}\right)$$

- On the plane of τ_{max}, value of normal stress σ_{n1} and σ_{n2}:

$$\boxed{\sigma_{n_1} = \sigma_{n_2} = \frac{\sigma_1 + \sigma_2}{2}}$$

- $\sigma_{Result} = \sqrt{\dfrac{\sigma_1^2 + \sigma_2^2}{2}}$ on plane of τ_{max}.

(ii) If σ_x and 'σ_y' are normal stress on vertical and horizontal plane respectively and this plane is accompanied by shear stress "τ_{xy}" then normal stress and shear stress on plane "a-a". Which is inclined at angle 'θ' from plane of σ_x.

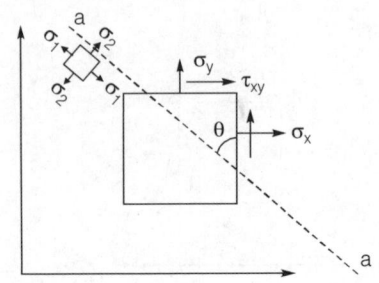

$$\sigma'_{1(a-a)} = \frac{\sigma_x + \sigma_y}{2} + \left(\frac{\sigma_x - \sigma_y}{2}\right) \cdot \cos 2\theta + \tau_{xy} \sin 2\theta$$

$$\sigma'_{2(a-a)} = \left(\frac{\sigma_x + \sigma_y}{2}\right) - \left(\frac{\sigma_x - \sigma_y}{2}\right) \cdot \cos 2\theta - \tau_{xy} \sin 2\theta$$

$$\tau_{(a-a)} = -\left(\frac{\sigma_x - \sigma_y}{2}\right) \sin 2\theta + \tau_{xy} \cos 2\theta$$

Remember: ....

- If θ occupies a position such that $\tau_{(a-a)}$ becomes zero, then such a plane is called principal plane and σ_1 and σ_2 become principal stress.

$$\tan 2\,\theta_p = \frac{2\,\tau_{xy}}{\sigma_x - \sigma_y}$$

θ_p = Angle of principal plane

(iii) Pure shear case

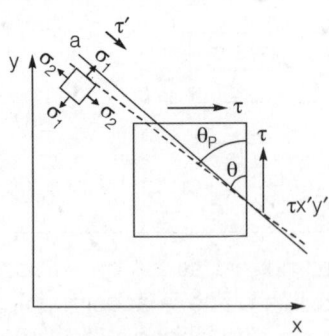

Thus normal stress on plane a – a

$$\sigma_{1(a\text{-}a)} = \tau_{xy} \sin 2\theta$$
$$\sigma_{2(a\text{-}a)} = -\tau_{xy} \sin 2\theta$$
$$\tau' = \tau_{xy} \cos 2\theta$$

4. Graphical Method of Analysis/Mohr Circle

Mohr circle is the locus of points representing magnitude of *normal* and *shear stress* at various plane in a given stress element.

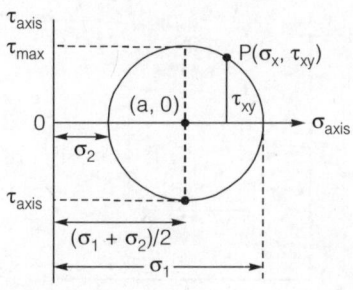

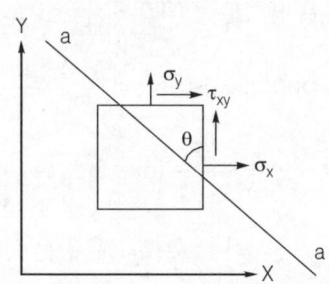

$$\sigma_1\!\Big/\!\sigma_2 = \frac{\sigma_x + \sigma_y}{2} \pm \sqrt{\left(\frac{\sigma_x - \sigma_y}{2}\right)^2 + (\tau_{xy})^2}$$

$$a = \frac{\sigma_1 + \sigma_2}{2}$$

- Radius of mohr's circle

$$r = \sqrt{\left(\frac{\sigma_x - \sigma_y}{2}\right)^2 + \tau_{xy}^2} = \frac{\sigma_1 - \sigma_2}{2}$$

Remember

- **Radius of Mohr circle** represent the value of **maximum shear stress.**
- **Normal stress** on the plane of maximum shear stress is represented by **coordinate of centre of Mohr circle.**
- Mohr circle reduce to *a point* in case of *hydrostatic* loading and *zero* shear.
- In case of *pure shear*, centre will fall at *origin*.

5. Analysis of Strain

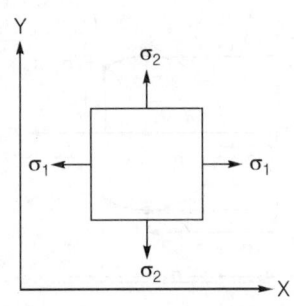

ϵ_1 = Major principal strain = $\dfrac{\sigma_1}{E} - \mu\dfrac{\sigma_2}{E}$

ϵ_2 = Minor principal strain = $\dfrac{\sigma_2}{E} - \dfrac{\mu\sigma_1}{E}$

$$\sigma_1 = \frac{E}{1-\mu^2}[\epsilon_1 + \mu\,\epsilon_2]$$

$$\sigma_2 = \frac{E}{1-\mu^2}[\epsilon_2 + \mu\,\epsilon_1]$$

Symbol has usual meaning:

Remember

- Plane stress does not lead to plain strains.

■ ■ ■ ■

Theory of Failure 5

1. Maximum principal stress theory (Rankines theory)

According to this theory, permanent set takes place under a state of complex stress, when the value of maximum principal stress is equal to that of yield point stress as found in a simple tensile test.

For design criterion, the maximum principal stress (σ_1) must not exceed the working stress 'f' for the material

$$\sigma_{1,2} \leq f$$

- Graphical representation

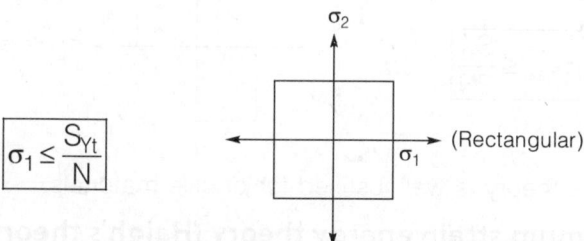

$$\boxed{\sigma_1 \leq \frac{S_{Yt}}{N}}$$

(Rectangular)

- For brittle material which do not fail by yielding but fail by brittle fracture, this theory gives satisfactory result.
- The graph is always square even for different of values of σ_1 and σ_2.

2. Maximum principal strain theory (ST. Venant's theory)

According to this theory, a ductile material begins to yield when the maximum principal strain reaches the strain at which yielding occurs in simple tension

$$\varepsilon_1 \leq \varepsilon_{Y.P.}$$

- Graphical Representation

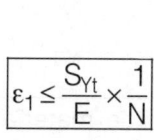

$$\boxed{\varepsilon_1 \leq \frac{S_{Yt}}{E} \times \frac{1}{N}}$$

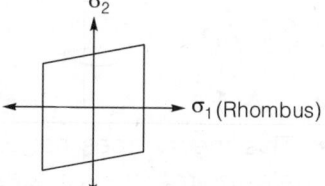

σ_1 (Rhombus)

- This theory over estimate the elastic strength of ductile material.

3. Maximum shear stress theory (Guest & Tresca's theory)

According to this theory, failure of a specimen subjected to any combination of loads when the maximum shearing stress at any point reaches the failure value (τ_f) equal to that developed at the yielding in an axial tensile or compressive test of the same material.

- Graphical Representation

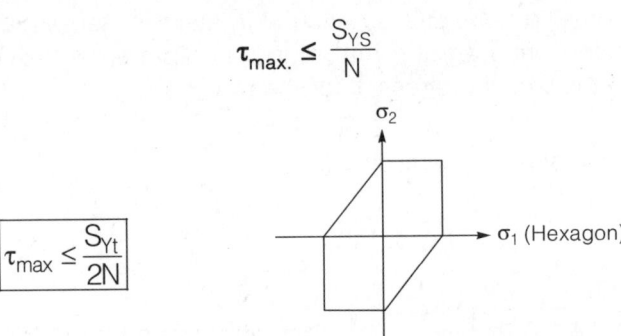

$$\tau_{max.} \leq \frac{S_{YS}}{N}$$

$$\boxed{\tau_{max} \leq \frac{S_{Yt}}{2N}}$$

- This theory is well justified for ductile materials.

4. Maximum strain energy theory (Haigh's theory)

According to this theory, a body under complex stress fails when the total strain energy on the body is equal to the strain energy at elastic limit in simple tension.

- Graphical Representation

$$\left[\left(\sigma_1^2 + \sigma_2^2 + \sigma_3^2\right) - 2\mu\left(\sigma_1\sigma_2 + \sigma_2\sigma_3 + \sigma_1\sigma_3\right)\right] \leq \frac{S_{Yt}^2}{N^2}$$

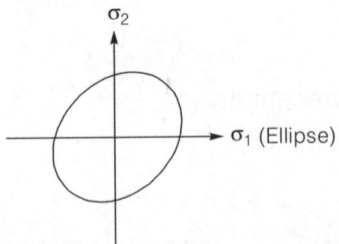

- This theory does not apply to brittle material for which elastic limit stress in tension and in compression are quite different.

5. Maximum shear strain energy/Distortion energy theory/ Mises-Henky theory

It states that inelastic action at any point in a body, under any combination of stress begins, when the strain energy of distortion per unit volume absorbed at the point is equal to the strain energy of distortion absorbed per unit volume at any point in a bar stressed to the elastic limit under the state of uniaxial stress as occurs in a simple tension/compression test.

$$\left[\left(\sigma_1-\sigma_2\right)^2+\left(\sigma_2-\sigma_3\right)^2+\left(\sigma_3-\sigma_1\right)^2\right]\le 2\left(\frac{S_{Yt}}{N}\right)^2$$

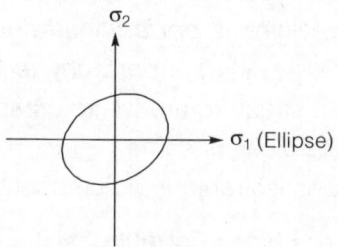

σ_1 (Ellipse)

Remember

- It can not be applied for material under hydrostatic pressure.

■ ■ ■ ■

Columns

6

1. STRUT

Structural member subjected to axial compressive load is called strut.

- **Column**

  Vertical structural member fixed at both ends and subjected to axial compressive load is called column.

2. Buckling Failure : Euler's Theory

- **Assumption in Euler's theory**

  (i) Axis of column is *perfectly straight* when *unloaded*.

  (ii) Compressive load is *perfectly axially* applied.

  (iii) Stress in structure are within *elastic limit*.

  (iv) Flexural rigidity is same.

  (v) Material is isotropic and homogeneous.

- **Limitation of Euler's Formula**

  (i) There is always *crookedness* in the column and the load may not be exactly axial.

  (ii) This formula does not take account the axial stress and the buckling load given by this formula may be much more than the actual buckling load.

$$P_e = \frac{\pi^2 \, EI_{min}}{l_e^2}$$

  P_e = Buckling load

  I_{min} = Moment of inertia about centroids axis

  l_e = *Effective* length

Remember: ....

- It is applicable for long column
- Effect of crushing is neglected.

3. Euler's load for different column with different end Condition

End condition	Both end hinged	One end fixed other free	Both end fixed	One end fixed and other hinged
Effective length(l_e)	L	2L	$\frac{L}{2}$	$\frac{L}{\sqrt{2}}$

4. Slenderness Ratio (S)

Slenderness ratio of a compression member is defined as the ratio of its effective length to radius of gyration.

$$S = \frac{L_e}{K}$$

L_e = Effective length
K = Least radius of gyration

$$K = \sqrt{\frac{I_{min}}{A}}$$

∴   Buckling stress $(\sigma_b) = \dfrac{P_e}{A} = \dfrac{\pi^2 E}{s^2}$

5. Rankine's Formula

$$\frac{1}{P_R} = \frac{1}{P_C} + \frac{1}{P_e}$$

Rankine load = P_R
Crushing load = $P_C = \sigma_c \times A$

Buckling load = $P_e = \dfrac{\pi^2 E I_{min}}{L_e^2}$

∴ $P_r = \dfrac{A\sigma_c}{1 + a\left(\dfrac{L_e}{k}\right)^2}$

Here,

     A = Area of column

     $a = \dfrac{\sigma_c}{\pi^2 E}$ = Rankine's constant

Remember

- This formula is applicable to any column.
- Effect of crushing and buckling is considered in this formula.

■■■■

Springs

7

- Spring are used to absorb energy and restore it slowly or rapidly.

1. Type of spring on the basis of helix angle

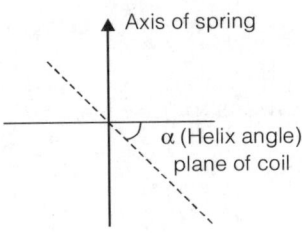

- If helix angle is *less than* or *equal* to 10° then it is called closed coil spring.
- If helix angle is *greater than* 10° then it is called open coil spring.

- The best form of spring absorbs *greatest amount* of energy for a given stress.
- Spring stored energy in the form of *resilience*.

Remember

2. Series and parallel arrangement of springs/Equivalent spring constant (k_{eq})

- In Series: $\dfrac{1}{k_{eq}} = \dfrac{1}{k_1} + \dfrac{1}{k_2} + \cdots \dfrac{1}{k_n}$

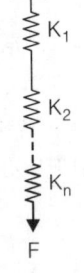

- In parallel: $k_{eq} = k_1 + k_2 + \cdots k_n$

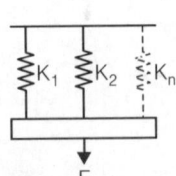

Remember: ....

- Stiffness of spring inversely proportional to number of coils in the spring. Therefore when a spring cut into two parts its stiffness become *double* for every *individual* part.

3. Closed coil helical spring under axial pull

- Axial deflection of spring (δ)

$$\text{Deflection } (\delta) = \frac{8FC^3N}{Gd}$$

- Shear stress in spring (τ_{max})

$$\tau_{max} = \frac{8K_wFC}{\pi d^2}$$

- Stiffness of Spring (K)

$$K = \frac{Gd}{8C^3N}$$

Here,   F = Applied load
C = Spring index
d = Spring wire diameter
G = Modulus of rigidity
N = Number of turn
k_w = Wahl's factor

Remember

- Spring index (C) = $\dfrac{D}{d}$.

- Wahl's factor is considered to consider the effect of direct shear stress and curvature effect.

---

4. Strain energy stored in spring (U)

$$U = \frac{1}{2}T \cdot \theta$$

T = torque applied
θ = angular deflection

5. Wahl's correction factor (k_w)/Stress concentration factor (k_c)

$$k_w = \frac{4C-1}{4C-4} + \frac{0.615}{C}$$

$$k_c = \frac{4C-1}{4C-4}$$

Here,   C = Spring index

Remember

- The *average* value of modulus of rigidity for *steel* used for spring equal to *79300 MPa.*
- *Shot peening*, result in raising the *fatigue* life of spring because it leave the surface in compression.

■■■■

Pressure Vessels

8

1. Types of Pressure Vessels

Pressure vessel are mainly of two type:

(i) Thin shells

If the thickness of the wall of the shell is less than 1/10 to 1/15 of its diameter, then shell is called thin shells.

$$t < \frac{D_i}{10} \text{ to } \frac{D_i}{15}$$

Remember: ....

- For thin shell, it is assumed that the normal stresses, which may be *either* tensile or compressive are *uniformly* distributed through the thickness of wall.

(ii) Thick Shells

If the thickness of the wall of the shell is greater than 1/10 to 1/15 of its diameter, then shell is called thick shells.

$$t > \frac{D_i}{10} \text{ to } \frac{D_i}{15}$$

2. Nature of stress in thin cylindrical shell subjected to internal pressure

(i) Hoop stress /circumferential stress will be *tensile* in nature.

(ii) Longitudinal stress/axial stress will be *tensile* in nature.

(iii) *Radial* stress will be *compressive* in nature.

- Radial compressive stress varies from a value at the inner surface equal to pressure 'P' to the atmospheric pressure at the outside surface.
- If internal pressure in thin cylinders is low, the radial stress is negligible compared with axial stress and hoop stress. This *radial stress* is neglected.

Remember

3. Analysis of thin cylinder

- Longitudinal Stress

$$\sigma_L = \frac{pd}{4t}$$

- Hoop Stress

$$\sigma_h = \frac{pd}{2t}$$

- Longitudinal Strain

$$\epsilon_L = \frac{pd}{4t\,E}(1-2\mu)$$

- Hoop Strain

$$\epsilon_h = \frac{pd}{4tE}(2-\mu)$$

Here,   p = Pressure of fluid

t = Thickness of cylinder

d = Inside diameter

μ = Poisson's ratio

- Ratio of Hoop Strain to Longitudinal Strain

$$\frac{\epsilon_h}{\epsilon_L} = \frac{2-\mu}{1-2\mu}$$

- Volumetric Strain (ϵ_v) of Cylinder

$$\epsilon_v = \frac{pd}{4t\,E}(5-4\mu)$$

Remember

- If fluid is compressible, volumetric strain will be

$$\epsilon_v = \frac{pd}{4t\,E}(5-4\mu) + \frac{p}{K}$$

k = Bulk modulus of fluid

P = Pressure of fluid

- Minimum thickness of cylinder required for a given pressure 'P' and diameter 'd' is

$$t \geq \frac{pd}{2\sigma}$$

4. Analysis of thin sphere

- Hoop stress/longitudinal stress

$$\sigma_L = \sigma_h = \frac{pd}{4t}$$

- Hoop strain/longitudinal strain

$$\epsilon_L = \epsilon_h = \frac{pd}{4tE}(1-\mu)$$

- Volumetric strain of sphere

$$\epsilon_v = \frac{3pd}{4tE}(1-\mu)$$

Remember

- Thickness ratio for cylindrical shell (t_c) and sphere (t_s), for *same strain* in both side.

$$\frac{t_c}{t_s} = \frac{2-\mu}{1-\mu}$$

- Thickness ratio for cylindrical shell (t_c) and sphere (t_s), for *same maximum stress* in both side.

$$\frac{t_c}{t_s} = 2$$

- Auto frottage is used for prestressing the cylinder. *Wire winding* is done for *strenghting thin shell*. *Compounding* is done for *thick* shell cylinders.

---

5. Analysis of Thick Cylinders/Lame's Theorem

- Lame's assumption
  - (i) Material of shell is homogeneous
  - (ii) Plane section of cylinder, perpendicular to longitudinal axis remains plane under pressure.

- Subjected to internal pressure

  (i) At $x = R_i$, $\sigma_h = \dfrac{P\left[R_o^2 + R_i^2\right]}{R_o^2 - R_i^2}$

  (ii) At $x = R_o$, $\sigma_h = \dfrac{2PR_i^2}{R_o^2 - R_i^2}$

- Subjected to external pressure

  (i) At $x = R_i$, $\sigma_h = \dfrac{-2PR_o^2}{R_o^2 - R_i^2}$

  (ii) At $x = R_o$, $\sigma_h = \dfrac{-P\left[R_o^2 + R_i^2\right]}{\left[R_o^2 - R_i^2\right]}$

Here, R_i = Inner radius

R_0 = Outlet radius

$$\boxed{(\sigma_h)_{max} = p + (\sigma_h)_{min}}$$

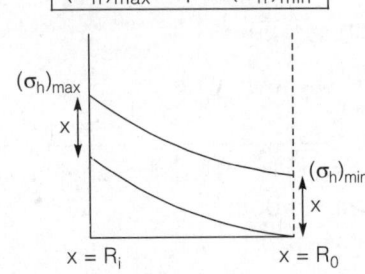

 • Radial and hoop compression vary *hyperbolically*.

■ ■ ■ ■

Deflection of Beam

For design purpose, a beam should be so designed that it has adequate stiffness so that the deflections are within permissible limits.

Stiffness of beam, inversely proportional to deflection.

1. Methods of Determining Deflection of Beam

- Double integration method.
- Moment area method
- Strain energy method
- Conjugate beam method

2. Deflection of Beam Under Different Loading/Support Condition

- Notation used

$$\theta_B^A = \text{Slope at B w.r.t A}$$

$$\Delta_B^A = \text{Deflection at B w.r.t A}$$

-

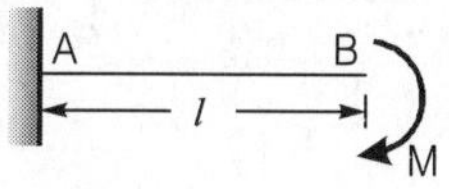

$$\text{Slope } (\theta_B^A) = \frac{Ml}{EI}$$

$$\text{Deflection} (\Delta_B^A) = \frac{Ml^2}{2\,EI}$$

-

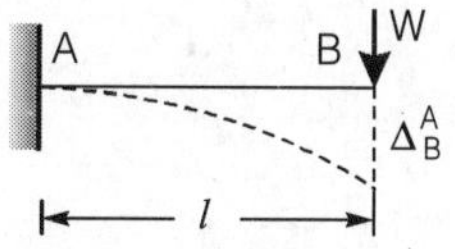

$$\text{Slope } (\theta_B^A) = \frac{Wl^2}{2\,EI}$$

$$\text{Deflection} (\Delta_B^A) = \frac{Wl^3}{3\,EI}$$

-

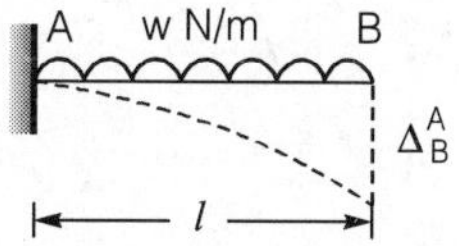

$$\text{Deflection} (\Delta_B^A) = \frac{wl^4}{8\,EI}$$

-

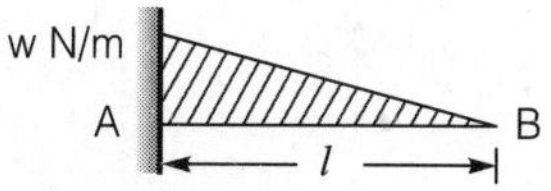

$$\text{Deflection} (\Delta_B^A) = \frac{wl^4}{30\,EI}$$

- 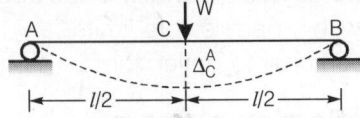   Deflection $(\Delta_B^A) = \dfrac{11}{120} \dfrac{wl^4}{EI}$

- **Simply supported beam**

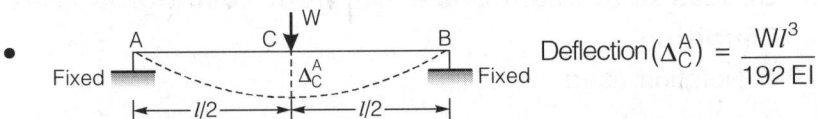

Deflection $(\Delta_C^A) = \dfrac{Wl^3}{48\,EI}$

Deflection $(\Delta_C^A) = \dfrac{5\,wl^4}{384\,EI}$

Deflection $(\Delta_C^A) = \dfrac{Wl^3}{192\,EI}$

Deflection $(\Delta_C^A) = \dfrac{wl^4}{384\,EI}$

■■■■

A Handbook on
Mechanical Engineering

Material Science &
Production Engineering

CONTENTS

- Change of phase means either there is a change in microstructure or there is change in the lattice structure.
- A phase diagram has temperature as its ordinate and alloy composition as abscissa.
- **Curie point (768°C):** There is no change in phase and only magnetic properties are disappearing.
- **Paramagnetic Materials:** Electrons are unpaired and such alloys exhibit colour.
- **Diamagnetic Materials:** Electrons are paired and such alloys are colourless.
- **Ferromagnetic:** Iron behave as paramagnetic as well as diamagnetic.
- **Iron Carbon Diagram**

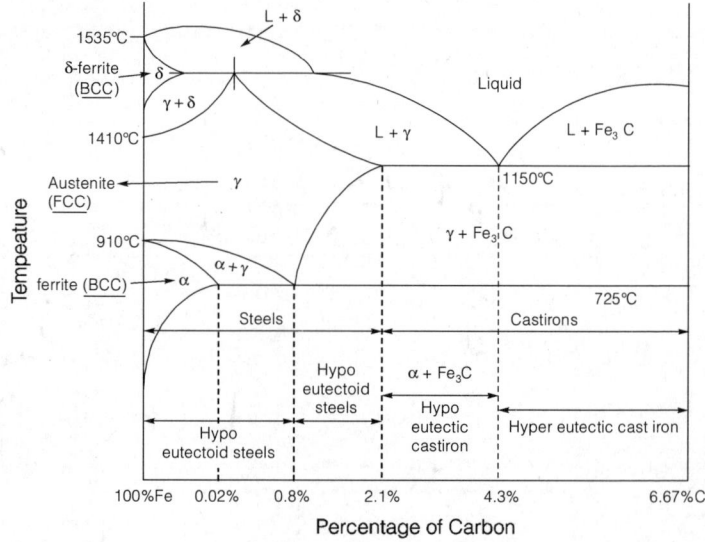

- $L \underset{4.3\%C}{\overset{1150°}{\rightleftharpoons}} \gamma + Fe_3 C$ [Eutectic reaction]

- $\gamma \underset{0.8\%C}{\overset{725°C}{\rightleftharpoons}} \alpha + Fe_3 C$ [Eutectoid reaction]

- $\delta + L \underset{0.18\%\,Carbon}{\overset{1495°C}{\rightleftharpoons}} \alpha(solid)$ [Peritectic reaction]

- Low C steel (mild steel) < 0.3% C
- High C steel > 0.7% C
- **Hot Shortness → Fe + S**
  - FeS has low melting point.
  - Upon warm working it melts out.
  - Mn is added to remove FeS because Mns form.
- **Binary Isomorphous System**
  Materials which are completely soluble in the liquid as well as solid.
  Cu-Ni phase diagram

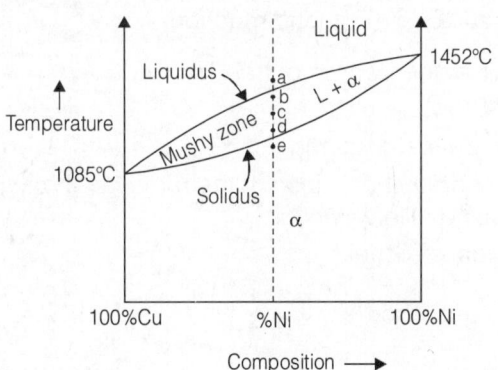

- Hume-Rothary condition has to be satisfied for the alloy formation.
  - Atomic size difference should be less than 15%.
  - Valency of both the material should be same.
  - Electronegativity difference should be minimum.
- **Grain Boundaries**
  The region where two solidification front meets there will be orientation mismatch.
  Some common grain boundary defect :

Hot shortness	Temper embrittlement	Grain boundary embrittlement
Local melting of impurity at grain boundary causes crack	Segregation of metal Impurity at grain boundary	Ductile metal when brought into contact with a low M.P. metal it undergoes crack at low stress.

- **Binary Eutectic System**
  Material which are completely soluble in the liquid but partially soluble in the solid.

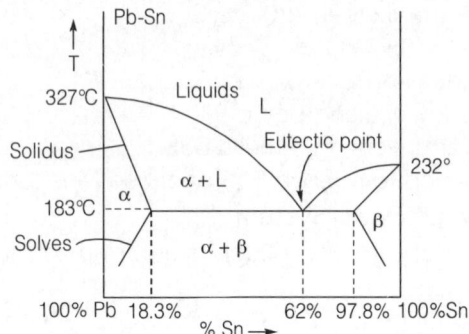

- $L \xrightleftharpoons[183°C]{62\% Sn} \alpha + \beta$ (Eutectic reaction).
- At eutectic point, degree of freedom = 0
- **Types of Steel**
  - Hadfield steel-12% Mn
  - By the addition of silicon, if the removal of oxygen is complete it is called killed steel.
- **Designation of Steel**

  $x \times \dfrac{x\,x}{100}\%$ of carbon

2 x x x	:	Ni	31 x x	:	Hadfield
3 x x x	:	Ni-Cr	4 x x x	:	Mo
3 2 1 3	:	Stainless steel	5 x x x	:	Cr

- **Gray C.I**
  Carbon is present in free or flake form (C = 2.5 – 4%).
- **White C.I**
  Carbon is present in combined form (C = 1.8 – 3.6%).
- **Chilled C.I**
  Normally freeze as gray but forced to freeze as white.
- **Ductile C.I**
  Nodular C.I, spherical C.I, malleable C.I.
- **KISH**
  Graphite is having lower density, due to which it comes over the surface of liquid Fe in the red hot condition and sparkles.

1. Effect of Alloying Element in Steel

- Sulphur: Machinability
- Molybdenum: Forms abrasion resisting particles/Improves creep properties.

- Phosphorous: Improves machinability in free cutting steel.
- Cobalt: Contributes to red hardness by hardening ferrite.
- Chromium: Corrosion resistance / Increase hardness.
- Nickel: Hardenability
- Tungsten: Heat resistance
- Silicon: Magnetic permeability/Increase resistance to high temperature oxidation.
- Vanadium: Hardenability/Increases strength.
- Manganese: Hardness.
- **Addition of Silicon to C.I**
  - Promotes graphite **flake** formation.
  - Increases **fluidity** of the molten metal.

Graphite flakes 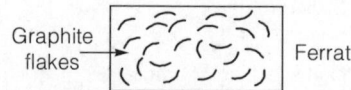 Ferratic gray C.I.

- **Addition of Mg to C.I**
  - It increases **ductility** and **strength** in tension.

Spheroidal graphite  Spheroidal (Ductile) C.I.

Alloy	Application
Monel metal $\begin{bmatrix} Ni+Cu \\ 70\%\ 30\% \end{bmatrix}$	Marine bearing
Invar (Nickel)	Thermal expansion
Gun metal $\begin{bmatrix} Cu,\ Sn,\ Zn \\ 88\quad 10\quad 2 \end{bmatrix}$	Journal bearing
Duralium (Al)	$\begin{bmatrix} Al, & Cu, & Mg & Si \\ 94.8 & 4.0 & 0.5 & 0.7 \end{bmatrix}$
Babbit alloyt (Cu base alloy having Sn, Pb)	Bearing, excellent embeddability
Low carbon steel (Mild steel)	Wire nails, crane hooks, rolled steel section
Medium carbon steel	Screw, shaft and axles
High carbon steel	Commercial beams, springs cold chisels
Hadfield Mn steel	Buldozer rolls
Constantan	Thermocouple
Chromal	Thermo couple wire

Nimonic alloy	GT blades
HSS ⎡T-series, W, Cr, V, Co M-Series Mo, Cr, V, C, W	Milling cutter
Tool steel	Making die, ball bearing
Silicon steel	Transformer lamination
Commercial bronze (10% zn)	Forging and stamping
Red brass (15% zn)	Radiator
Al. brass (22% Zn, 2% Al) [Admirality brass]	Condenser tubes, heat exchanger
P-bronze (11% Sn, small amount of P)	Spring metal
Spring steel	Non sparking character

- ## TTT Curve (Bain's Curve)

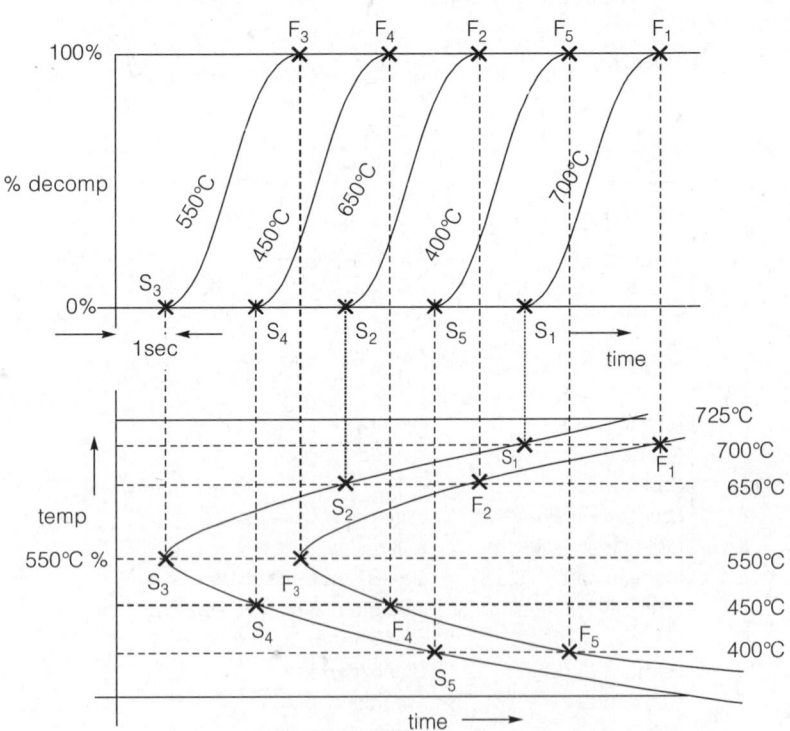

- **Incubation Period**
  - While quenching austenite below 725°C, it was observed that for the substantial period of time there was no change in the microstructure this period is called incubation period.
  - Incubation period decreases as the temperature of quenching decreases.

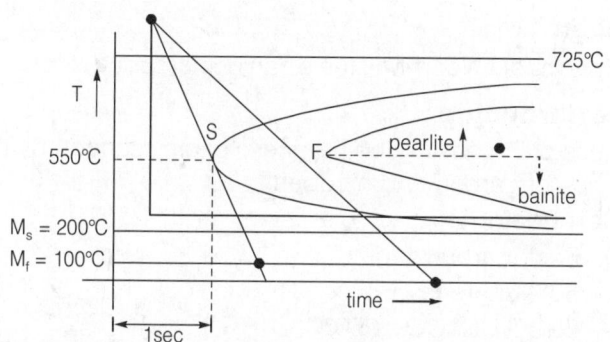

Upon cooling austenite at a rate greater than CCR. Carbon will freeze at its location and the microstructure will be like colloidal solution of cementite into ferrite called martensite.

- **Hardenability**

  Ease of martensite formation and it can be improved when martensite can be formed at slower cooling rates. Achieved by switching TTT towards right.

HEAT TREATMENT

1. Annealing
- **Quenching Medium: Furnace**
- Full annealing → Ductility inc. and toughness inc.
- Process annealing → stress relieving
- Spherodise annealing → machinability increases
- Diffusion annealing → To homogenize the chemistry of material.

2. Normalizing
- Quenching medium : AIR
- Produce hard surface and tough core.

- Normalizing gives more homogenized and refine grain structure than annealing.
- Normalizing is fast, efficient and less costly.

Remember

3. Hardening

The objective of hardening is to produce the martensite.

- Quenching medium
  - Water → Vapor blanket forms (HT ↓ 2) Non uniform cooling.
  - Salt bath → Decreases the tendency of vapor blanket formation.
  - Oil bath → uniform cooling

Remember: ....

- Only medium carbon steels can be hardened.

4. Case Hardening

Process of increasing the hardness of only surface by the process of diffusion of carbon and nitrogen.

- Carburizing
  - (a) Pack carburizing
    50% charcoal
    $BaCO_3$, $CaCO_3$, 950°C
  - (b) Liquid carburizing
    20% NaCN
    950°C
  - (c) Gas carburizing
    CH_4 gas
    950°C
- Cyaniding
  30% NaCN, NaCl, Na_2CO_3, 850°C
- Flame hardening
  - Carburizing flame (2900°C)
  - Neutral flame (3150°C)
  - Oxidizing flame (3500°C)
- Induction hardening
  - Fastest method of case hardening.
  - **Medium carbon steel** can be hardened by this technique.
  - **Copper Induction coil** is used in Induction hardening for application over camshaft, crankshaft.
- Tempering
  - Process of introducing toughness.
  - High temperature tempering (500-650°C): SORBITE
  - Medium temperature (350-500°C): TROOSTRITE spring and dies.
  - Low temperature (on 250°): Matensite is heated at (in 250°C) 250° and coded slowly to relevane internal speed.

5. Defect in Material

- Point Defect
  - Vacancy defect
  - Intertial defect
  - Substitutional defect
  - Frankle defect
  - Schottky defect
- Surface Defect
  - Grain boundary defect
  - Twin boundary defect
  - Tilt boundary defect
  - Staking faults
- Line Defect
  - Edge dislocation defect
  - Screw dislocation defect

■■■■

Welding

2

Welding is a process in which localized coalescence (permanent joint) is produced by heating the material upto suitable temperature with or without application of filler material.

If filler material is different from base material it is heterogenous welding.

1. Advantage/Disadvantage of welding:

- Advantage
  - *Different* material can also be welded
  - Welding can be done anywhere
- Disadvantage
- In the heat affected zone properties of base material is also affected
- To dismantle, we have to break the weld
- Initial investment is more
- Highly skilled operator is required.

Remember

Weldability : It is the ability of material to weld a material.

$$\% \text{ Weldability} = \frac{\text{Resistivity} \times 100}{K_{\text{Relative constant}} \times T_{\text{Melting point}}}$$

2. Type of Welded Joints

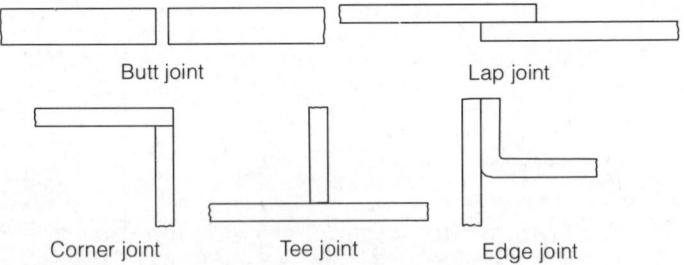

Butt joint   Lap joint

Corner joint   Tee joint   Edge joint

3. Type of Welded Joints according to Position

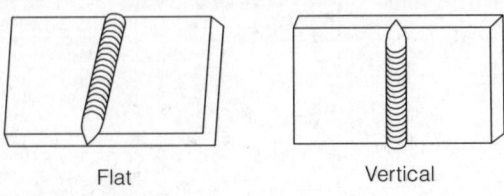

Flat   Vertical

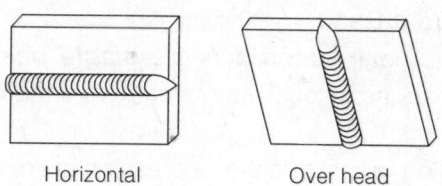

Horizontal   Over head

4. Welding Terms

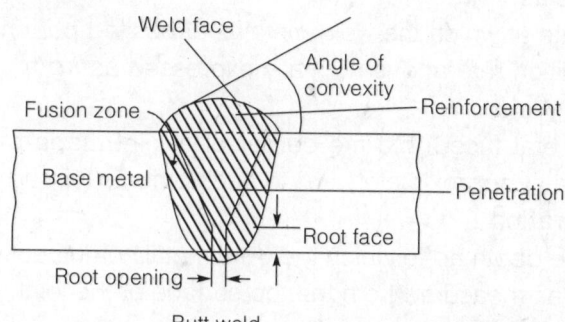

Butt weld

$$\text{Dilution} = \frac{A_P}{A_P + A_R}$$

A_p = Area of penetration
A_R = Area of reinforcement

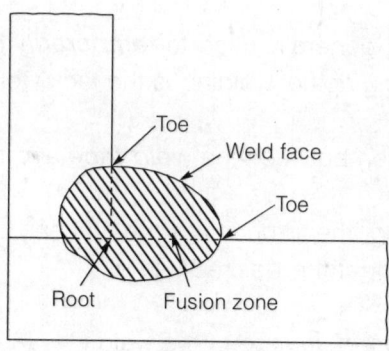

Filled weld

- **Backing**
  It is the material support provided at the ***root side*** of a weld to aid in the ***control of penetration***.
- **Base Metal**
  The metal to be joined or cut is termed the base metal.

- **Bead or Weld Bead**

  Bead is the metal added during a *single pass* of welding. The bead appears as a separate material from the base metal.

- **Crater**

  In arc welding, a crater is the depression in the weld-metal pool at the point where the arc strikes the base metal plate.

- **Deposition Rate**

  The rate at which the weld-metal is deposited per unit time is the deposition rate and is normally expressed as *kg/h*.

- **Filled Weld**

  The metal fused into the corner of a joint made of two pieces placed at approximately 90° to each other is termed fillet weld.

- **Penetration**

  It is the depth up to which the weld metal combines with the base metal as measured from the top surface of the joint.

- **Puddle**

  The portion of the weld joint that is melted by the heat of welding is called puddle.

- **Root**

  It is the point at which the two pieces to be joined by welding are nearest.

- **Tack Weld**

  A small weld, generally used to *temporarily* hold the two pieces together during actual welding is the tack weld.

- **Toe of Weld**

  It is the junction between the *weld face* and the *base metal*.

- **Torch**

  In gas welding, the torch mixes the fuel and oxygen and controls its delivery to get the desired flame.

- **Weld Face**

  It is the *exposed* surface of the weld.

- **Weld Metal**

  The metal that is solidified in the joint is called weld metal. It may be only a base metal or a mixture of base metal and filler metal.

- **Weld Pass**

  A single movement of the welding torch or electrode along the length of the joint which results in a bead is a weld pass.

5. Types of Welding

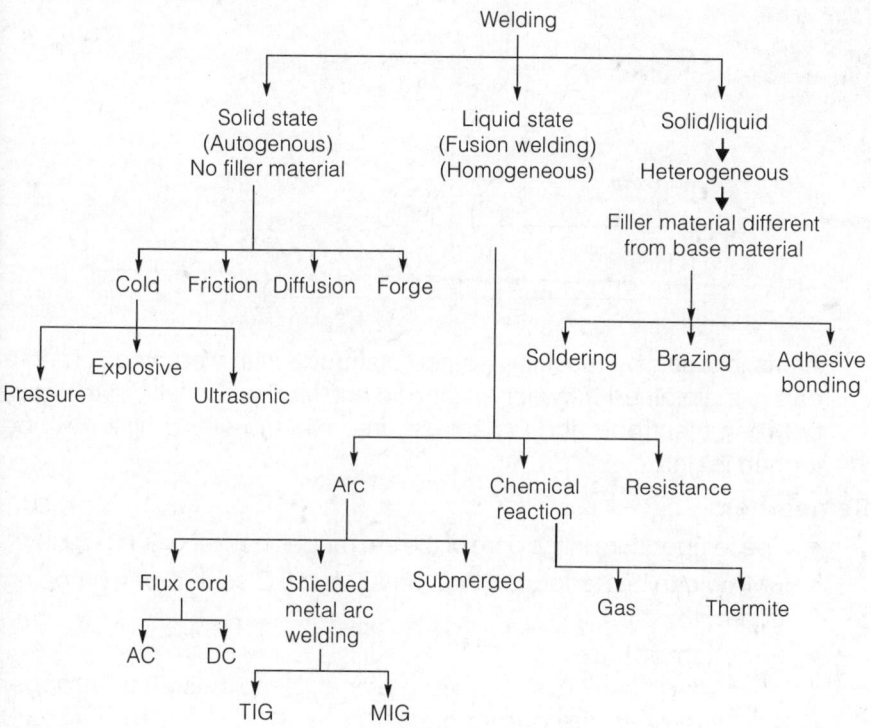

ARC WELDING

- **Principle of Arc**
  - An arc is generated between two conductors of electricity, cathode and anode (considering direct current, DC), when they are thouched to establish the flow of current and then separated by a *small distance*. An arc is a sustained electric discharge through the *ionized gas column,* called *plasma*, between the two electrodes.
  - In order to produce the arc, potential difference between the two electrodes (voltage) should be sufficient to allow them to move *across the air gap*. The *larger air gap* requires *higher potential* differences. If the air gap becomes too large the arc may be extinguished.

6. Shielded Metal Arc Welding (Manual Metal Arc Welding)

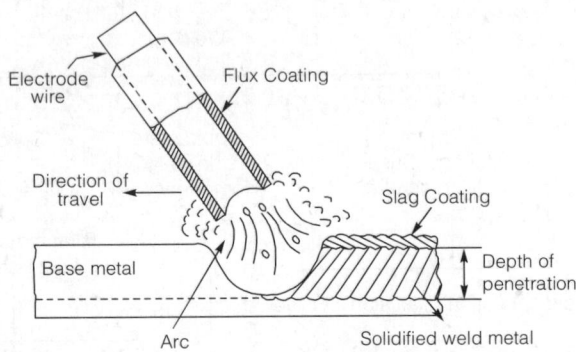

Arc is initiated by touching coated electrode with workpiece. Then a gap is maintained between electrode and workpiece with in the gap, between electrode and workpiece, free electrons and holes will be formed in gas.

Remember: ....

- The temperature in the core of the arc ranges between 6000 -7000°C.
- SMAW can be performed both on AC and DC source with drooping characteristics.
- Function of Flux
  - Flux gives alloys to bead material, bead becomes *stronger than even the parent metal.*
  - It provide heat treatment to bead.
  - It protect the bead from the attack of atmospheric gases.
  - Arc stabilization is good
  - It controls weld pool viscosity.
- Flux Coating Material

Deoxidizing element	Alumina, Graphite, Manganese, Silicon
Slag formation compound	Iron oxide, sio_2
Arc stabaliser	Na_2O, CaO, TiO_2, MnO
Weld strength	Cr, W, Ni, Co, V

- Limitation of FLux Coating
  - Slug inclusion
  - Hydrogen embrittlement
- Short Circuit Current
  It is that current which flow when we touch workpiece by electrode. This is the current required *during* arc generation.

- **Open Circuit Current**
  It is that current, which flow when we maintain a gap between workpiece and electrode.

  $$V = A + Bl$$

  Here, V = Voltage
  l = Arc length (mm)
  A, B = Constant

- **Open Circuit Voltage**
  It is the amount of voltage required to generate the arc under no load conditions. Generally open circuit voltage is around 80 volts. (Voltage at time of short-circuit current)

- **Duty Cycle**
  It is the amount of time during which the transformer will be used for welding under normal loading conditions.

Remember

- Normally, 60% duty cycle is accepted worldwide in 10 minute. While in India we consider the cycle time of 5 minute.

- **Automatic and Semiautomatic Welding**
  During welding if the movement of electrode with respect to the workpiece will be controlled automatically these welding machines are called automatic welding machines. If this movement of electrode controlled by *automatic* machines and the relative movement of the electrode is controlled *manually* this is known as semiautomatic machines.

TRANSFORMER

- **Constant current type transformer**

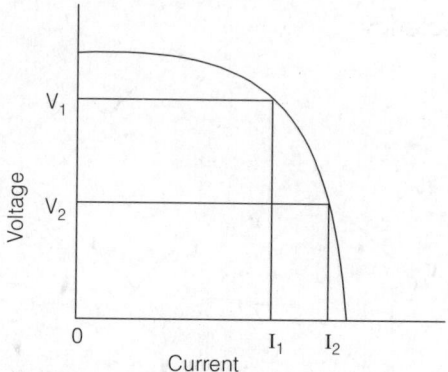

This type of transformer is used in manual arc welding because arc length can not be controlled, So arc current is controlled by the transformer. It has the drooping volt-ampere characteristic curve, shows that welding power source produces maximum output voltage with no load.

- **Constant Voltage Transformer**

  This transformer is used in automatic welding as well as in semiautomatic welding

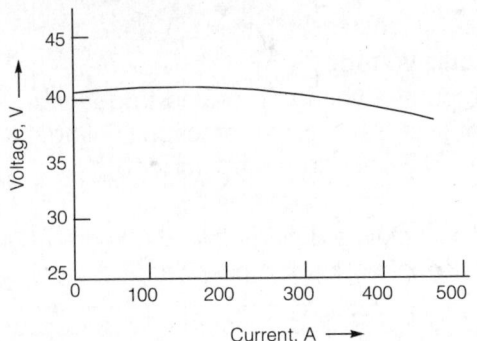

It has essentially a **flat volt-ampere** characteristics curve through usually with a slight droop.

$$V_{arc} = (OCV) - \left[\frac{OCV}{SCC}\right]I$$

This is for semiautomatic arc welding

OCV = Open circuit voltage

SSC = Short-circuit current

For stable arc in constant voltage transformer

$$V_{arc} = V_{transformer}$$

For stable arc in constant current transformer

$$I_{arc} = I_{transformer}$$

- **Melting Efficiency**

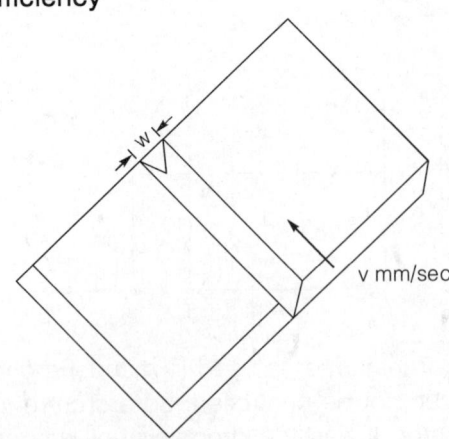

H_s(Heat input supplied) $= \dfrac{V \cdot I}{v \cdot a}\left(J/mm^3\right)$

a = weld bead area (mm²)
v = velocity mm/sec

$$\eta_t = \frac{H_a}{H_s}$$

where, H_a = Actual heat input (J/mm³)
η_t = **heat transfer efficiency**
H_m = heat required to melt unit volume of material.

$$\eta_m = \frac{H_m}{H_a}$$

Where, $H_m = m\, cp\Delta T + mL$ (J/mm³)

Melting efficiency; $\quad \eta_m = \dfrac{H_m}{(VI/v \cdot a)\,\eta_t}$

In most case $\eta_{nm} \approx 30 - 40\%$

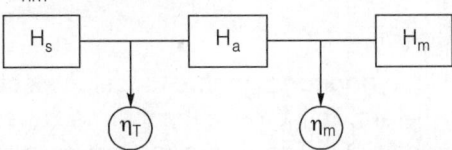

COMPARISON OF DIFFERENT ELECTRODE POLARITY

	DC, electrode positive	DC, electrode negative	AC,
1. Penetration	Shallow	Deep	Intermediate
2. Heat generation	2/3rd at electrode 1/3rd at workpiece	1/3rd at electrode 2/3rd at workpiece	50% on both
3. Metal deposition rate	High	Low	Intermediate
4. Thickness of work to be welded	Thin sheets	Thin sheets	Intermediate
5. Stable smaller arc	Easier	Easier	Difficult
6. Arc blow	Severe	Severe	Insignificant

- **Arc Blow**

  While welding near the edges, since most of the magnetic lines will be concentrated in the material, arc will be deflected towards the workpiece. This phenomenon is called arc blow which results in spatter.

7. Tungsten Inert Gas Welding (TIG)/Gas Tungsten Arc Welding (GTAW)

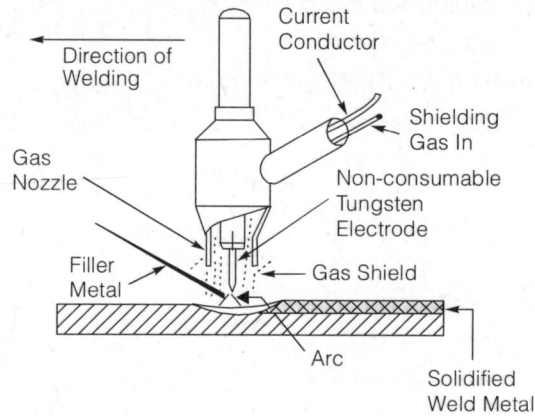

- **TIG Welding**

  In this technique *nonconsumable* tungsten electrode will be used to generate the arc. To increase the thermal resistance of tungsten alloying elements like *Thorium and Beryllium* are added.

  This will be used for *less thickness* material without using filler material.

  - Molten weld pool is protected by inert gas atmosphere (Helium, Ne, Ar, N_2 and CO_2 etc.). N_2 is preferred for welding copper.

  - The power supply will be *AC or DC* depending on workpiece material for all other material except Al, Mg alloys *'Direct Current Straight Polarity' (DCSP)* will be used so that heat concentration will be *more on workpiece* side and depth of penetration is more.

  - For Al and Mg alloys if we use straight polarity due to high temperature oxide formation is severe and this will not allow electrons from the electrode. To overcome this *'Direct Current Reverse Polarity" (DCRP)* is used but the melting rate of the electrode is more when compared to melting of the workpiece due to this depth of penetration is reduced. To overcome this AC welding will be used for welding of Al and Mg alloys.

- TIG uses DSCP for all material except Al and Mg alloys (AC power source is used for Al and Mg alloys).
- DCRP in TIG causes more distortion and large HAZ.

Remember

- Applications
  - Welding of Al and Mg alloys in Aerospace industries and automobile industries.
  - Since weld bead thickness is less it can be used for *welding in any position*.
- Limitation
  For welding of high thickness materials if we use filler material movement of the filler material with respect to electrode is difficult.

8. Metal Inert Gas Welding (MIG)/Gas Metal Inert Gas Welding (GMAW)

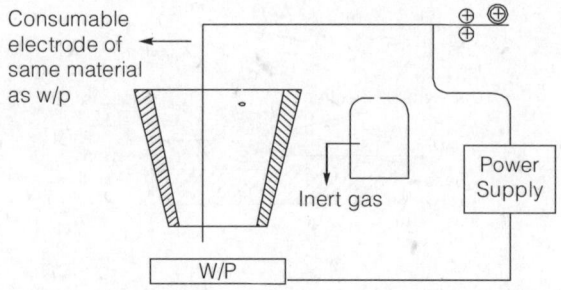

- MIG
  In this technique *consumable electrode* in the form of spool wire will be supplied.
  - DCRP will be used for welding of all materials increase the weld deposition rate.
  - By increasing the current supply (100 to 300 amps) we can increases the depth of penetration and we can weld high thickness material also.
  - Constant voltage type of power source is used since the control of movement of electrode and feeding of the electrode is done automatically.

- DCRP(DCEP) in MIG result into smooth metal transfer and stable arc with high metal deposition rate.
- DCSP will cause unstable arc that result into large spatter.

Remember

- **Application**

  Used for welding of stainless steels, Aluminium, Mg, Cu and Ni alloys in aircraft and automobile industries.

- **Limitation**

  High initial cost.

9. Submerged Arc Welding

It is semiautomatic version of shielded metal arc welding. It is used for long weld run.

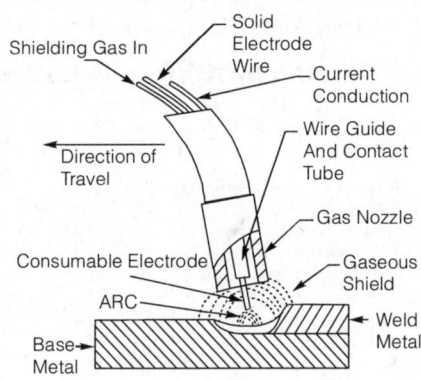

- In this technique arc will be submerged inside the molten weld pool any length of welding with high thickness of materials can be done by using this technique in a single run. High welding current will be used to increase the melting rate of electrode and workpiece (200 – 2000 amps).

- Best suited for *high deposition rate,* high **welding speeds** and *high depth* of penetration.

- The flux material contains **CaO** and **CaF$_2$.**

- **Application**

  ▪ Welding of **large dia pipes**, boilers, **pressure vessel** fabrication and ship building etc.

- **Limitation**

  ▪ *Only flat position* welding is possible.

  ▪ Heat affected zone is high.

10. Plasma Arc Welding

- Plasma is the pool of ionized gas. Initially the arc is being created between Tungsten electrode and the work piece, plasma gases by taking energy from the spark converts into plasma.

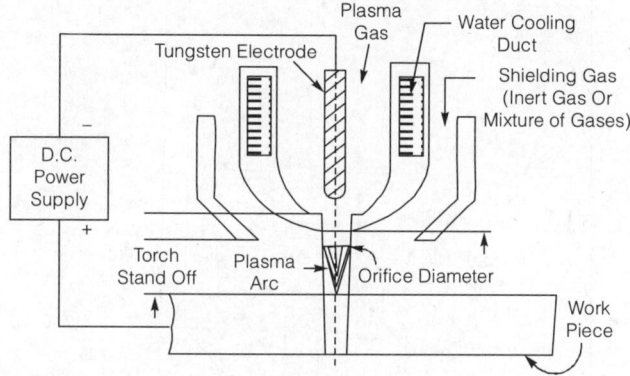

- The temperature of plasma is around 50000°C but for the welding process it is restricted to 20000°C. This high temperature arc when impinges upon the work piece results in reuniting of electrons and ions to form atomic and then molecular gas, releasing heat in the process which is thus utilized for welding.
- The major application of the process is in welding stainless steel, titanium, metals having very high melting points and super alloys. Commercially it is used in aeronautical industry, precision instrument industry and jet engine manufacturing.
- The main drawback is that intense arc will produce extensive ultra-violet and infrared radiation, which is harmful for skin.

11. Electro Slag Welding/Electro Gas Welding

- Electro Slag Welding

  Welding is started by generating the electric arc and completed by resistance heating effect of slag material.

  After getting sufficient amount of molten metal power supply will be stopped and heat will be maintained due to resistance effect of slag material. After sometime if the heat is reduced power supply will be given, we can use automatic welding machines to control the feeding of consumable electrode and power supply.

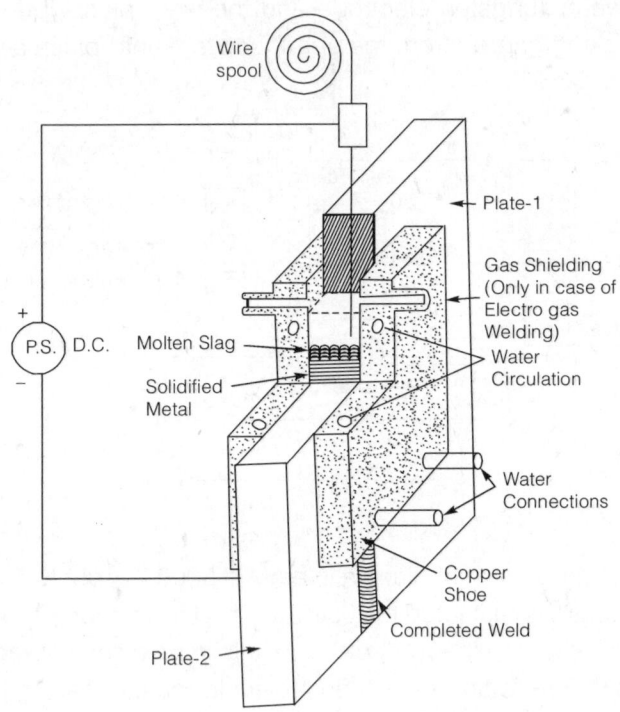

- More no. of consumable electrodes are also used to increase the deposition rate.
- High rate of current with DCRP will be used.
- It is used for welding of thick plates upto 450 mm thickness in single pass.

- Application
  - Fabrication of press frames
  - Pressure vessels
  - Heavy plates fabrication, etc.
- Limitation
  - Welding is done in vertical position only.
  - Heat affected zone is very high. Coarse grain structure will be developed in the welded portion. (strength is reduced)

CHEMICAL WELDING

12. Gas Welding

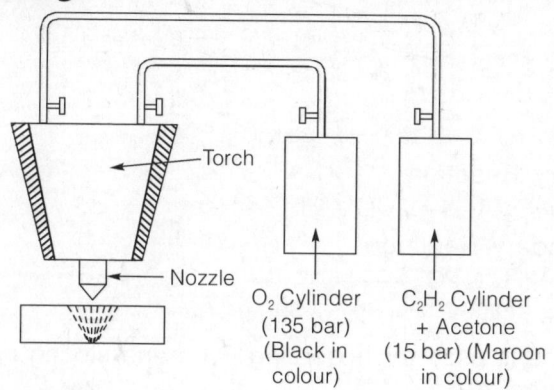

TYPES OF FLAME

- **Neutral Flame**

  For this flame, we use both the gases in the ratio of 1 : 1 by *volume*. This flame is used for welding of cast iron, mild steel, low carbon steel, aluminium. ($C_2H_2 =. O_2$)

    - **Hissy** and **little smoke**.

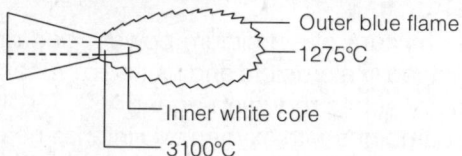

- **Carburizing Flame**

  For this flame, excess fuel gas (C_2H_2) is used. It is used for welding medium carbon steel, nickel etc. ($C_2H_2 > O_2$)

    - **Smoky** and **quiet flame**.

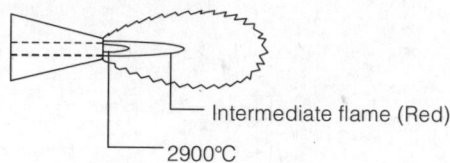

- **Oxidizing Flame**

  For this type of flame excess oxygen (O_2) is used. This type of flame is used for welding oxygen free copper (Cu) alloys, brass, bronze and zinc base materials. ($C_2H_2 < O_2$).

- **No smoke** and **noisy.**

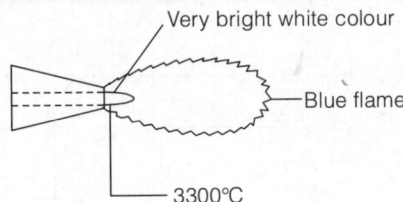

Very bright white colour

Blue flame

3300°C

- **Primary Reaction**

$$C_2H_2 + O_2 \longrightarrow 2CO + H_2\uparrow + \text{Heat}$$

- **Secondary Reaction**

$$2CO + 2H_2 + 2O_2 \longrightarrow \text{'}O \quad {}_2O + \Delta h$$
$$C_2H_2 + 5/2O_2 \longrightarrow 2CO_2 + H_2O$$

Acetone is used for absorbing C_2H_2 and storing in cylinders at high pressure.

  - Cast iron is best welded by using gas welding with neutral or slightly oxidizing flame.
  - If acetone comes out and participate in reaction colour of flame will be purple.
  - For complete combustion of 1 mole of C_2H_2, 2.5 moles of oxygen is required.

13. Thermit Welding

- Thermit is a mixture of aluminium powder and metal oxide.
- Mixture is placed in a crucible and ignited by means of a fire cracker.
- The aluminium act as reducing agent.
- Aluminium combines with oxygen and intense heat will be released.
- Temperature generated near about 3000°C.
- It is used for repair of railway track.

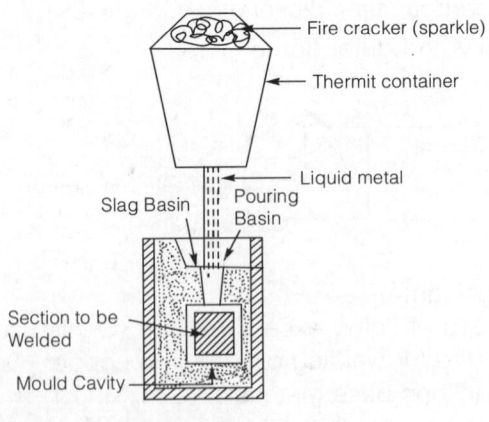

Fire cracker (sparkle)

Thermit container

Liquid metal

Slag Basin   Pouring Basin

Section to be Welded

Mould Cavity

$$8Al + 3Fe_3O_4 \rightarrow 4Al_2O_3 + 9Fe + \text{Heat}$$

- Heat released is 242×10^3 kJ/mole
- Small percentage of manganeese is added as alloying to increase strength of joint.

RESISTANCE WELDING

The resistance welding is produced by means of electrical resistance across the two components to be joined. The heat generated in this process is given by

$$H = I^2RT$$

     H = heat generated in joules

     I = current in amperes.

     R = resistance in ohms.

     t = time of current flow in seconds

The welds produced by resistance welding are normally without the addition of any filler material, electrodes and shielding gases.

14. Spot Welding

- Copper electrodes are used.
- Identation is produced by applying pressure.
- Step down transformer are used.
- High current is used.
- Maximum resistance will be at the interface between the two sheets due to presence of air gap.
- Because of heat, metal liquified and nugget will be formed.

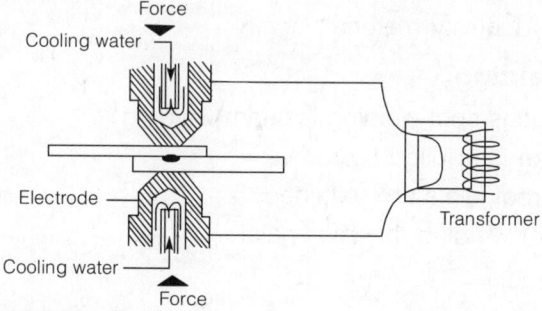

Current $\approx$ 10,000 amp

Time Ω 0.01 sec

Here,   S = Squeeze time

      W = Welding time

      H = Hold time

      O = Off time

(Current is passed during time period 'W.')

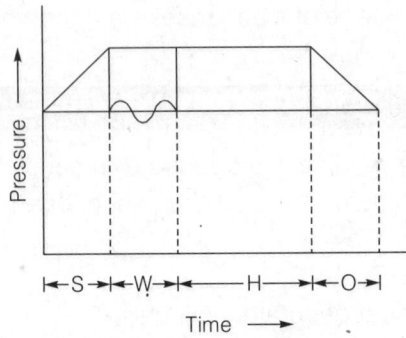

Time ⟶

The diameter of the nugget is $d_n = 6\sqrt{t}$

Height of the nugget is $h_n = 2(t - \text{indentation})$ per sheet.

If sheets of different thickness then, $t = \dfrac{t_1 + t_2}{2}$

- **Application**

  Spot welding is mainly used for lap welding of thin sheets particularly in the welding of automobile and refrigerator bodies and high quality work in automobile engines.

- **Heat Balance in Resistance Welding**

  - For welding two sheets which differ in thickness always use smaller electrode tip diameter on the side of thicker work material.

  - For welding the metals which differ in composition always use smaller electrode tip area on the side of high electrical conductivity material.

15. Seam Welding

- Continuous spot welding is seam welding.
- Pressure is applied by rollers.
- Leak proof joints are obtained.
- Used for welding thin materials.

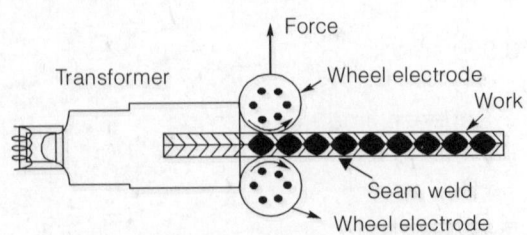

16. Projection Welding

The welding of two sheets without any indentation can be done by using this technique.

- On one of the sheets some projection will be provided by using embossing operations.
- By applying the pressure through large area of electrodes, nuggets will from at the cross-section of projection.

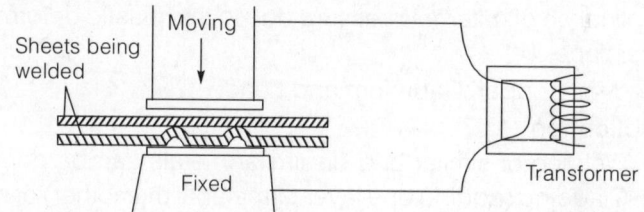

- Projections are provided in the thicker sheet or the sheet which has higher thermal conductivity.
- Good bond between the two pieces.
- Less distribution around the weld zinc.

17. Flash-Butt Welding

- Two workpiece which are to be welded will be clamped in the electrode holders, high pulsed current in the range of 1×10^5 amps is supplied to the workpiece material.
- In this two electrode holder are used of which one is fixed and other is movable.
- Initially current is supplied and movable clamp is forced against the fixed clamp due to contact of these two workpieces at high current, flash will be produced.
- Current is stopped and axial pressure is increased to make the joint. Weld is formed through plastic deformation.

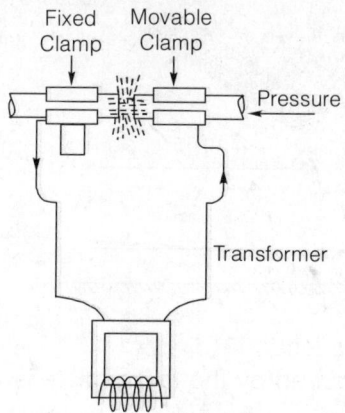

- Used for mild steel, alloy steel, titanium.
- Used for end to end or edge to edge joining.

SOLID STATE WELDING

18. Pressure Welding

- Solid state welding.
- Done at room temperature.
- Application of external pressure due to this plastic deformation will take place.
- Used for Copper, Cadmium and Lead.
- Application
  - Welding of similar and dissimilar metals can be done.
  - Cladding (adding one layer of material over other) of materials.
  - Used in welding of wires which break during wire drawing operation.
  - Used in highly explosive areas.
- Limitations
  - Cannot be used for brittle and high strength materials.
  - In pressure welding, strength of weld is less.

19. Explosive Welding

- Low intensity explosive are used.
- Plastic deformation is used for joining.
- Two plates which are to be welded are maintain at a distance (stand off) then movable plate hit the target and plastic deformation take place.

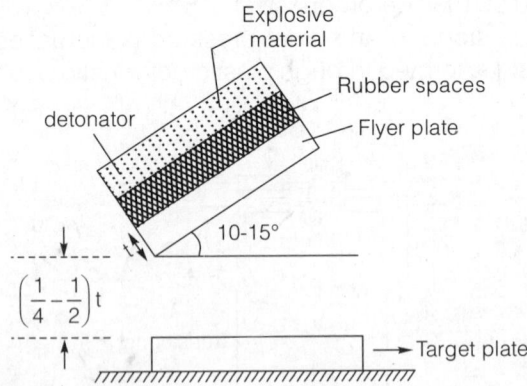

- Limitation
  - Accuracy is less.
  - Rigidity required by the target plate is very high.

20. Ultrasonic Welding

Ultrasonic vibrations combined with static clamping force produces dynamic shear stresses between the contact of 2 workpiece materials. Then local plastic deformation will take place and joint formation will take place at the interface.

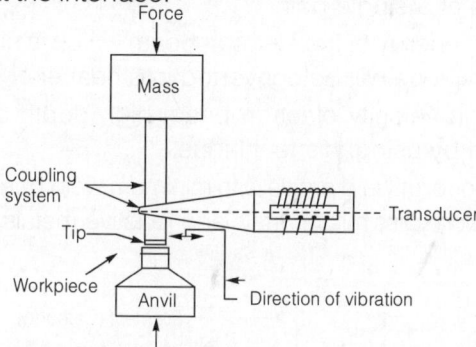

Transducers changes high frequency electrical power into ultrasonic vibrating energy. The bonding will take place in solid state without applying external heat, filler material and max. pressure.

- In this welding, frequency applied is 20 kHz to 60 kHz.
- It can be used for welding less than 1 mm thickness sheets heat affected zero is minimum in this welding.
- fabrication of nuclear reactor components.
- Armature winding, in electronic industry.

21. Friction Welding

It will be used for welding of similar and dissimilar metals with different cross-section.

Of the two objects to be welded, one will be having axial movement and another will be having rotation movement @6000 rpm.

- If we make contact between two workpiece due to friction heat generation will take place and material will become soften, then apply the axial pressure because of plastic deformation joint will be formed.

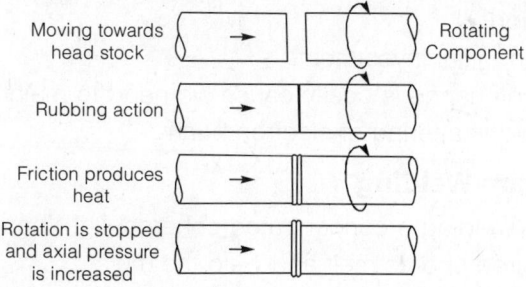

RADIANT ENERGY WELDING

22. Electron Beam Welding

- In this technique high velocity electron beam is to curved on the workpiece at a single point.
- The kinetic energy of the electron beam which is hitting the surface of the workpiece will be converted into heat energy.
- Due to high velocity electron beam high depth of welds can be performed by using this technique.
- Precise control of weld profile with minimum weld bead and high depth of penetration. Dissimilar metals and reactive metals can be welded.

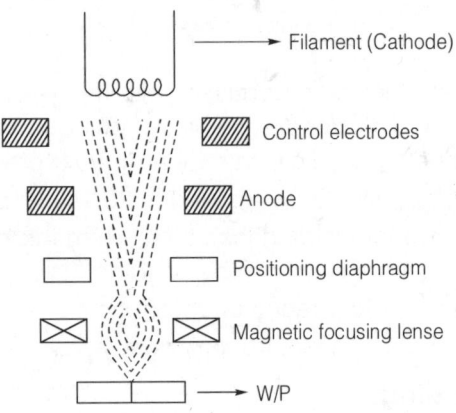

In this potential difference of 20-200 kV, and current range of 50-500 mA is used. Velocity of electron will be 50,000-2,00,000 m/sec.

- **Application**

  Welding of titanium, nickel and stainless steel and its alloys used in fabrication of gas turbine parts space industries, aircraft industries defence sector (missile etc.)

- **Limitation**
  - High initial investment.
  - Maintenance is costly cause we need to maintain vacuum to avoid scattering of electron beam.

23. Laser Beam Welding

- In laser welding a concentrated coherent light beam impinges at the desired spot to melt and weld the metal.

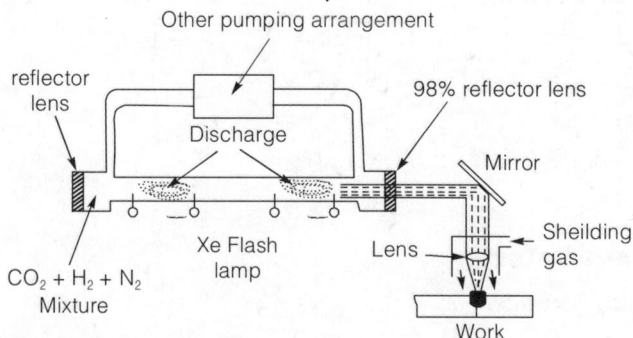

- The most useful laser for welding is the laser in which the lasing medium is mixture of CO_2, nitrogen, and helium in the ratio of 1:1:10 at a pressure of 20 to 50 mm of mercury.
- Application
  - Welding Cu and Al alloys in electronic industries.
  - Heat generated in this is less than electron welding.
- Limitations
  - High reflective materials cannot be welded.

24. Difference between Soldering and Brazing

	Soldering		Brazing
1.	The melting point temperature of filler material is less than 430° and it is also less than melting pt. temperature of base material.	1.	The melting point temperature of filler material is more than 430°C but less than the melting pt. temperature of base material.
2.	Filler material is an alloy of lead and tin it is known as solder.	2.	Filler material is an alloy of copper and zinc, Cu and silver; Cu and aluminium. This is knows as spelter.
3.	The flux used in soldering is zinc chloride (ZnCl2) and HCl.	3.	The flux used is borax and boric acids.
4.	The strength of joint is less when compared to brazing.	4.	Strength is more.
5.	Used in electronic industry.	5.	Used in pipe fitting where leak proof joints are required for intricate light weight components.

In both the cases the filler material will enter into the gap between workpiece by means of capillary action.

25. Type of Solders

		Pb : Sn
1.	Soft solder	40% : 60%
2.	Medium	50% : 50%
3.	Electricians	60% : 40%
4.	Plumbers	70% : 30%

26. Welding Detects

* **Weld Porosity**

  This is due to entrapment of atmospheric gasses inside the molten weld pool due to this the strength of the weld is reduced.

  Remedies: This can be reduced by proper selection of filler metals, by preheating the weld area, by proper cleaning the weld zone, and by reducing the welding speeds.

* **Slug Inclusion**

  Inclusions may be caused by compounds such as oxides, fluxes, and electrode coating materials, which are trapped in the weld zone.

  Remedies: Use inert gases to protect the molten weld pool instead of flux coatings. Submerged arc inside the molten metal.

* **Incomplete Fusion**

  It is usually caused by insufficient heat and too fast travel of torch or electrode.

  Remedies: This can be avoided by raising the temperature of the base metal, cleaning the weld area, providing enough shielding gas.

* **Weld Spatter**

  This is due to high welding current and too low welding speed and arc blow.

* **Cold Cracking**

  This is due to different in shrinkage rates and hydrogen entrapment and variation of weld composition.

  Remedies: Pre heating, constant weld pool composition.

* **Hot Cracking**

  This is due to hydrogen embrittlements and high temperature gradients and high heat effected zone.

Remedies: By preheating.

- Cold crack occurs in weld pool and HAZ both. But cold crack mainly occurs in HAZ.
- Hot crack occurs mainly in weld pool and occurs along centreline of the weld bead thus called centreline crack.

Remember

- **Under Cutting:** It is the melting or burning away the base metal at the toe of the weld as sharp recess or notch. Undercut can be as stress raiser and can reduce the fatigue strength of the joint. In such cases it may lead to premature failure. These undercuts are generally due to excessive wearing of electrode.

27. Representation of Weld

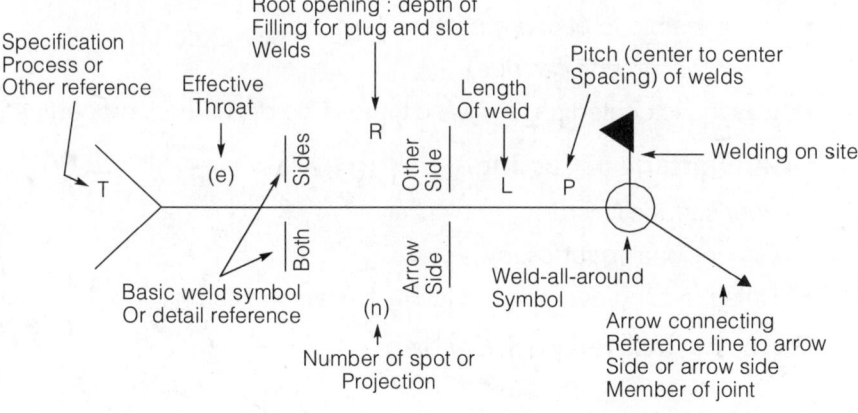

■■■■

Casting means pouring molten metal into a refractory mould with a cavity of the shape to be made, and allowing it to solidity. The solidified object is called casting. Application of casting are cylinder blocks, *piston*, *piston ring*, wheels, housings.

1. Advantage of Casting

- Molten metal can flow in any *small* section.
- Any *intricate* shape can be made.
- It is possible to cast *almost* any material.
  (Ferrous or non-ferrous)
- Casting is cooled uniformly, so there is *no directional* properties.

2. Disadvantage of Casting

- Poor surface finish
- Dimensional *inaccuracy*
- Defect arising out of the moisture present in casting.

3. Terms Associated With Casting

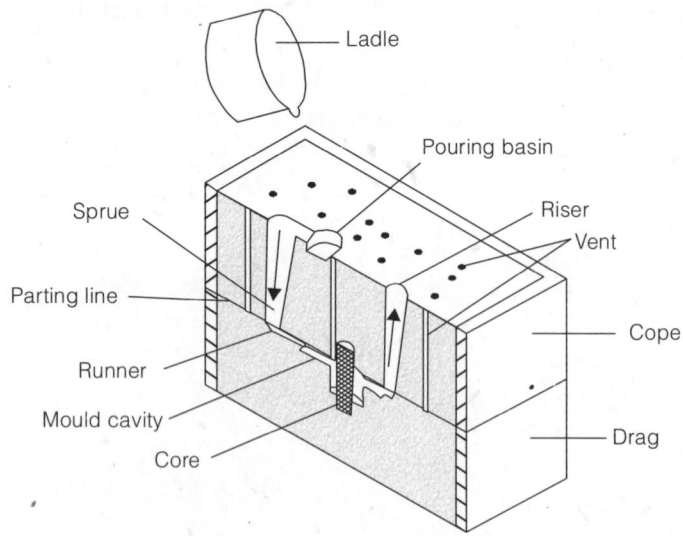

Cross section of a sand mould ready for pouring

- **Flask**

  A moulding flask is one which holds the sand mould intact. Depending upon the position of the flask in the mould structure, it is referred by various names such as drag, cope and cheek. It is made up of wood for temporary applications or more generally of metal for long-term use.

- **Core**

  It is used for making hollow cavities in castings.

- **Pouring Basin**

  A small funnel-shaped cavity at the top of the mould into which the molten metal is poured.

- **Sprue**

  The passage through which the molten metal from the pouring basin reaches the mould cavity. In many cases it *controls* the flow of metal into the mould.

- **Runner**

  The passage ways in the parting plane through which molten metal flow is regulated before they reach the mould cavity.

- **Gate**

  The *actual* entry point through which molten metal entries the mould cavity.

- **Chaplet (Metallic Spacings)**

  Chaplets are used *to support cores* inside the mould cavity to take care of its own weight and over-come the *metallostatic* forces.

- **Chill**

  Chills are metallic objects, which are placed in the mould to *increase the cooling rate* of castings to provide *uniform or desired* cooling rate. Chills are metal of high thermal conductivity.

Remember: ....

- Padding is extra metal of casting (material) to provide uniform cooling rate.

- **Riser**

  It is a reservoir of molten metal provided in the casting so that hot metal can flow back into the mould cavity when there is *a reduction in volume* of metal due to solidification.

4. Factor Governing Selection of Pattern Material

- *Shape and size* of component to be produced.
- *Surface finish* of components be produced.
- Number of component produced.

- Moulding method we are going to use.
- Cost of the pattern material.

5. Type of Pattern Allowances

- Shrinkage Allowance

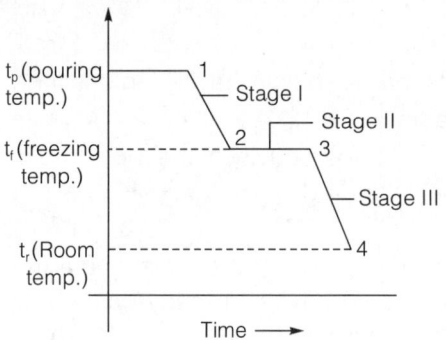

- 1-2   liquid shrinkage
- 2-3   Solidification shrinkage
  (Phase transformation shrinkage)
- 3-4   Solid shrinkage

Remember

- Stage I and Stage II shrinkage are compensated by *riser*.
- Stage III are compensated by giving **shrinkage allowance** i.e., oversize pattern.
- Liquid shrinkage > Solid shrinkage > Phase transformation shrinkage

---

- Shrinkage Allowances for important material

Material	Allowance
Bismuth	Negligible
Cast iron	10 mm/ metre length
Aluminium alloys	12-15 mm/metre length
Bronze	15 mm/metre length
Pure aluminium	17 mm/metre length
Plane Carbon steels	20 mm/metre length
Grey C.I.	*Negative allowance*

- Draft Allowances / Taper Allowance
  Draft/Taper allowances provided for easy removal of pattern without damaging mould.
  $x = h \tan \theta$

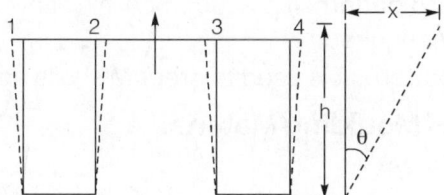

The average value of the *taper allowance* is (1/2) – 2°.

Internal surface require *more* taper when compared to *external surfaces* cause for the external surfaces the mould strength is more compared to internal surfaces. (since dimesion on internal side are less hence less strength).

- **Machining Allowance**

  The machining or finish allowance is provided on the pattern in order to remove some amount of material after the casting has been produced in order *to get smooth surface finish.*

- **Rapping Allowance**

  (i) Provide to make clearance between pattern and mould walls.

  (ii) Depends on the skill of operator and it is negative in nature.

- **Distortion Allowance or Camber Allowance**

  Some typical application like V-shape, U-shape and flat objects with *very less thickness* can distort, keep it in there shape. In order to arrest that distortion opposite to the direction of distortion we need to provide sum excess amount of material in the form of distortion allowance.

6. Type of Pattern

- **Single piece pattern:** It is used for very simple casting.
- **Split pattern/Two piece pattern:** It used when complexity of part is more and the withdrawal of pattern is difficult.
- **Gated pattern :** It is used when small components in mass production is required.
- **Cope and drag pattern :** It is used for produced of big size castings.
- **Match plate pattern :** It is used for production of small size precision casting in mass production e.g., *Piston ring*.
- **Loose piece pattern:** It used when parts with internal webs are produced.
- **Sweep pattern:** It is used when 2-D pattern used to produce symmetrical 3-D casting for eg., Cone, Bells of temples.

- **Follow board pattern:** It is used when thin or overhanging sections in casting is required.
- **Skeleton pattern:** It is used to prepare shells and drums.

7. Properties of Moulding Material

- **Refractoriness**
  It is the ability of material to withstand *the high temperature* of molten metals. It should be *high*.

- **Green Strength**
  The moulding sand that contains *moisture* is termed as green sand. The green sand should have enough strength so that the constructed mould retains its shape.
  Green compressive strength → 130-160 kPa.
  Green shear strength → 10-40 kPa.

- **Dry Strength**
  When molten metal is poured into a mould, the sand around the mould cavity is quickly converted into dry sand as the moisture in the sand immediately evaporates due to the heat in the molten metal. At this stage, it should retain the mould cavity and at the same time withstand the metallostatic forces.
  Dry compressive strength →120-140 kPa.
  Dry shear strength → 30-80 kPa.

- **Hot Strength**
  It is the strength of the sand that is required to hold the shape of the mould cavity, after all the moisture is eliminated.

- **Permeability**
  The gas evolving capability of moulding sand is known as permeability. This will be expressed by **permeability number (p_n)**.

$$p_n = \frac{VH}{PAT} = \frac{3007}{T}$$

Here,   V = Volume of air passing through the specimen
         H = Height of standard specimen
         P = Pressure of the air passing through the specimen
         A = cross–sectional area of cylindrical specimen
         T = Time required to pass through the specimen

Remember: ....

- *Universal testing machine* is used to test green strength.
- *Sand muller* is used to mix and prepare moulding sand.

- **Grain Fineness Numbers**
  GFN will indicate the average grain size distribution of a given moulding.
  Greater the GFN lower the grain size.
- **Flowability**
  The ability of the sand to flow over and around the pattern when the mould is *rammed*.
- **Adhesiveness**
  The bond formation between two different material i.e., moulding sand and mould flask. Between moulding sand and pattern.
- **Cohesiveness**
  Bond formation between 2 similar materials i.e., between 2 sand grains.
- **Toughness**
  Ability to *resist impact and shock loads* by the moulding sands. *Shatter index test* is done for toughness testing shocker observed when molten metal is poured.
- **Collapsibility**
  The properties of the moulding sand which creates *resistance to metal contraction* is called collapsibility. During the solid contraction of the casting part if the mould creates resistance cracks will appear over the casting. High collapsibility is preferred to improve collapsibility. Saw dust or wood powder is added to moulding sand. Since when molten metal is poured wood powder burns to ash due to heat and hence shrinks in size causing the mould near casting to easily collapse and provide resistance less shrinkage.

8. Types of Sand

Composition of Moulding Sand	
Silica sand	70-85%
Clay	10-20%
Water	3-6%
Additive	1-6%

- **Green Sand**
  If the sand contains *2-6% of moisture* then the sand is called green sand.
- **Dry Sand**
  The moisture available in the moulding sand evaporates cause of high temp. of molten metal then the sand is called dry sand.
- **Facing Sand**
  The sand which is used near the mould cavity with more clay and fine silica sand is called facing sand.

- **Parting Sand**

  It is a pure silica sand without clay and water which is used to avoid striking of the moulds with other surfaces.

- **Backing Sand**

  The sand which is used near the mould flask away from the mould cavity with less moulding sand properties to support the mould is known as backing sand.

9. Type of Moulding Techniques

- Bench moulding
- Floor moulding
- Pit moulding
- Machine moulding

10. Additive Used in Moulding Sand

- **Wood flour/saw dust**

  To improve *green strength* and *collapsibility* of moulding sand.

- **Starch and Dextrin**

  These are organic binders used to improve *resistance to deformation* of the mould and improves the *skin hardness* of the mould.

- **Iron oxide and aluminium oxide**

  Used to improve *hard strength* of green sand.

- **Coal dust, sea coal, silica flour**

  These are carbonous materials which are used in the moulding sand to improve *surface finish* and *resistance to metal penetration*.

11. Design of Gating System

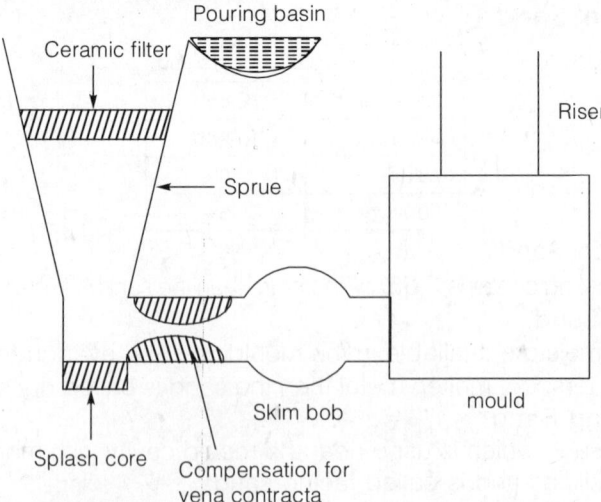

- **Objective of Gating System Design**
  - We need to design gating system elements such that molten metal can enter into the mould cavity without increase in velocity and turbulence of the molten metal within a specified time.
  - Molten metal has to enter into the mould cavity **without eroding** gating elements and mould cavity.
  - Molten metal has to enter into the mould cavity with full of molten metal through all the gating elements in order to **avoid air aspiration effect.**
  - We need to design the gating elements such that casting **yield is maximum.**
  - The molten metal has to enter in the mould cavity without any **slag particles and impurities.**
- **Design of Top Gate**

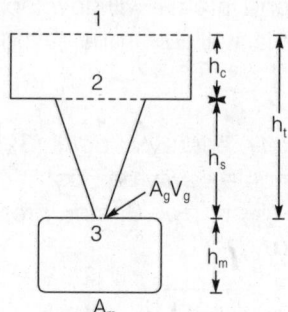

- **Assumption**

  The molten metal is entering into mould cavity directly from the end of the sprue at atmospheric pressure.

  Top gating system is used for ferrous materials of large size castings.

- **Velocity at Gate (V_g)**

$$V_g = V_3 = \sqrt{2gh_t}$$

- **Filling Time (t_f)**

$$t_f = \frac{A_m \times h_m}{A_g \times V_g} = \frac{A_m \times h_m}{A_g \times \sqrt{2gh_t}}$$

Here, h_c = height of pouring cup
h_s = height of spure
h_m = height of the mould
A_m = Area of mould
A_g = Area of gate

- **Design of Bottom Gate**

  This is used for non ferrous materials.

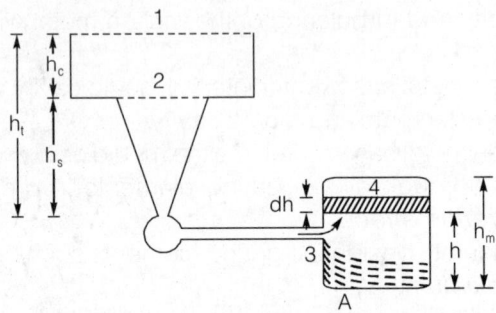

By using bottom gating system the molten metal will enter the mould at the bottom of the mould cavity. With this the velocity of molten metal will reduce and intern it will develop the back pressure in the molten metal which is available in the gating elements thus is reduces the air aspiration effect.

- **Assumption**

  $V_4 \approx 0$ and the kinetic energy at point (3) of the molten metal after entering into the mould cavity has lost.

  $\therefore V_3 = 0$, $h_3 = 0$, $h_4 = h$, $p_4 = 0$ (atm. pressure)

- **Velocity at Gate (V_g)**

$$V_g = V_3 = \sqrt{2g(h_t - h)}$$

- **Filling Time**

$$t_f = \frac{2\,A_m}{Ag \cdot \sqrt{2g}} \cdot \left[\sqrt{h_t} - \sqrt{h_t - h_m} \right]$$

12. Solidification Time

$$t_s = K\left(\frac{V}{SA}\right)^2$$

Here,

V = Volume of casting

SA = Surface area of casting

K = Solidification factor $\left(\dfrac{Sec}{m^2}\right)$

Volume represents amount of **heat content**.

SA represents amount of *heat transfer*.

- For sphere

$$\frac{\text{Vol.}}{\text{SA}} = \frac{\frac{4}{3}\pi R^3}{4\pi R^2} = \frac{R}{3} = \frac{D}{6}$$

- For cylinder (h = d)

$$\frac{\text{Vol.}}{\text{SA}} = \frac{\frac{\pi}{4}d^2 \times d}{2 \times \frac{\pi}{4}d^2 + \pi d \times d} = \frac{\frac{\pi}{4}d^3}{\frac{3\pi}{2}d^2} = \frac{d}{6}$$

- For cube (side)

$$\frac{\text{Vol.}}{\text{SA}} = \frac{a^3}{6.a^2} = \frac{a}{6}$$

13. Type of Riser/Riser Design

- **Type of Riser**
  - Side riser
  - Top riser
  - Blind riser
- **Riser Design Methods**

$$\left(\frac{\text{SA}}{\text{Vol}}\right)_{\text{Riser}} \leq \left(\frac{\text{SA}}{\text{Vol}}\right)_{\text{Casting}}$$

- $t_s = \dfrac{1}{\beta^2 \alpha}\left(\dfrac{V}{\text{S.A}}\right)^2$

- where, β = Geometry constant α = Thermal diffusivity

(a) Caine's Formula

$$\text{F.R.} = \frac{\left(\dfrac{V}{\text{SA}}\right)_R}{\left(\dfrac{V}{\text{SA}}\right)_C}$$

Here,

R → Riser

C → Casting

$y = \dfrac{V_R}{V_C}$ Sound casting

Defective casting

$x = F_R$ = Freezing ratio

$X = \dfrac{a}{Y-b} + c$

a & b are constants.

- Naval Research Laboratory Method

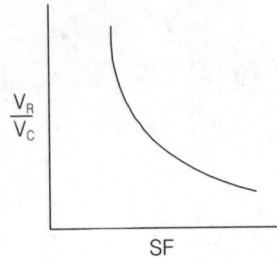

$$SF = \frac{L + W}{T}$$

L = Length of casting

W = Width of casting

T = Average section thickness of casting

- Modulus Method

$$Modulus = \frac{1}{\left(\frac{SA}{V}\right)}$$

$$M_R = 1.2 \, M_C$$

- Distance between two riser ≤ 4t
- Distance between a riser and end of the casting ≤ 4.5t

  t = thickness of casting

14. Gating Ratio

- For non-pressurized gating system
  Sprue : runner : ingate = 1 : 4 : 4
- For pressurized gating system
  Sprue : runner : ingate = 2 : 2 : 1

Non-pressurised G.S	Pressurized G.S.
Useful for nonferrous example	Useful for ferrous example
Al, Mg, Alloys	C.I., Steel, Gray C.I.

15. Types of Solidification

- Skin Forming

  It take place either in *pure* material or the *alloys* having *eutectic* composition. It move towards centre and solidification appear *layer by layer.*

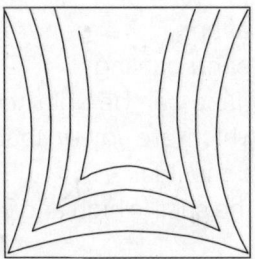

- **Dendritic Growth**

  It take place when mushy zone appear during solidification of alloy. In this solidification is *not* uni-directional. To provide uni-direction solidification chills are used.

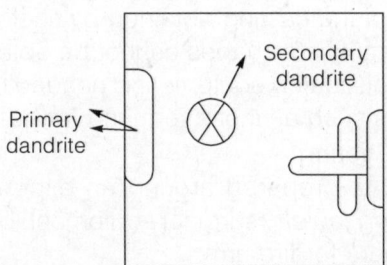

16. Classification of Casting

- **Expendable moulds (Temporary Moulds)**
  - Sand casting
  - Shell moulding
  - Investment casting
  - Full moulding
  - CO_2 moulding
- **Permanent Moulds (Metallic Moulds)**
  - Centrifugal casting

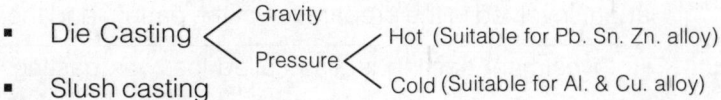

  - Die Casting ⟨ Gravity / Pressure ⟨ Hot (Suitable for Pb. Sn. Zn. alloy) / Cold (Suitable for Al. & Cu. alloy)
  - Slush casting

CONTINUOUS CASTING

- **Shell Moulding**

  It is a process in which the sand mixed with a *thermosetting resin* is allowed to come into contact with a heated metallic pattern plate, so that a *thin and strong* shell of mould is formed around the pattern.

Advantage:

- Shell-mould castings are generally more *dimensionally accurate* than sand castings.
- A *smoother* surface can be obtained.
- *Draft* angles, which are *lower* than the sand castings, are required in shell moulds.
- Permeability of the shell is high and therefore no gas inclusions occur.
- Very small amount of sand needs to be used.
- Mechanization is readily possible because of the simple processing involved in shell moulding.

- **Limitations**
  - The patterns are very *expensive.*
  - The *size* of the casting obtained by shell moulding is limited.
  - Highly complicate shapes cannot be obtained.
  - More sophisticated equipment is needed for handling the shell mouldings such as those required for heated metal patterns.
- **Investment Casting**
  In this mould is prepared around an expendable pattern. It is used for making *jewelry* surgical equipment, blade for gas, turbine bolt and triggers for fire arm.
  Advantage:
  - *Complex shapes* which are difficult to produce by any other method are possible since the pattern is withdrawn by melting it.
  - Very *fine details* and *thin sections* can be produced by this process.
  - Very *close tolerances* and *better surface* finish can be produced.

Limitations:

- The process is normally limited by the size and mass of the casting.
- This is a more expansive process because of larger manual labour involved in the preparation of the pattern and the mould.

- Investment casting is also called lost wax casting.
- Binder used is (Ethyl Silicate + Silica flour + water).

- **Diecasting**
  Diecasting involves the preparation of components by injecting molten metal at *high pressure into a metallic* die.

The diecasting machines are of two types:

- hot chamber diecasting
- cold chamber diecasting

The main difference between these two types is that in the hot chamber casting, the holding furnace for the liquid metal is *integral with the diecasting machine*, whereas in the cold chamber machine, the metal is melted in a *separate furnace* and then poured into the diecasting machine with a ladle for each casting cycle which is also called shot.

- When pressure die casting is applied over plastic (casting material) called **Injection moulding.**

  Advantages:

  - Because of the use of the movable cores, it is possible to obtain fairly complex castings than that feasible by permanent mould casting.
  - *Very small thicknesses* can be easily filled because the liquid metal is injected at high pressure.
  - Very high production rates can be achieved.
  - Because of the metallic dies, very good surface finish of the order of 1 micron can be obtained.
  - Closer dimensional tolerances of the order of +0.8 mm for small dimensions can be obtained compared to the sand castings.
  - It is very economical for *large-scale production.*

  Limitations:

  - The *maximum* size of the casting is *limited*.
  - This is not suitable for all materials because of the limitations on the die materials. Normally, zinc, aluminium, magnesium and copper alloys are diecast.
  - The air in the die cavity gets trapped inside the casting and is therefore a problem often with the diecastings.

Remember
- Gravity die casting is used for C.I.
- Hot chamber: Low m.p. alloy like, Pb, Sn, Zn, alloy.
- Cold chamber: High mP alloy like Al and Cu alloy.

- **Centrifugal Casting**

  This is a process where the mould is *rotated rapidly about its central axis* as the metal is poured into it. Because of the centrifugal force, a *continuous pressure* will be acting on the metal as it solidifies.

The slag, oxides and other inclusions being lighter, gets separated from the metal and segregates toward the centre.

There are three main types of centrifugal casting processes. They are

(i) True centrifugal casting

This is normally used for the making of *hollow pipes, tubes, hollow bushes,* etc., which are axi-symmetric with a concentric hole.

Advantages:

- The mechanical properties of centrifugally cast jobs are better compared to other processes, because the inclusions such as *slag and oxides segregates towards the centre* and can be easily removed by machining.

  Also, the pressure acting on the metal throughout the solidification, causes the porosity gets eliminated giving rise to dense metal.

- Up to a certain thickness of objects, proper directional solidification can be obtained.

- **No cores** are required for making concentric holes in the case of **true centrifugal casting.**

- There is no need for gates and runners, which increases the casting yield, reaching almost 100%.

- The axis of rotation can be either horizontal vertical or any angle in between.

Limitations:

- Only certain shapes which are *axi-symmetric* and having concentric holes are suitable for true centrifugal casting.

- The equipment is expensive and thus is suitable only for *large scale production.*

(ii) Semi-centrifugal casting

In this process mould is placed on the horizontal plane and it is rotated along the vertical axis. The *outer* portions of the mould will be filled *by purely* centrifugal action and as the liquid metal approaches towards centre, the *centrifugal component decreases* and *gravity component increases.* The *central portion* is purely filled by *gravity.*

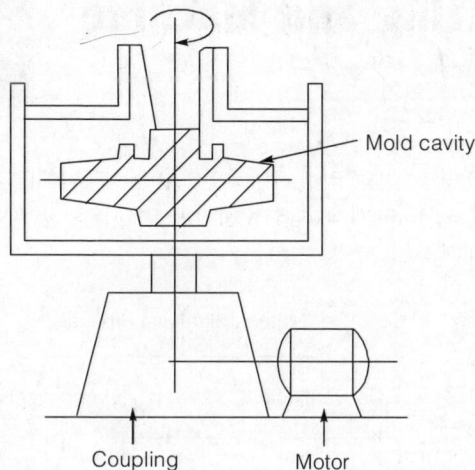

Coupling   Motor

(iii) Centrifuging

The centrifuging process is used in order to obtain higher metal pressures during solidification when casting shapes are not axisymmetrical. This is suitable only for small jobs of any shape.

Remember
- Centrifuging casting consist of central sprue to feed metal info cavity through radial gates.
- Casting yield: True centrifugal > semicentrifugal > centrifuging

17. Casting Defect

- **GAS Defect**
  - Blow holes and open blow (due to moisture)
  - Pin hole porosity
- **Moulding Material Defect**
  - Cuts and washes ▪ Metal penetration ▪ Fusion ▪ Runout
- **Pouring Metal Defect**
  - Misruns and cold shuts ▪ Slag inclusion
- **Metallurgical Defect**
  - Hot tear ▪ Hot spot

Remember
- Pin hole porosity is the path traced by H_2 gas.
- Hot tear is the crack occurs due to variation in solidification rate while hot spot is due to high temperature point.

■■◆■■

Metal Cutting and Machine Tools

4

Machining is the process of removing unwanted material from workpiece. The major drawback of this process is *loss of* material in the form of chips.

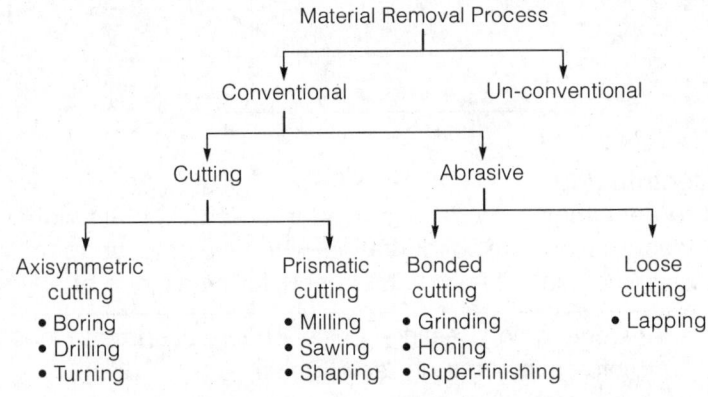

1. Terminology of Cutting Tool

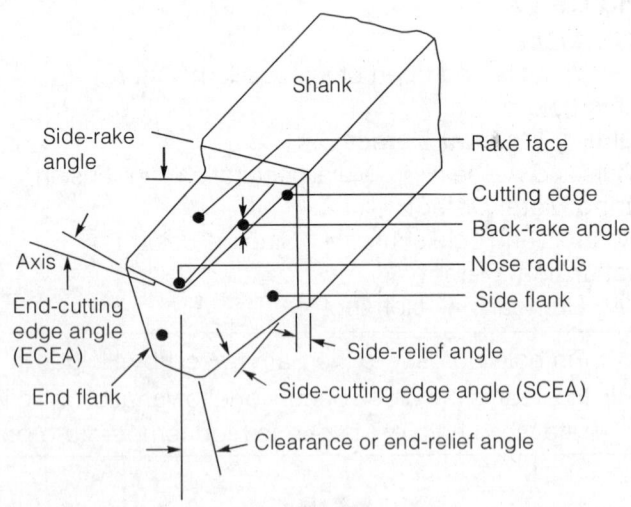

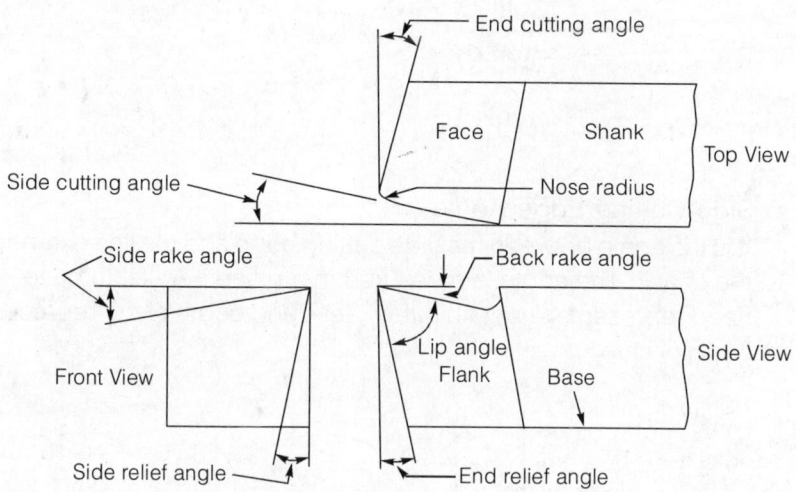

- ## Back Rake Angle

  It is the angle between the line parallel to the tool axis passing through the tip and the rake face and angle is measured in a plane perpendicular to the base.

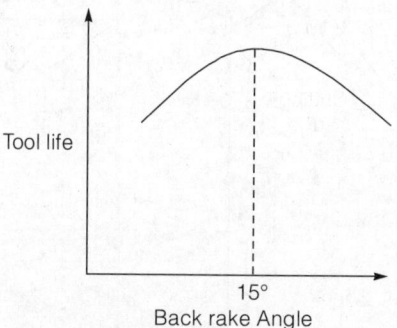

For **brass**, back rake angle will be **zero**.

Positive rake angle is used for soft and ductile material and negative rake angle is used for hard and brittle material.

- Negative, back rake angle shift the point where cutting force (F_c) is acting i.e. away from tool tip and thus body will bear the brunt of F_c.

Remember

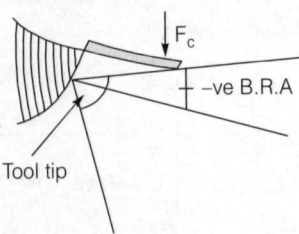

- ## Side Cutting Edge Angle (ψ)

  It is the angle between the side cutting edge and the line extending the shank. The angle is measured in a plane parallel to base. Feed represents uncut chip thickness and depth of cut represents width of chip.

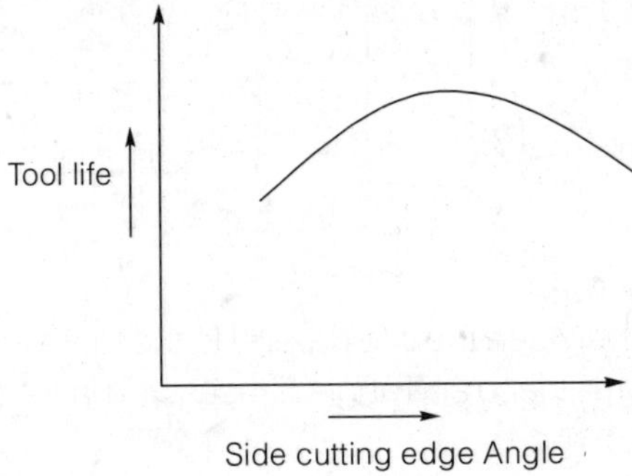

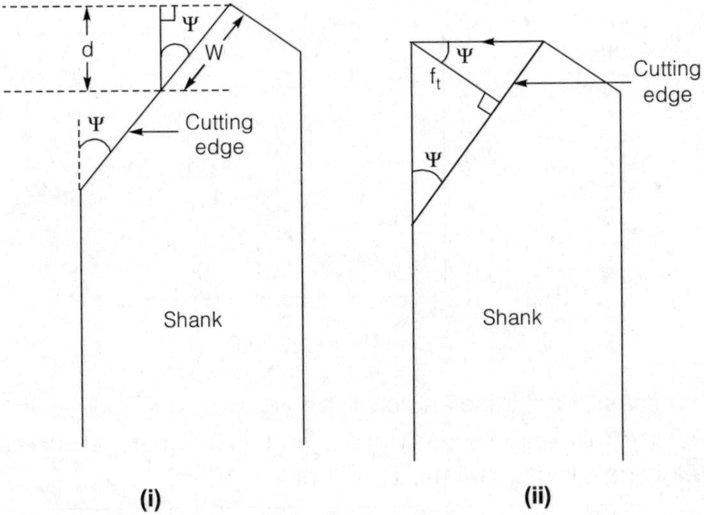

From Fig. (i)

$$W = \frac{d}{\cos \psi}$$

d = depth of cut

w = width of cut

From Fig. (ii)

$$\frac{f_t}{f} = \cos \psi \Rightarrow \boxed{t_1 = f \cos \psi}$$

True feed = $f_t = f \cos \psi$

$f_t = t_1$ = uncut chip thickness

- **Side Rake Angle**

  It is the angle between the rake face and the line passing through the tip perpendicular to the tool axis and the angle is measured in a plane perpendicular to the base. Normally this angle varies between 5-15°.

  Large positive side rake angle result in less chip formation, less cutting force, good surface finish, low heat dissipation rate.

- **Side Relief Angle**

  It is the angle between the side flank and the line passing through the tip perpendicular to the base and the angle is measured in a plane perpendicular to the tool axis. This angle varies in the range of 5-15°.

  This angle is provide to avoid side rubbing due to elastic recovery of work piece material.

- **End Cutting Edge Angle (ECEA)**

  It is the angle between the end cutting edge and the line passing through the tip perpendicular to the tool axis and the angle is measured in a plane parallel to base.

  Due to small end cutting edge angle, tool may chatter. Normal value of ECEA is 8-15°.

- **End Relief Angle / Clearance Angle:** It is the angle between the end flank and the line passing through the tip perpendicular to the base and angle is measured in plane parallel to the tool axis.

  In ductile material elastic recovery is more, so large clearance angle is provided.

- **Nose Radius:** It provide good surface finish and longer tool life because a sharp point at the end of tool is highly stressed, short lived and leaves a groove in the finished workpiece.

  Large nose radius will result in more power consumption, better heat dissipation, higher tool life.

2. Tool Signature

- **ASA Tool Signature**

  Back rake angle - Side rake angle - End relief angle -Side relief angle - End cutting edge angle - Side cutting edge angle - Nose radius.

  ASA (α_b, α_s, γ_e, γ_s, Ψ_e, Ψ_s, r_n)

- **Orthogonal Rake System (ORS)**

  Angle of inclination - normal rake angle - side relief angle - end relief angle - end cutting edge angle - approach angle.

  ORS (i, α, γ_s, γ_e, c_e, λ, r_n)

- **Surface Roughness**

  Let,   f = feed

  ψ = SCEA, ψ_1 = ECEA

  R = Nose radius

  $$\text{Peak to valley height}(H_{max}) = \frac{f^2}{8R}$$

  $$H_{max} = \frac{f}{\tan\psi + \cot\psi_1}$$

  $$\text{Centre line average value}(R_a) = \frac{H_{max}}{4}$$

4. Metal Cutting Process

- **Orthogonal Cutting Process (Two dimensional cutting)**

  In this process, the cutting edge of tool is oriented at 90° to tool feed. In this, radial force is zero.

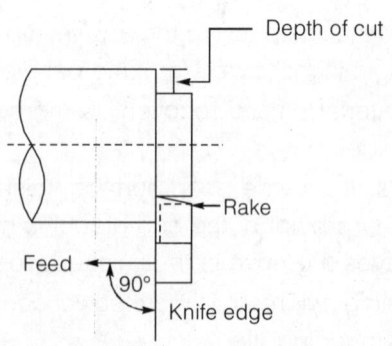

- ## Oblique Cutting Process (Three-dimensional cutting)
  In this, cutting edge of tool is oriented at an acute angle to tool feed.

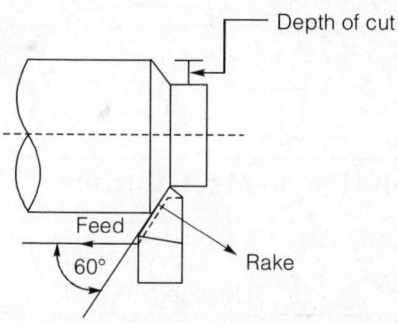

5. Difference Between Orthogonal and Oblique Cutting

	Orthogonal Cutting		Oblique Cutting
(i)	Chip flow in a direction perpendicular to cutting edge.	(i)	Chip flow at an angle to cutting edge.
(ii)	Chip flow angle is zero.	(ii)	Chip flow angle is more than zero.
(iii)	Chips get coiled in a spiral fashion.	(iii)	Chips flow in a side way direction in a wider area thus less concentration of heat.
(iv)	Tool life is less.	(iv)	Life is more.
(v)	There are two component of forces (no radial force).	(v)	There are three component of force.
(vi)	Surface finish is poor.	(vi)	Surface finish is good.
(vii)	It is used in slotting, parting, grooving, pipe cutting.	(vii)	It is used in turning milling, drilling, grinding.

6. Type of Chip

Continuous Chip	Discontinuous Chip	Chips with built-up edge
• Ductile workpiece	• Brittle work	• Ductile
• High speed	• Low speed	• Low speed
• Low feed	• High feed	• High feed
• Low depth of cut	• High depth of cut	• High depth of cut
• High back rake angle	• Low back rake angle	

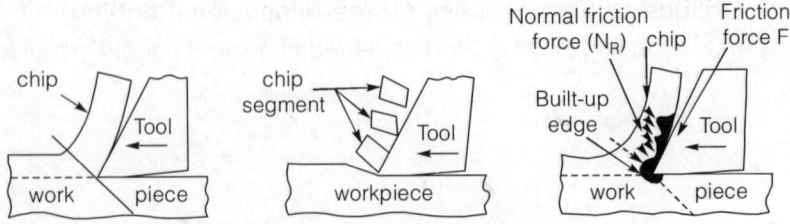

7. Distribution of Heat in Metal Cutting

Zone	Fraction of Heat generation	Reason
1. Primary deformation zone	70-80%	Due to plastic deformation.
2. Secondary deformation zone	20-25%	Due to drag between chip and rake face.
3. Work-tool contact-zone	2-5%	Due to burnishing action

8. Merchant's Analysis

- **Assumption**
    - (i) The tool is perfectly sharp and has no contact along the clearance face.
    - (ii) The surface, where shear is occuring is a plane.
    - (iii) The cutting edge is a straight line extending perpendicular to the direction of motion, and generates a plane surface as the work moves past it.
    - (iv) The chip does not flow to either side or no side spread.
    - (v) Uncut chip thickness is constant.
    - (vi) Width moves with a uniform velocity.
    - (vii) A continuous chip is produced without any built-up edge.
    - (viii) Work moves with a uniform velocity.
    - (ix) The stresses on the shear plane are uniformly distributed.

- **Chip Thickness Ratio**

    Let,

    t_1 = Chip thickness before cutting (uncut chip thickness)

    t_2 = Chip thickness after cutting

    ϕ = Shear plane angle

    α = Back rake angle

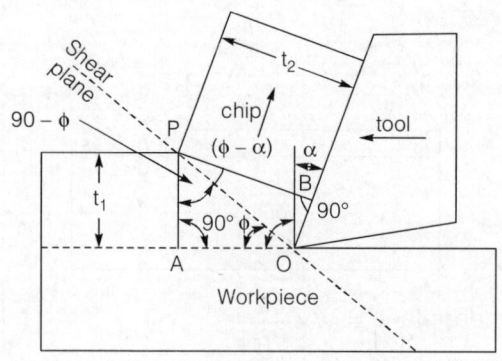

$$\tan\phi = \frac{\cos\alpha}{\dfrac{t_2}{t_1} - \sin\alpha} = \frac{r\cos\alpha}{1 - r\sin\alpha}$$

$$r = \frac{t_1}{t_2} = \frac{\sin\phi}{\cos(\phi - \alpha)}$$

= Chip thickness ratio

- **Cutting Force**
  Here,

$$F_T = \text{Thrust force}$$
$$F_C = \text{Cutting force}$$

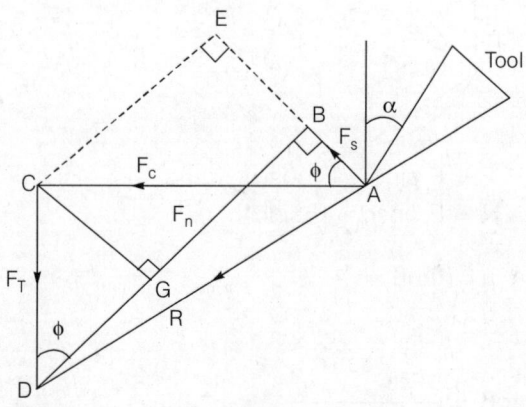

F_s (Shear force) $= F_c\cos\phi - F_T\sin\phi$
F_n (Normal shear force) $= F_c\sin\phi + F_T\cos\phi$
F_c (Cutting force) $= F_s\cos\phi + F_n\sin\phi$
F_T (Thrust force) $= F_n\cos\phi - F_s\sin\phi$

- ## Area of shear plane, Shear strength

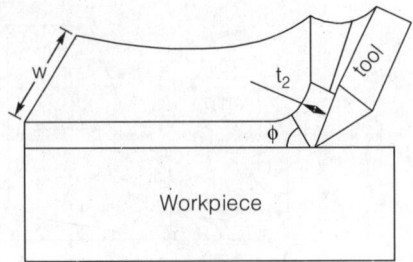

Area of shear plane = $w \times \dfrac{t_1}{\sin \phi}$

$\dfrac{F_S}{\dfrac{w.t_1}{\sin \phi}} = \tau_s$ = Shear strength of the material

- ## Cutting force with friction and normal friction forces

Here,   N = Normal friction force

F = Friction force

β = Friction angle

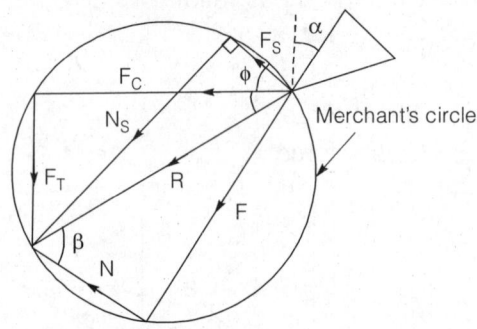

Merchant's circle

$F = F_c \sin\phi + F_T \cos\alpha$

$N = F_c \cos\alpha - F_T \sin\alpha$

$\mu = \tan\beta = \dfrac{F}{N}$

$\tan\beta = \dfrac{\dfrac{F_T}{F_C} + \tan\alpha}{1 - \dfrac{F_T}{F_C} \cdot \tan\alpha}$

- ## Shear Strain(γ)

$\gamma = \tan(\phi - \alpha) + \cot\phi$

Work done in shearing per volume of material removed(vmr) $= \tau_s \times \gamma$

- $\boxed{\left[W_{specific} / vmr\right] = \text{Shear stress} \times \text{shear strain}}$

- $\boxed{\text{Shear strain rate}(\gamma) = \dfrac{V_{shear}}{t_1 / \sin\phi}}$

9. Velocity Triangle

Let,

V_s = Shear velocity

V_c = Chip velocity

V = Velocity of un-cut chip

$$\frac{V}{\sin\left(90^\circ - \phi + \alpha\right)} = \frac{V_C}{\sin\phi} = \frac{V_S}{\sin\left(90^\circ - \alpha\right)}$$

Specific cutting power $= \dfrac{F_C}{w \cdot t_1}$

10. Merchant's Circle Assumptions

1. Cutting edge is straight and sharp.
2. Material is homogeneous.
3. Cutting is orthogonal.

11. Ernest and Merchant Theory

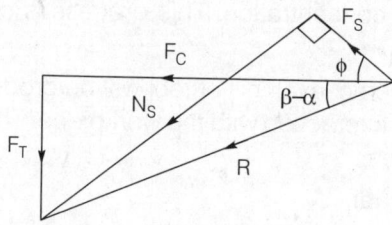

$$F_C = F_S \frac{\cos\left(\beta - \alpha\right)}{\cos\left(\phi + \beta - \alpha\right)}$$

For minimum power consumption.

- First analysis:

$$\boxed{2\phi + \beta - \alpha = \frac{\pi}{2}}$$ For synthetic plastic

- Second analysis:

$$\boxed{2\phi + \beta - \alpha = \cot^{-1}(k)}$$ For metal   Here, k = machining constant

$$\boxed{2\phi + \beta - \alpha = C_m}$$

12. Slip Line Field/Lee and Shaffer

Assumption

1. The work material ahead of the tool behaves as ideal plastic mass.
2. There exists a shear plane which separates the chip and work piece.
3. No hardening in chip occurs

$$\phi + \beta - \alpha = 45°$$

13. Mechanism of Tool Wear

- **Abrasion**

  It one of the surface contain very hard particles then these particles during the process of sliding may dislodge material from the other surface by abrasine action.

- **Adhesion**

  Due to high temperature and pressure between chip and tool metallic bonding takes place at the contact point in the form of spot welds. Sliding causes fracture of these welded joint and material is lost from the surface.

- **Diffusion**

  Atoms in a metallic crystal move from a region of high concentration to that of low concentration. This process is called "Diffusion".

- **Fatigue Wear**

  On a microscopic level, hill of tool will be eroded and fresh hill will be formed by interaction with the workpiece. This process is known as Fatigue wear.

- **Oxidation Wear**

  After machining operation, oxide layer will be formed over tool which will be removed in the next cut. This type of wear is called oxidation wear.

14. Type of Tool Wear

- **Flank Wear**

  It is due to work hardening. The main reason is abrasion and adhesion.

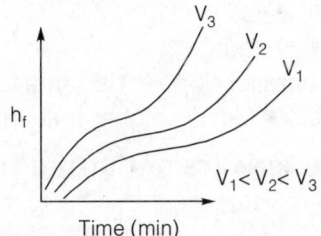

Time (min)

Flank wear is measured by height of wear land (h_f) in mm.

- **Crater Wear**

  It takes place on the face of tool at a short distance from the cutting edge by the action of chip particles flowing over the face at very high temperature. Its main reason is diffusion along with abrasion.

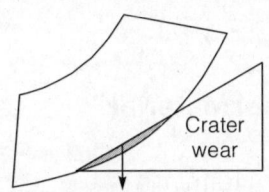

Depression tooks place at rake face

15. Taylor's Tool Life Equation

The tool life is mainly affected by cutting speed. Higher the cutting speed the smaller the tool life. Taylor gave the relation between cutting speed and tool life that is

$$VT^n = C$$

Here,

   V = cutting speed

   T = tool life.

   C = machining constant.

   n = Tool life exponent (depends only on tool material)

For HSS,   n = 0.08 – 0.2
For carbides   n = 0.2 – 0.6
For ceramics   n = 0.5 – 0.8

C depends upon both tool and work piece.

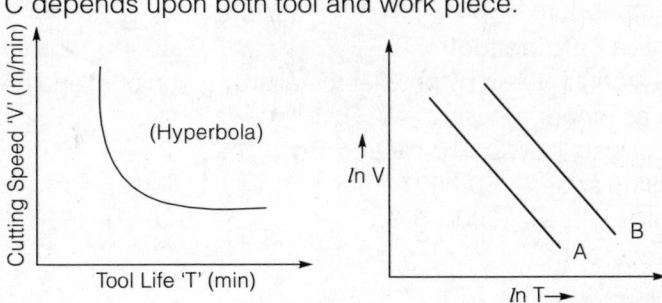

$$VT^n = C$$
$$\log V + n \log T = \log C$$

As the two lines are parallel slope (n) is same for both. B makes higher intercept on the axis. 'C' for B is higher than that for A.

16. Important Characteristics of Cutting Tool Material

- Hot hardness
- Wear resistance
- Toughness
- High thermal conductivity
- Low thermal expansion
- High young and elastic modulus
- Low coefficient of friction
- Ease of Fabrication
- Chemically stable

17. Some Important Tool Material

- Carbon steel tool
  - Cutting speed – 10 m/min
  - Temperature range – 200-250°C
  - Used for making hand drills, narrow blade, taps, dies, chisels
  - It is used for matching soft materials like, Magnesium, Aluminium, Wood.

- High speed steel (HSS)
  - Speed range 30-50 m/min
  - Temperature – 650°C
  - Toughness of HSS is highest among all cutting tool material.
  - Tungsten series of HSS contain 18% Tungsten, 4% Chromium and 1% Vanadium.

- Satellite (cast alloy)
  - It is non-ferrous alloy of cobalt (40-55%), chromium (30-35%), Tungsten (10-20%) along with carbon, molybdenum and boron.
  - Cutting speed 50-80% higher than HSS.
  - Temperature – 900°C

- Cemented carbide tool
  - It is manufactured by powder metallurgy technique, and cobalt act as binder.
  - It consists of **WC = 94%** and **Co = 6%**
  - Cutting speed – 60-200 m/min
  - Temperature – 1000-1200°C

- **Ceramics or Cemented oxides**
  - It consists of fine grained powder of Al_2O_3 along with SiC, chromium oxide, Magnesium oxide, Nickel oxide, Titanium carbide.
  - It is manufactured by powder metallurgy.
  - Cutting speed – 300-600 m/min.
  - Temperature – 1400°C
  - It can not cut Aluminium and Titanium due to great affinity.
- **Diamond**
  - It is hardest substance among all known material.
  - It has good thermal conductivity, Low friction coefficient, Low thermal expansion and Brittle.
  - It is used for machining very hard non-ferrous material like hard carbide, nitrides etc.
  - It is not used for maching steel.
- **Cubic Boron Nitride (CBN)**
  - It is second hardest element after diamond and not a natural material.
  - Cutting speed – 600-800 m/min.
  - It is also used as abrasive in grinding wheels.
- **UCON**
  - It consist of **50%** Colombian, **30%** Titanium and **20%** Tungsten.
  - It is produced by rolling.

18. Effect of Parameters on Tool Life

- **Cutting Speed**

  Higher the cutting speed less will be the tool life.

- **Feed and Depth of Cut**

  By increasing the feed and depth of cut, tool life will decrease because it increases the cutting forces.

  The empirical formulae is given by

  $$V = \frac{257}{T^{0.19} f^{0.36} t^{0.8}}$$

  V = cutting speed in m/min

  T = tool life in min

  f = feed rate in mm/min

  t = depth of cut in mm

- **Size and Structure**

  When the workpiece has coarse grain structure, tool life will decrease and when the tool material has fine grain structure, tool life will increase.

- By increasing the intermittent cuts, tool life decreases.

19. Machinability Index

$$\text{Machinability Index} = \frac{V_t}{V_s} \times 100$$

Here,

V_t = Cutting speed of metal for 1 min. tool life

V_s = Cutting speed of standard

Free-cutting steel for 1 min. tool life

Material	Machinability index
Stainless steel	25
Low carbon steel	55-65
Copper	70
Red brass	180
Aluminium alloys	300–1500
Magnesium alloys	500–2000
C–45 (0.45%C)	100
Mn steel	60
Free cutting steels	200
C–14 (0.14%C)	160

20. Economics of Machining

Optimum cutting speed is chosen to optimize tool life, minimize production cost, maximize, production rate.

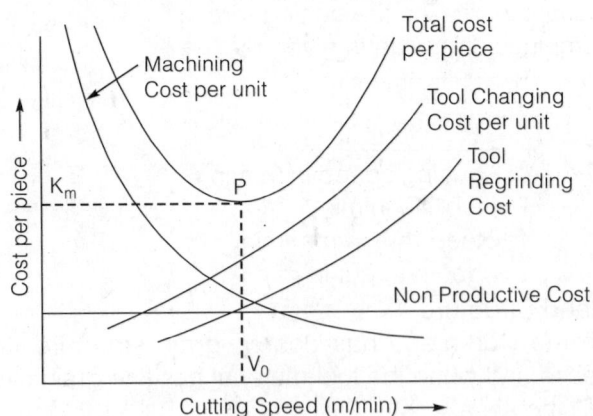

Let, C_m = Machining cost/time

T_m = Machining time (in minute)

L = Length of job to be machined (mm)

D = Diameter of work-piece (mm)

F = Feed rate (mm/rev.)

N = R.P.M.

V = Cutting speed (m/min.)

C_e = Cost of tool regrind per grind

T_i = Idle time (Job loading/Un-loading time)

T_c = Tool changing time per failure

T = Tool life

f = feed

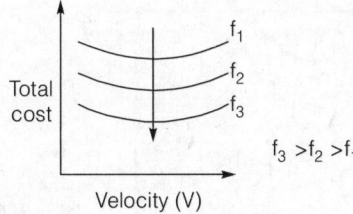

- **Maching cost/unit (C_1)**

  ∴ $C_1 = C_m \cdot T_m$

$$= C_m \cdot \frac{\pi DL}{1000 \cdot f \cdot V}$$

- **Idle cost/unit (C_2)**

$C_2 = C_m \cdot T_i$

- **Tool changing cost/unit (C_3)**

$$C_3 = C_m \times \frac{\text{Tool failure}}{\text{Unit}} \times \frac{\text{Tool changing time}}{\text{Failure}}$$

$$= \frac{\pi DL V^{\left(\frac{1-n}{n}\right)}}{1000 \times f \cdot C^{1/n}} T_C \times C_m$$

- **Tool regrainding cost/unit (C_4)**

$$C_4 = C_e \times \frac{T_m}{T} = \frac{C_e \cdot \pi DL(V)^{\frac{1-n}{n}}}{1000 \times f \cdot C^{1/n}}$$

- Total cost per unit (C_P)

$$C_P = C_1 + C_2 + C_3 + C_4$$

- Total machining time per unit (T)

$$T = T_m + T_i + \frac{T_m}{T} \cdot T_C$$

- Rate of production (R_P)

$$R_P = \frac{1}{T} = \frac{1}{T_m + T_i + \frac{T_m}{T} \cdot T_C}$$

- Minimum cost criteria for cutting speed, (V_{mc}) and Tool life (T_{mc})

$$V_{mc} = \frac{C}{\left[\left(\frac{1}{n} - 1 \right) \left(T_C + \frac{C_e}{C_m} \right) \right]^n}$$

$$T_{mc} = \left(\frac{1}{n} - 1 \right) \left(T_C + \frac{C_e}{C_m} \right)$$

- Maximum production rate criteria for cutting speed (V_{mp}) and Tool life (T_{mp})

$$V_{mp} = \frac{C}{\left[\left(\frac{1}{n} - 1 \right) T_C \right]^n}$$

$$T_{mp} = \left[\left(\frac{1}{n} - 1 \right) T_C \right]$$

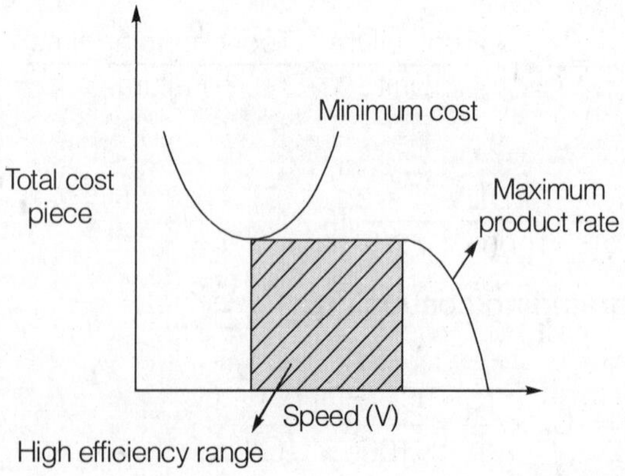

- For maximum profit rate

$$\text{Profit rate} = \frac{R - C_0}{T_m + T_h + \dfrac{T_m}{T} \times T_c}$$

R = selling price

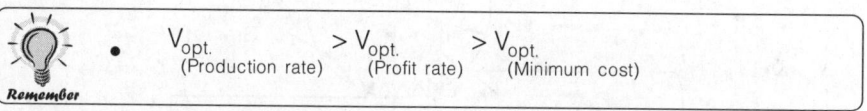

- $V_{opt. \text{(Production rate)}} > V_{opt. \text{(Profit rate)}} > V_{opt. \text{(Minimum cost)}}$

Remember

When V is increased beyond max productivity, tool life decreases. Frequent grinding is required when cutting speed increases and hence productivity decreases.

21. Properties of Cutting Fluid
- High thermal conductivity
- Low viscosity
- Large specific heat
- Small molecular size to help rapid penetration between tool and chip.
- Low surface tension and chemically stable.
- Non-toxic, Non-flammable and odourless.

22. Selection of cutting fluid based on work material

Material	Cutting fluid to be used
Gray cast iron	Soluble oils and thinner neat oils are satisfactory for flushing swarf and metal dust.
Aluminium alloys	It is necessary to have the tool surface highly polished. Generally, a soluble oil to which an online agent has been added. For more difficult application, straight neat oil, fatty oil, or kerosene could be used.
Copper alloys	Water based fluids could be conveniently used. Tougher alloys, a neat oil blended with fatty or inactive EP additive, is used.
Mild steel and low to medium carbon steels	Milky type soluble oil or mild EP neat cutting oil could be used.
Alloy steels	EP cutting oil is used. In some applications, milky soluble can be used.
Stainless steels and heat resistant alloys	Use high performance neat oils with high concentration of chlorinated additives. Sulphur additives are to be avoided. Some high performance EP soluble oils may sometimes be useful.

23. Grinding

It is used to get desired surface finish, correct size and accurate shape of product.

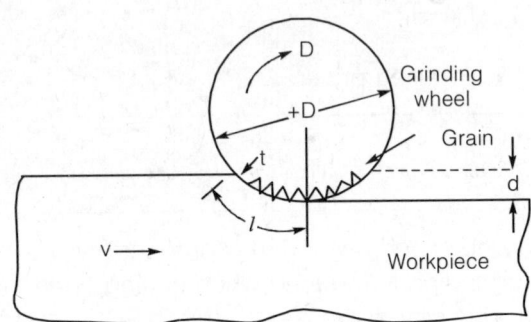

Grinding wheel consists of abrasive particles, bonding material and voids. The projecting abrasive particles act like cutting tool tips and remove metal. A properly selected grinding wheel exhibits self sharpening action. As cutting proceeds the abrasive particles, at cutting edge become dulled, and eventually these becomes plane and which resists the cutting action. Thus new cutting edge points are produced which carry out the further cutting action. This process continues till the abrasive grains get worn down till the level of bond. At this position new cutting edges will do further cutting action.

- **Structure**

  It is the distance between the two cutting edges.
- **Open structure**
  - Space is more in open structure .
  - It is used for ductile material.
- **Close structure**
  - Space is less in close structure.
  - It is used for brittle and hard material.

24. ISO Designation of Grinding wheel

$$\overset{①}{51} - \overset{②}{C} - \overset{③}{30} - \overset{④}{M} - \overset{⑤}{5} - V - 20$$

- **First and last number**

  Manufacturer has to give special information about the grinding wheel.

  According to this system the various elements and characteristics of grinding wheel are represented on all the wheels in a definite sequence as follows.

(i) **Type of abrasive**
- C - carbide
- A - Aluminium oxide (Al_2O_3)
- D - Diamond
- Emery ($Al_2O_3 + Fe_3O_4$)
- Artificial SiC (Carborundum)

(ii) **Grain size**
- 10 - 24 roughening
- 30 - 60 Medium
- 70 - 180 Finishing
- 220 - 600 Superfinishing operation
  Per square inch opening in the mesh. These are number of opening per square inch area.

(iii) **Hardness**
- A - H soft wheels
- I - P medium wheels
- Q - Z hard wheels

(iv) **Structure**
- (0 – 7) - close structure
- (8 – 16) - open structure

(v) **Type of bond**
- V - Vitrified bond (glass material which produces strong wheel)
- S - Silicate
- B - Resinoid (Thermosetting materials which are flexible than vitrified bond)
- R - Rubber (Most flexible bond material)
- M - Metal bond

25. Drilling

It is a process of creating a hole by oblique cutting. Drills are made by forging and it is smaller in size so that there is some margin for reaming.

26. Boring

It is the process of enlarging already existing hole to bring it to the required size, and have a better finish. Expected accuracy is 0.125 mm.

27. Reaming

It is the process of exacting the hole and finishing it. Expected accuracy of finish is ±0.005 mm.

28. Trepanning

Drill is in the form of a tube at the periphery of which there are cutting edges. Complete material is not removed and initially this process was used in Gun Barrel manufacturing process.

29. Counter Boring

Counter boring is making the hole little larger. Counter boring is done by end milling. It is seating place for bolt heads. The enlarged hole forms a square shoulder with the original hole. This is necessary in some cases to accommodate the heads of bolts, studs and pins. The cutting edges have straight or spiral teeth.

30. Counter Sinking

It is the process of making the holes slightly tapered in the beginning. It is a seating place for screws.

NON-CONVENTIONAL MACHINING

These are known as unconventional method because compared to conventional method there is no contact between tool and workpiece, specific power consumption is very large, metal removal rate is low and is used for machining very hard materials for precision work.

1. Ultrasonic Machining (USM)

Ultrasonic means those vibrational waves having a frequency above the normal hearing range (i.e., 20 K – 30 K Hz). In this method a slurry of abrasive grains are hammered on to the work surface by a vibrating tool normal to work surface removing the w/p material in the form of extremely small chips. Tools is feed gradually towards work by feed mechanism. The concentrator is used to increase the amplitude to the level needed for machining.

- Abrasives

  Commonly used abrasives are Boron Carbide, Silicon Carbide, Al_2O_3

  In USM the abrasives are suspended in a fluid medium to form a slurry water is the most commonly used fluid other liquids used are Bengene, Oils, etc.

- Advantage of USM
  - Operate is noiseless
  - Equipment is easy to operate
  - Little or no heat generation
  - Good finish and high accuracy

- Hard materials can be easily machined irrespective of conductivity.
- **Disadvantage**
  - Low MRR (Metal Removal Rate)
  - Higher specific cutting energy
  - Softer materials are difficult to machine
- **Application**
  - Normally used for making dies
  - Machining hard carbides, glasses and precious stones
  - Also used for dental application.

2. AJM (Abrasive Jet Machining)

- **Principle**

  The fundamental principle of abrasive jet machining involves the use of high velocity stream of abrasive particles carried by a high pressure gas through a nozzle on the workpiece. The metal removal occurs due to erosion caused by the abrasive particles impacting the work surface at high speed. Abrasives normally used are Al_2O_3 SiC or B_4C for machining and sodium bicarbonate for polishing work.

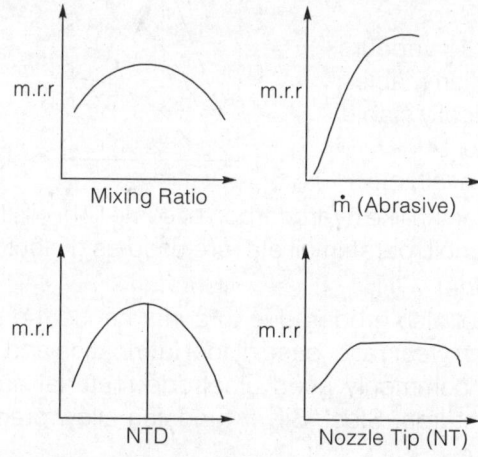

- **NTD (Nozzle Tip Distance)**

  Distance between workpiece and tip of nozzle is called nozzle tip distance. It varies from 7 to 13 mm.

  Nozzles are subjected to high degree of wear so day are made of hard materials like tungsten carbide or ceramics. Gases used are N_2, CO_2 or clean dry air.

- **Advantages**
  - Low cost and ease of operation.
  - No heat generation

- Disadvantage
  - Poor surface finish
  - Very low MRR

3. Electric Discharge Machining (EDM)

In EDM the metal is removed due to erosion caused by rapidly occurring discharge between the tool and work. The mechanism of material removal is melting and evaporation. The basic principle is that when a discharge takes place between two points intense heat is generated near the zone having temperature range of 8000-12000°C which evaporates the materials in the sparking zone. Positive terminal erodes out faster so w/p is made anode.

- Dielectric Fluid

  The use of liquid dielectric provides following main advantages.
  - It acts as a vehicle to drive away the chips and thus preventing them from sticking to the surface of tool.
  - It helps in increasing the MRR by promoting spark between tool and work.
  - Acts as coolant medium.

- Requirement of Dielectric
  - Low viscosity
  - Nontoxic vapours
  - Non inflammable
  - Chemically stable
  - Low cost
  - Easily available

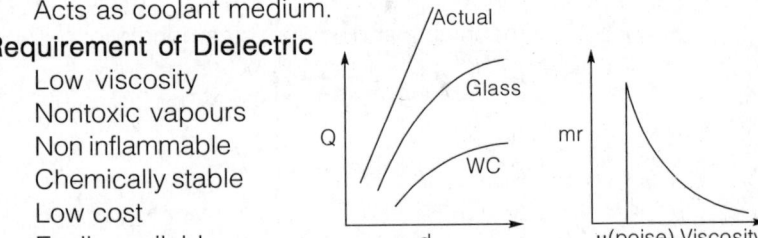

Liquid dielectric like hydrocarbon or some light oils like kerosene oil, transformer oil, paraffin oil etc. are used as dielectric fluid.

- Tool Material

  In EDM tool also erodes due to spark hence the selection of tool depends on wear ratio, ease of tool fabrication and cost of material. The most commonly used electrode material are Cu, Tungsten alloys, Cast Iron, Steel, Silver tungsten alloy, graphite.

- Workpiece

  In EDM only good conductors of electricity can be machined irrespective of hardness.

- Advantage
  - High MRR
  - Good surface finish
  - Complex cavities can be cut

- Limitation
  - High tool wear

- Tool wear limits accuracy
- Only good conductors can be machined
- Application
  - Blind, complex cavities
  - Hard material dies
  - Making holes in dies

4. ECM (Electrochemical Machining)

This process is an extension of already known process of electroplating with modification in a reverse direction. A shaped tool or electrode is used in the process which forms cathode and workpiece forms anode. A small gap is maintained between the tool and workpiece and an electrolyte is pumped through it. Low voltage direct current is used which in the presence of electrolyte enables a control removal of metal from workpiece by anodic dissolution. The tool is provided with a constant feed motion towards workpiece. The electrolyte is pumped at a high pressure through the small gap between tool and work.

- Main Function of Electrolyte
  - Complete the electrical circuit between the tool and workpiece and allow large current to pass through it.
  - To carry away the heat and waste product of reactions.
  - Allow electrochemical reactions at a faster rate.
- Properties of Electrolyte
  - High electrical and thermal conductivity
  - Low viscosity
  - High specific heat
  - Non corrosive and nontoxic
  - Chemically stable, cheap and easily available.
- Normally used Electrolyte are:
  - $NaCl + H_2O$
  - HCl
  - H_2SO_4, HNO_3
  - Strong alkaline sol.

MRR — 3000-4000 mm³/min
Which is very high

$$Q = \frac{AI}{\rho ZF} = \frac{eI}{\rho F}$$

$$e = \frac{A}{Z} = \text{equivalent wt.}$$

Q = MRR cm³/sec

A = Gram atomic wt. of w/p metal

I = Current in (amp.)

ρ = Density of anode (i.e., w/p) in gm/cm^3

Z = Valency

F = Faradays constant $\approx$ 96,500 columbs

5. Laser Beam Machining

In this method focussing of laser on the workpiece result into a very high surface temperature, resulting in metal removal due to melting and evaporation.

The rays of laser are perfectly parallel and they can be focussed to a very small diameter.

It is a costly method and is used only when it is not feasible to machine with other process.

It is used to drill microholes, cutting very narrow slot in very hard material like diamond.

Used for making wire drawing dies. Heat affected zone in this process is very less.

6. Electron Beam Machining (EBM)

EBM is the process of machining material with the use of very high velocity beam of electrons. The workpiece is held in vacuum chamber and the electron beam focused on it when strikes the surface of material, it results into generation of very high temperature. Leading to metal removal due to melting and evaporation.

The setup consists of electron gun that is the source for electron. A high DC power source, electromagnetic focussing lens, enclosed vacuum chamber and deflecting coils.

Used for drilling very fine holes, cutting, contours and very narrow slots.

■■■■

Metal Forming

Forming is the process in which the desired size and shape are obtained through the plastic deformation of a material. The stresses induced during the process are greater than the yield strength, but less than the fracture strength, of the material.

1. Classification of Metal Forming

Type of metal forming	Working temperature
Cold working	less than $0.4\ T_m$
Warm working	$0.4\ T_m - 0.6\ T_m$
Hot working	Greater than $0.6\ T_m$

- Cold Working

  When the material is deformed *below recrystallization temperature*. The recrystallization temperature generally varies between one-third to half the melting point of most of the metals.
- Advantage
  - Cold working increases the strength and hardness of the material due to strain hardening. Further, there is no possibility of decarburization of the surface.
  - Since the working is done in the cold state, no oxide would form on the surface and consequently, good surface finish is obtained.
  - Better dimensional accuracy is achieved.
  - It is far easier to handle cold parts and also is economical for smaller sizes.
- Disadvantage
  - Since the material has higher yield strength at lower temperatures, the amount of deformation that can be given to it is limited by the capability of the presses or hammers used.
  - Since the material gets strain hardened, the maximum amount of deformation that can be given is limited. Any further deformation can be given after annealing.
  - Some materials which are brittle cannot be cold worked.
- Hot Working

  In hot working, process is carried *above the recrystallization* temperature with or without actual heating. In hot working, the temperature at which the working is completed is important because any extra heat left after working will aid in grain growth, thus giving poor mechanical properties.

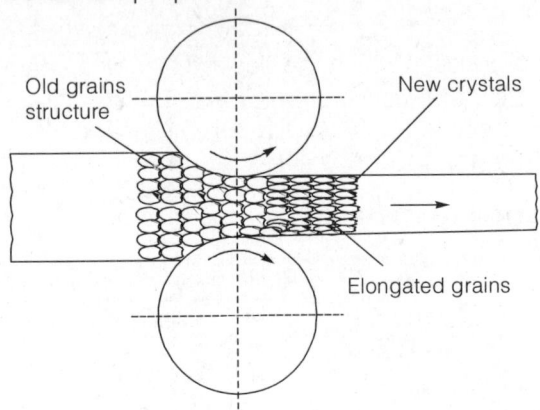

Grain reformation in hot rolling process.

- **Advantages**
  - As the material is above the recrystallization temperature, any amount of working can be imparted since there is no strain-hardening taking place.
  - At high temperature, the material would have higher amount of ductility and therefore there is no limit on the amount of hot working that can be done on a material. Even **brittle materials** can be hot worked.
  - Since the shear stress gets reduced at higher temperatures, hot working requires much less force to achieve the necessary deformation.
  - It is possible to continuously reform the grains in metal working and if the temperature and rate of working are properly controlled, a very favorable grain size could be achieved giving rise to better mechanical properties.

- **Disadvantages**
  - Some metals cannot be hot worked because of their brittleness at high temperatures.
  - Higher temperatures of metal give rise to scaling of the surface and as result, the surface finish obtained is poor. Also, there is a possibility of the decarburization of skin in steels due to the high temperature.
  - Because of the thermal expansion of metals, the dimensional accuracy in hot working is hard to achieve since it is difficult to control the temperature of workpieces.

- Handling and maintaining of hot metal is difficult and troublesome.

2. Forging

Forging is the operation where the metal is heated and then a force is applied to manipulate the metal in such a way that the required final shape is obtained. Forging is generally a hot working operation though cold forging is also used some times.

3. Type of Operations in Forging

- **Drawing**

  This is the operation in which the metal gets elongated with a reduction in the cross-sectional area. For this purpose, the force is to applied in a direction, perpendicular to the length axis.

- **Upsetting**

  This is applied to increase the cross-sectional area of the stock at the expense of its length. To achieve the upsetting, force is applied in a direction parallel to the length axis.

4. Type of Forging

- **Smith Forging**

  This is the traditional forging operation done openly or in open dies by the village blacksmith or modern shop floor by manual hammering or by power hammers.

- **Drop Forging**

  This is the operation done in closed impression dies by means of drop hammers. Here, the force for shaping the component is applied in a series of blows.

- **Press Forging**

  Similar to drop forging, press forging is also done in closed impression dies with the exception that the force is a continuous squeezing type applied by the hydraulic presses.

- **Machine Forging**

  Unlike the drop or press forging where the material is drawn out, in machine forging, the material is only upset to get the desired shape.

5. Forging Defect:

- **Unfilled Sections**

  In this, some sections of the die cavity are not completely filled by the flowing metal. The causes of this defect are improper design of forging die or using faulty forging techniques.

- **Cold Shut**

  This appears as a small, crack at the corners of the forging. This is caused mainly by the improper design of the die wherein the corner and fillet radii are small, as a result of which the metal dies not flow properly into the corner and ends up as a cold shut.

- **Scale Pits**
  This is seen as irregular depression on the surface of the forging. This is primarily caused because of the improper cleaning of the stock used for forging.

- **Die Shift**
  This is caused by the misalignment of the two die halves, making the two halves of the forging to be of improver shape.

- **Flakes**
  These are basically internal ruptures caused by the improper cooling of the large forging. Rapid cooling causes the exteriors to cool quickly causing internal fractures. This can be remedied by following proper cooling practice.

- **Improper Grain Flow**
  This is caused by the improper design of the die, which makes the flow of metal not following the final intended directions.

6. Wire Drawing

The process of wire drawing is to obtain wires from rods of bigger diameter through a die wire drawing is always a cold-working process.

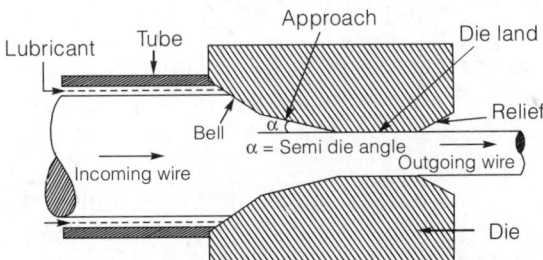

Relationship between drawing force and semi-die angle (α)

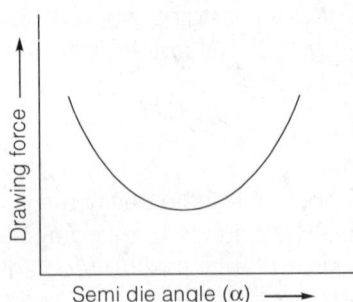

- **Defect in wire drawing**
    - Oxide produce centreline cracks. So removal of oxide from wire is required.
    - There are longitudinal scratches or folds in the material (seams). It may open up during operation.
    - In cold drawn products there will be residual stresses. These residual stresses can be significant in causing stress corrosion cracking of the part over a period of time.

7. Tube Drawing

It is similar to wire drawing except it requires a mandrel of the requisite diameter to form the internal hole.

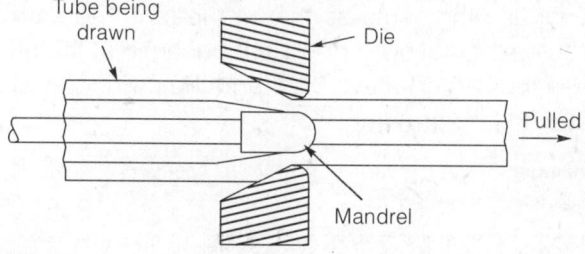

8. Swagging

Swagging is a mechanical deformation technique of reducing or shaping the cross-section of rods or tubes by means of repeated impacts or blows. The swagging process consists of dies which are given the requisite external shape. These dies intermittently hammer the stock to produce the deformation. This hammering action, besides producing the necessary shape, ensures good surface qualities, better grain structure and higher tensile strength.

9. Extrusion

Extrusion is the process of confining the metal in a closed cavity and then allowing it to flow from only one opening so that the metal takes the shape of the opening. The operation is identical to the squeezing of toothpaste out of a toothpaste tube.

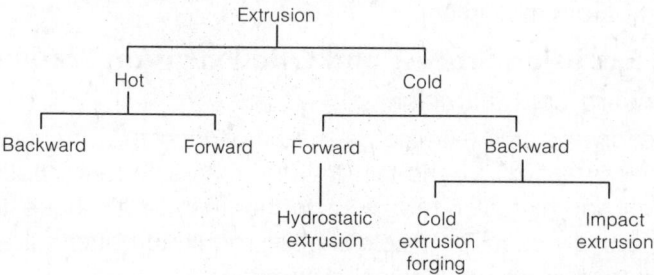

10. Forward Hot extrusion/Backward Hot Extrusion Process

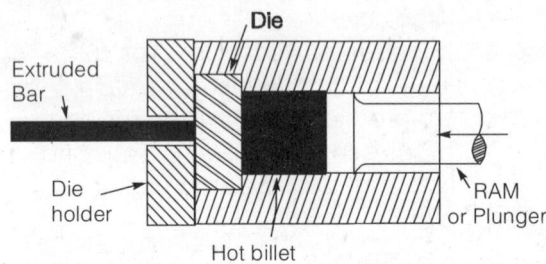

Forward hot extrusion, signifying the flow of metal in the forward direction, i.e., the same as that of the ram. In forward extrusion, the problem of friction is prevalent because of the relative motion between the heated metal billet and the cylinder walls.

- **Backward Hot Extrusion**

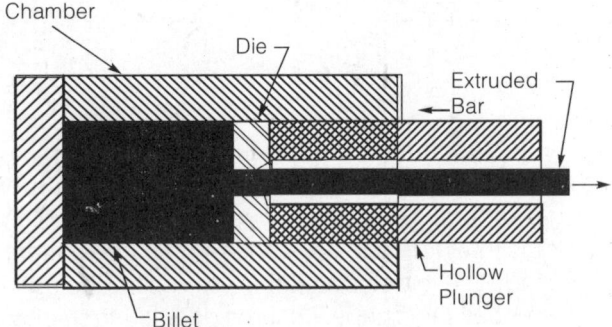

In order to completely overcome the friction, the backward hot extrusion, as shown in above figure used. In this, the metal is confined fully by the cylinder. The ram which houses the die, also compresses the metal against the container, forcing it to flow backwards through the die in the hollow plunger or ram. It is termed backward because of the opposite direction of the flow of metal to that of ram movement.

11. Cold Extrusion Process and Cold Extrusion Forging

- **Forward cold extrusion**

  The forward cold extrusion is similar to that of forward hot extrusion process except for the fact that the extrusion ratios possible are lower and extrusion pressures higher than that of hot extrusion. It is normally used for simple shapes requiring better surface finish and to improve mechanical properties.

- **Backward Cold Extrusion/Impact Extrusion**

  In impact extrusion a heavy punch is allowed to fall over the material and material takes the shape of the die by flowing in the clearance between punch and die. Collapsible tubes can be made by this. This is used to make collapsible tubes of soft alloys such as tooth paste containers. This process is limited to soft and ductile materials.

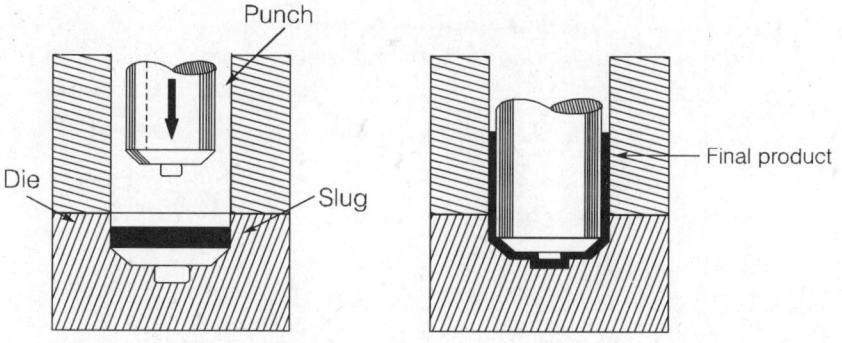

- **Cold Extrusion Forging**

  Cold extrusion forging is similar to impact extrusion but with the main difference that the side walls are much thicker and their height is smaller.

12. Hydrostatic Extrusion

In Hydrostatic extrusion, the billet in the container is extruded through the die by the action of a liquid pressure medium rather than by direct application of the load with a ram. The process of pure hydrostatic extrusion differs from conventional extrusion in that the billet is completely surrounded by a fluid, which is sealed off and is pressurized sufficiently to extrude the billet through the die.

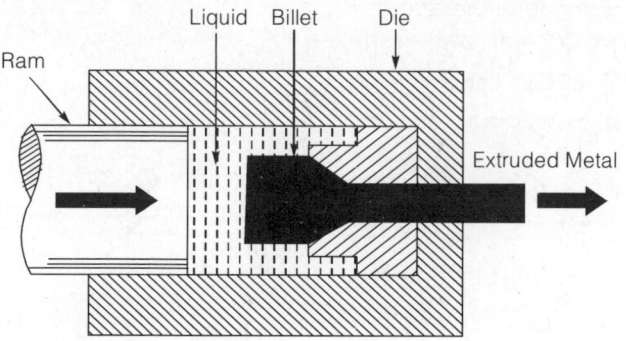

Some common extrusion defect and its reason:

Defect Type	Reason
Pipe defect/fish Tail/Tail pipe	High friction
Centreburst / Internal Crack	Hydrostatic Tensile
bamboo/Surface	Due to hot shortness

13. Analysis of Wire Drawing and Extrusion

- Johnson's Equation

$$\sigma_d = \sigma_0 \, [a + b \ln R]$$

where a, b are Johnson's constant, σ_0 is nominal stress and R is extrusion ratio.

$$\sigma_0 = \frac{K \in^n}{n+1}$$

K = Strength coefficient

$\in$ = Extrusion strength

    = $\ln[1 + e]$

e = True strain

n = Strain hardening coefficient

R = Extrusion ratio = $\dfrac{A_i}{A_f}$ (A_i is initial area and A_f is final area)

- Slab Method

$$\sigma_0 = \frac{K \in^n}{n+1}$$

$$(\sigma_d)_{r = R_0} = 2K'\left(\frac{1+B}{B}\right)\left[1 - \left(\frac{R_f}{R_i}\right)^{2B}\right] + \sigma_b \left(\frac{R_f}{R_i}\right)^{2B}$$

Here,

α = Semi die angle

μ = Coefficient of friction

σ_b = Back pull

B = $\mu \cot\alpha$

$K' = \dfrac{\sigma_0}{\sqrt{3}}$ (for plane strain)

$K' = \dfrac{\sigma_0}{2}$ (for plane stress)

- ## For hot extrusion

$$\sigma_d = K_1 \ln R$$

K_1 = Extrusion constant which depends upon temperature.

14. Rolling

When a metal passes through the rolls, grain structure of metal changes. As a result of squeezing, velocity of the material at exit is higher than at entry.

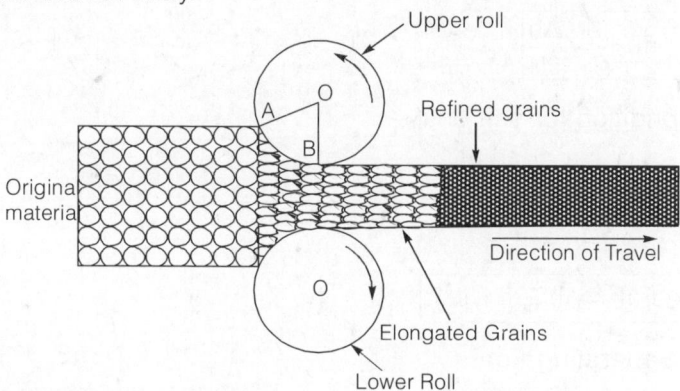

15. Mechanism of Rolling

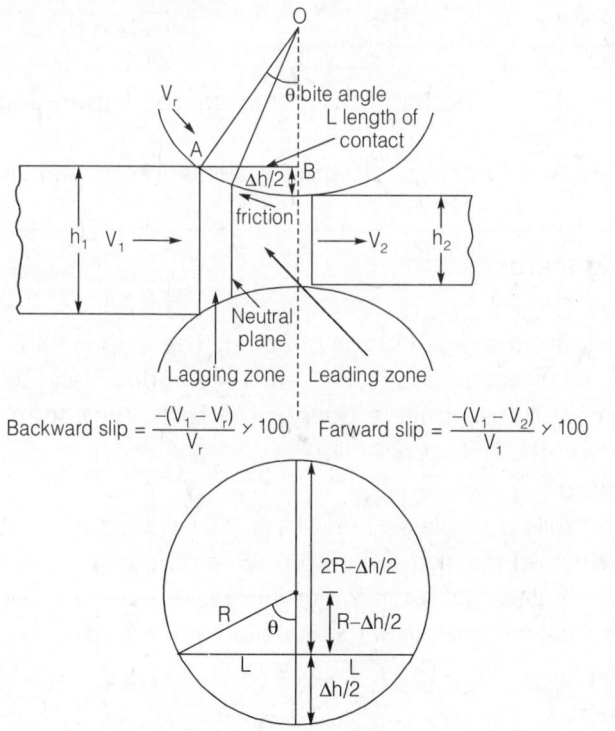

$$\text{Backward slip} = \frac{-(V_1 - V_r)}{V_r} \times 100 \qquad \text{Farward slip} = \frac{-(V_1 - V_2)}{V_1} \times 100$$

Here,

θ = Bite angle

L = Length of contact

Δh = $h_1 - h_2$

μ = coefficient of friction

- **Bite angle**

$$\cos\theta = 1 - \frac{\Delta h}{2R}$$

- **Condition for self entry**

$$\mu \geq \tan\theta$$

- **Reduction (Draft) Δh**

$$\Delta h \leq \mu^2 R$$

$$\text{Max draft} = \Delta h_{(max)} = \mu^2 R$$

- **Roll Separating Force**

$$F_{sep.roll} = LW\left(1 + \frac{\mu L}{2h}\right)$$ Where $$\frac{L = \sqrt{R\Delta h}}{W = \text{width}}$$

- As the **radius** of roll increases, **roll separating force** will increase.
- **Roll forming** is stagewise bending of metal strip under a series of roll.

16. Rolling Defect

- **Wavy Edges**

  These are the result of roll bending. The strip is thinner along its edges than at its centre. Because the edges elongate more than the centre they buckle because of restraining from expanding freely in the rolling direction.

- **Spread**

  In the rolling of plates and sheet having high width to thickness ratio the width of the material remains constant during rolling. With smaller ratios the width increases considerably in the roll gap. This increase in width is called spreading.

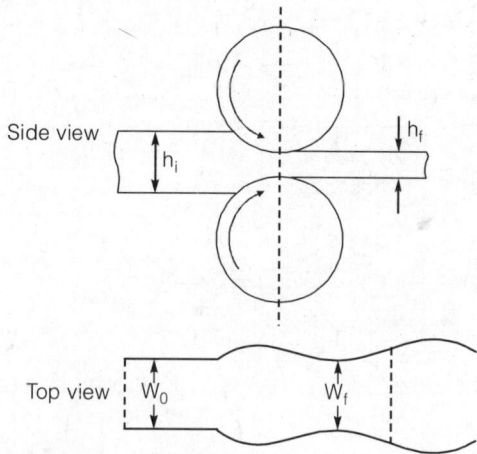

- ## Crocodile crack
  Due to weakness at the center sheet bifurcates into two parts. This is called crocodile crack.

17. Blanking and Piercing

This is done by shearing action. In this the metal is brought to the plastic stage by pressing the sheet between two shearing blades so that fracture is initiated at the cutting point. The fractures on either sides of sheet further progressing downwards with the movement of upper shear, finally result in the separation of the slug from the parent strip.

- ## Clearances
  The clearance between two shears is one of the principal factors controlling a shearing process. This clearance depends essentially on the material and thickness of the sheet metal. This clearance can be approximated per side C as

$$C = 0.0032 \times t \times \sqrt{\tau} \text{ mm}$$

Where, t = sheet thickness in mm

τ = material shear stress in MPa

- ## Blanking
  It is a process in which the punch removes a portion of material from the stock which is a strip of sheet metal of the necessary thickness and width. The removed portion is called a blank and is usually further processed to be of some use, e.g., blanking of a padlock key.

- ## Piercing
  Sometimes also called punching, piercing is making holes in a sheet. It is identical to blanking except for the fact that the punched

out portion coming out through the die in piercing is scarp. Normally, a blanking operation will generally follow a piercing operation.

A typical set-up used for blanking or piercing operation is shown in fig.

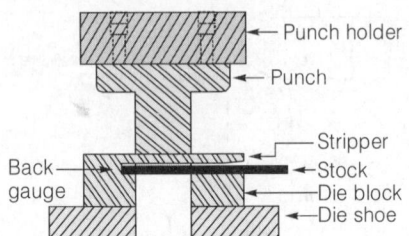

In blanking, clearance is provided to punch **(dp = dsize – 2C)** while.

In piercing clearance is provided to die **(d$_{Die}$ = d$_{size}$ + 2C)**

Remember

- **Angular Clearance**

  After final breaking, due to release of stored elastic energy, the blank cling to the die face unless the die opening is enlarge. This enlargement is called angular clearance.

  Angular clearance depends upon material, thickness and shape of stock used.

  High angular clearance is provided for thicker and softer materials. Its normal rule is 0.25-0.75 degree per side.

- **Stripping Force (P$_s$)**

  Stripper is used to separate the punch from the stock. The force required for stripping depends on material, position and size of punched hole. Thicker material or small hole in the middle of a strip require more stripping force. It vary from 2.5 to 20% of punch force

  $$P_s = kLt$$

  Here   L = perimeter of cut (mm)

          t = stock thickness (mm)

          k = stripping constant

- **Punching Force (P)**

  It is the force required to be exerted by the punch in order to shear out the blank from the stock

  $$P = Lt\tau$$

  Here τ = Shear strength

  Sometimes the tensile strength may replace the shear strength in the above expression because shear is not the only active force the press has to overcome.

  The punching force for holes which are smaller than the stock thickness, may be estimated as follows:

$$P = \frac{dts}{\sqrt[3]{\dfrac{d}{t}}}$$

where 　　d = diameter of the punch, mm

　　　　　s = tensile strength of the stock, MPa

- **Shear**

  The shear is relieved of the punch or die face so that it contacts the stock over a period of time rather than instantaneously. It may be noted that providing the shear only reduce the maximum force to be applied but not the total work done in shearing the component. Maximum shear force (P) when shear is applied to the punch or the die.

  $$P = \frac{Lt(tf)}{t_1}$$

  Here, 　　f = Penetration of punch as a fraction

  　　　　　t_1 = Shear on the punch or die.

- **Trimming**

  In operations such as drop forging and die casting, a small amount of extra metal gets spread out near the parting plane, which is termed flash. The flash is to be trimmed before the casting or forging is to be used.

- **Shaving**

  In a blanking or piercing operation, the edge of the blank or the hole is not perfectly clean because of the burr generated in the shearing process, which is equal to the clearance on the die. For close tolerance work, the blanking or piercing process is followed by shaving which removes the burr left on the product.

- **Nibbling**

  Nibbling is removing the metal in small increments. When specific contour is to be cut in a sheet metal, a small punch is used to punch repeatedly along the necessary contour, generating the required profile.

- **Notching**

  Notching is a method to cut a specified small portion of metal towards the edge of the stock.

18. Deep Drawing

In deep drawing, which is also called cup drawing or radial drawing, flat thin sheets (blanks) are formed into cup shaped components by pressing the central portion of the sheet into die opening using a punch to draw the metal into the desired shape.

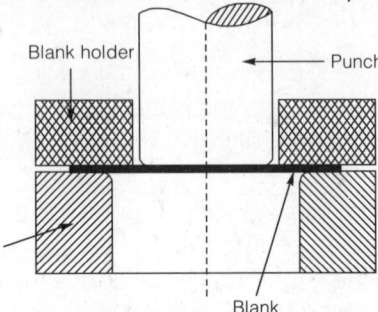

The blank may be circular or rectangular, or of a more complex outline. In deep drawing operation, bending of sheet takes place over the die curvature. The bend zone experiences localized strains which are tensile on the outside of the neutral axis and compressive on the inside.

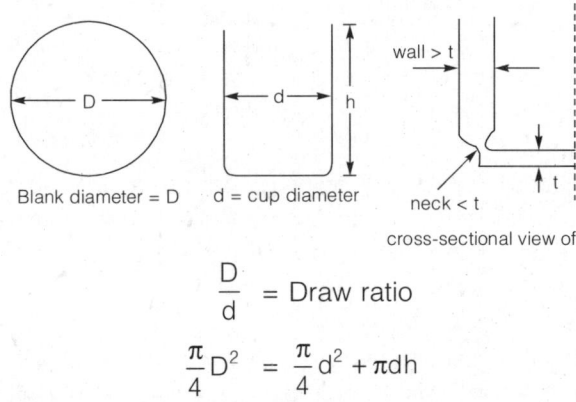

Blank diameter = D   d = cup diameter   cross-sectional view of cup

$$\frac{D}{d} = \text{Draw ratio}$$

$$\frac{\pi}{4}D^2 = \frac{\pi}{4}d^2 + \pi dh$$

$$\text{Limiting draw ratio} = \frac{D_{max}}{d}$$

$$\text{Die-corner radius} = 10\,t$$
$$\text{Punch corner radius} = 6\,t$$
$$\text{Clearance} = (1.2 \text{ to } 1.4)\,t$$
$$\text{Blank Holding Pressure} = 2\% \text{ of yield strength}$$

19. Bending

It is the operation of deforming a flat sheet around a straight axis where the neutral plane lies.

Bending is the cold working process involving plastic deformation in which the total surface area remains constant. In bending the outer fibers of the metal are in tension while the inner fibers are in compression. After bending when the load is removed from work piece, the extent of bending will decrease because of elastic recovery which is called Spring back.

- **Bend Allowance**
  The length of neutral axis in the curvature region of bend is called bend allowance.

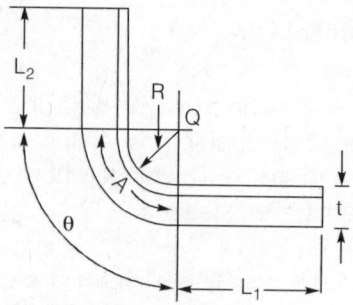

$$\text{Bend allowance} = \theta (R + Kt)$$
$$t = \text{sheet thickness}$$
$$\theta = \text{angle in radians}$$
$$R = \text{bend radius}$$
$$K = \text{stretch factor}$$
$$= 0.33 \text{ when } R < 2t$$
$$= 0.5 \text{ when } R \geq 2t$$

20. Embossing

Embossing is the operation used in making raised figures on sheets with its corresponding relief on the other side. The process essentially involves drawing and bending of the metal.

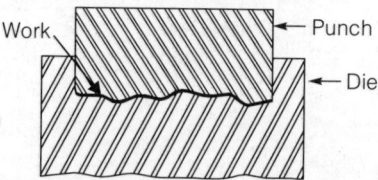

Embossing operation is generally used for providing dimples on sheets to increase their rigidity and for decorative sheet work used for panels in houses and religious places.

21. Coining

Coining is essentially a cold-forging operation except for the fact that the flow of the metal occurs only at the top layers and not the entire volume.

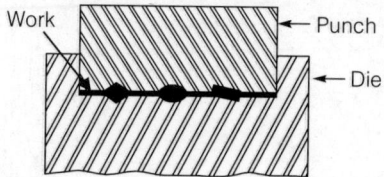

The type of impression obtained on both sides would be different unlike the embossing. Coining is used for making coins, medals and similar articles, and for impressions on decorative items.

22. Type of Sheet Metal Die

- **Progressive Dies**

  The progressive dies perform two or more operations simultaneously in a single stroke of a punch press, so that a complete component is obtained for each stroke. The place where each of the operations are carried out are called stations.

- **Compound Die**

  In a compound die, as distinct from a progressive die, all the necessary operations are carried out at a single station, in a single stroke of the ram. To do more than one set of operations, a compound die consists of the necessary sets of punches and dies.

  For blanking operation, the punch used for piercing becomes a die i.e., blanking is done in a direction opposite to that of piercing. It is slower than progressive die in operation.

- **Combination Die**

  A combination die is same as that of a compound die with the main difference that here non-cutting operations such as bending and forming are also included as part of the operation.

■■■■

Metrology

1. Normal Distribution

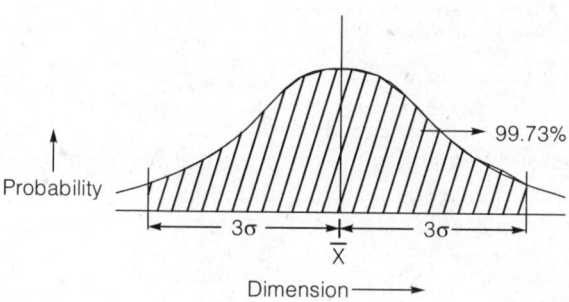

- No two dimensions match.
- Desired tolerance is the limit as specified by the consumer.
- Hole is internal feature of the assembly.
- Shaft is external feature of assembly.
- If process capability of both hole and shaft is equal then randomly a hole is selected from whole lot and shaft is selected from a shaft lot. Assembly can be mode.

2. Selective Assembly

- Process capability range of both hole and shaft is divided into different sub group in such a way that variation in dimension within the sub group is equal to the desired tolerance.

  To do that each and ever best is to be inspected and place in different sub group. Now the randomly a hole is selected from sub group A (capital) and a shaft is selected from a (small) and assembly can be made.

  The process of achieving fully interchangeable system is called *"Selective Assembly"* (Although machines are not capable).

  - Basic size
  - Width of tolerance zone
  - Fundamental deviation

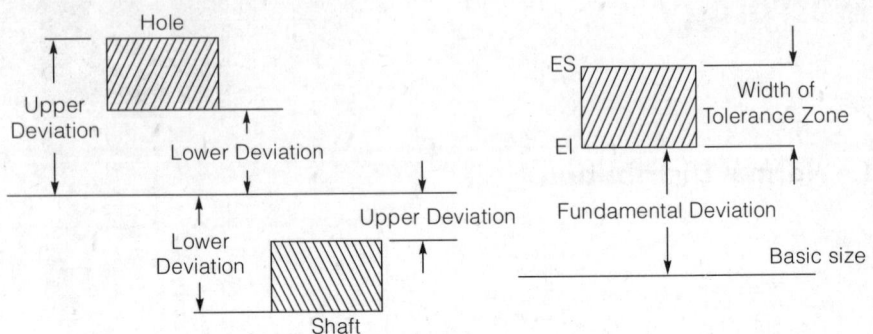

Fundamental derivation is the distance between the zero line and limit closer to it. Either upper limit or lower limit.

Hole Basis System

Shaft Basis System

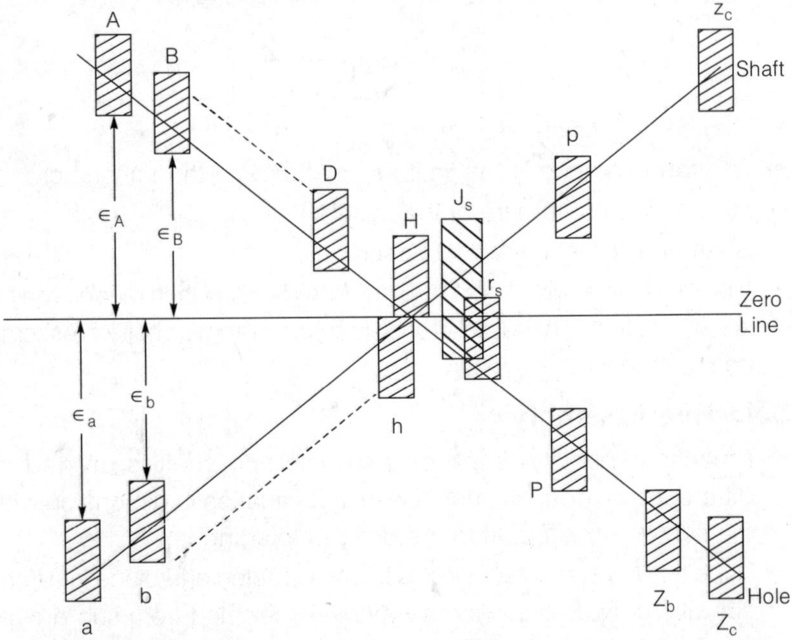

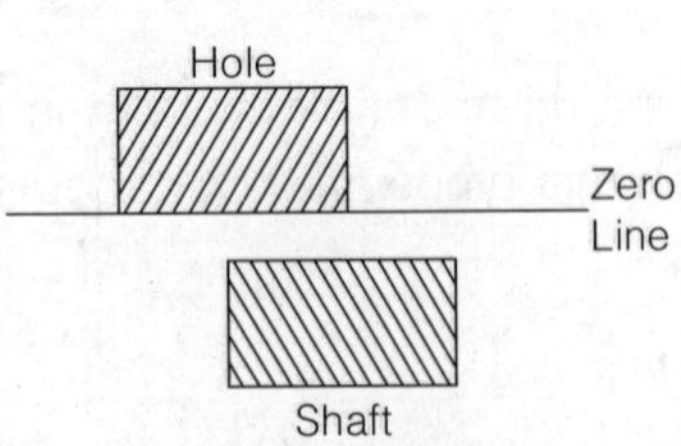

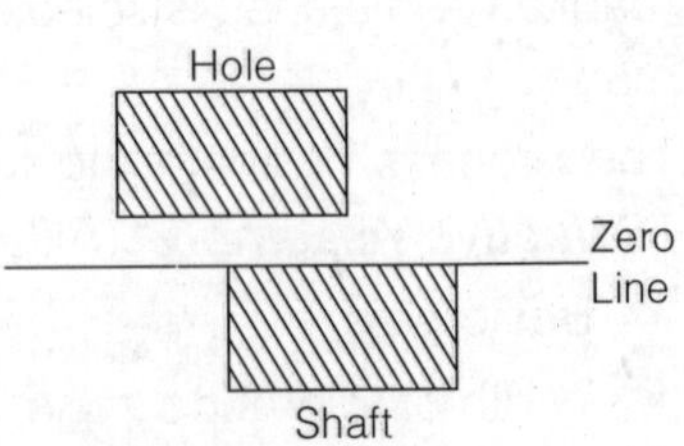

- If H type of hole appears in the assemble it is called hole basis assembly.
- If h type of shaft appears, it is called shaft basis system.
- If the tolerances are only one side it is called unilateral tolerances.
- If the tolerances are both the sides it is called bilateral tolerances.
- A consumer has to select any one of these 25 type fundamental deviation.

$$D = \sqrt{D_{max} \times D_{min}}$$
$$i = 0.45\ D^{1/3} + 0.001\ D$$

  - Five step preferred series
  - IT01 - IT1 → Empirical formula
  - IT1 - IT5 → Exact geometric mean
  - IT5 - IT16 → Preferred series

4. Width of Tolerance

IT01	IT0	IT1	IT2
0.3 + 0.008 D	0.5 + 0.012 D	0.8 + 0.02D	ar
		a	
IT3	IT4	IT5	IT6
ar^2	ar^3	$\approx 7i = ar^4$	10i
IT7	IT8	IT9	IT10
$10(10)^{1/5}i$	$10(10)^{2/5}i$	$10(10)^{3/5}i$	$10(10)^{4/5}i$
= 16i	= 25i	= 40i	= 64i
IT11	IT12	IT13	IT14
100i	160i	250i	400i
IT15	IT16		
640i	1000i		

- Fit is the relationship between hole and shaft before assembly.
- Fit and tolerance is always refers to a group and not the individual component.
- Maximum size of shaft and minimum size of hole are called maximum material limit.
- The different maximum material limit of hole and shaft is called allowance. Allowance is either minimum clearance or maximum interference.
- **Clearance fit**
  If lower limit of hole is larger than the upper limit of shaft. It is considered as clearance fit.

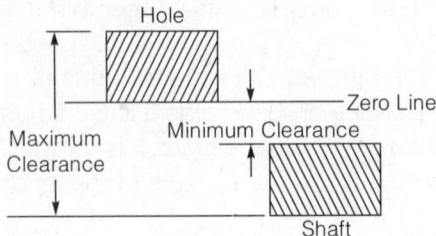

- **Transition Fit**

  This type of fit appears when there is overlap in the tolerance zones. Physically it indicates that when a part is selected randomly from hole lot and a shaft lot, some of the assemblies can be made without the application of force and for some of the assemblies force will be required.

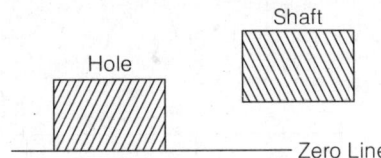

- **Interference Fit**

  If maximum size of hole is smaller than the minimum size of shaft then force has to be applied to make the assembly. Such fits are called interference fits.

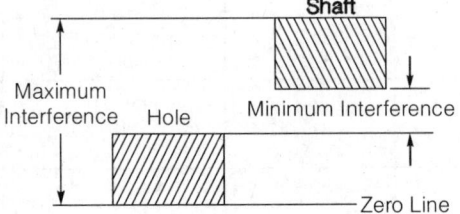

5. Limit Gauges

- **Plug gauge is used to check the hole.**
  - GO plug gauge - LLH (Lower limit hole)
  - NO-GO plug gauge - ULH (Uppers limit hole)

- **Snap (Ring) gauge is used to check the shaft.**
  - GO-snap gauge - ULS (Upper limit shaft)
  - NO-GO snap gauge - LLS (lower limit shaft)

- GO: Design for maximum material limit.
- NO-GO: Design for minimum material limit.

Limit gauges are used to check weather holes and shaft are coming within the acceptable range or not.

Gauges are designed according to Taylor principle which has two statements.

- Go gauges are designed at maximum material limit condition and NO GO gauges are design at minimum material limit condition.
- Go gauges are designed to check shape as well size. And has to be in the full form.

$$\text{Gauge Tool} = \frac{1}{10} \text{(Work tolerance)}$$

$$\text{Wear allowance} = \frac{1}{10} \text{(G.T)} = \frac{1}{100} \text{(Work tolerance)}$$

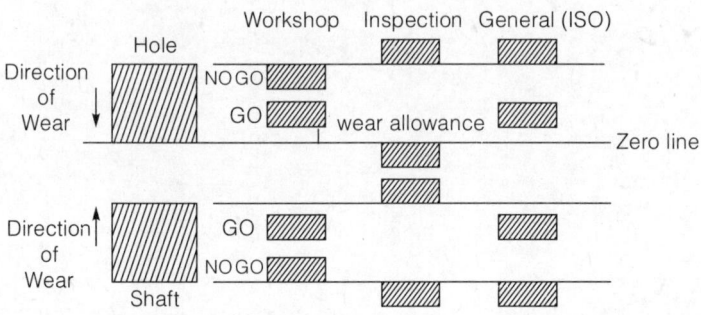

- In the work shop gauges, tolerances are towards work tolerance.
- In the inspection gauges, tolerance are away from the work tolerance.
- In ISO gauge, GO is taken from the work shop and NO-GO is taken from the inspection.

Remember: ....

- When nothing is mention problem design ISO gauge.
- In objective (Gate), maximum size of Go and NO GO should be given for plug gauges.

6. Tolerance Sink

When a design engineer present a design, one detail must be left blank without any tolerances. The tolerance of the sink is algebraic summations of all other tolerances. Only like minded tolerances are added.

7. Types of Error

- Instrument Error
  - **System Error:** Since no two products will match, hence error in the measurement follow a pattern. This error can be limited by calibrating the instrument.
  - **Short Period Error:** Short period error appears due to change in environment like unexpected increase in temperature and vibrations. This error can be taken care of neglecting the data.
  - **Erratic Error:** This error appears due to improper relation between different links that has to be avoided from metallurgical instruments.

■■■■

A Handbook on
Mechanical Engineering

11

Industrial Engineering

CONTENTS

◀ ☐ ▶

Break-Even Analysis

Break-Even analysis establishes the relationship among the factors affecting profit. It is a simple method of presenting to management the effect of changes in volume on profit.

1. Fixed Cost / Variable Cost / Mixed Cost

- **Fixed Cost**
  The cost which do not change for a *given period*, with *change in volume* of production e.g. rent, taxes, insurance etc.

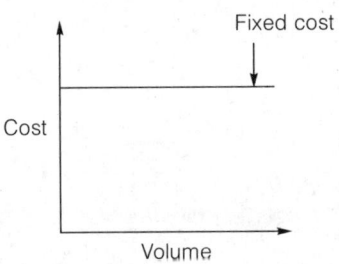

Remember: ....

- Fixed cost does not mean they never change. They are constant upto a specific volume or range of volume.

- **Variable cost**
  These cost vary directly with output volume.

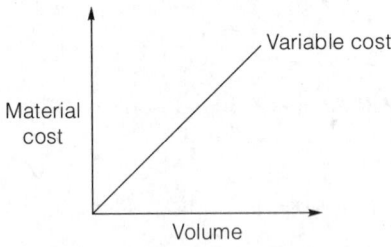

Remember: ....

- The ratio between *change in cost* and change in the level of *output* remain constant.

- **Mixed Cost**

This cost is a combination of semi-variable and semi-fixed cost.

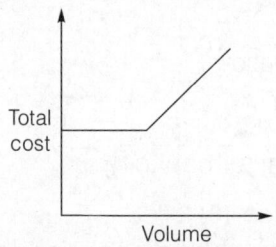

2. Assumption in Break-Even Analysis

- Selling price will remain constant.
- *Linear* relationship between sales volume and cost.
- Production and sales quantity are equal.
- No other factor, *except quantity*, will affect the cost.

3. Break-Even Analysis

Lets assume

    F = Fixed cost

    v = Variable cost per unit

    s = Sales price per unit

    P = Profit

    Q = Quantity produced and sold

    Q^* = Break-even quantity (No loss, no profit point)

$$\boxed{Q^* \cdot S = F + Q^* \cdot V}$$

∴ $$\boxed{Q^* = \frac{F}{S - V}}$$

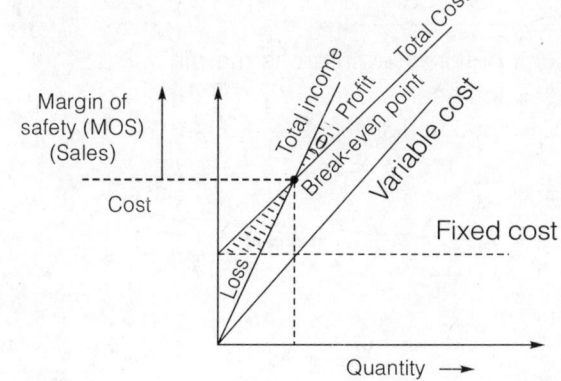

$(MOS)_{sales}$ = Sales income at *present capacity* – Sales income at break even point

$$= \frac{P}{(P/V)\,ratio}$$

$$(MOS)_{sales\%} = \frac{(MOS)_{sales}}{Sales\ at\ full\ capacity} \times 100$$

- **Angle of Incidence (θ)**

  It is the angle at which total sales lines cuts the total cost line.

- Large angle of incidence (q) indicate that profit are being made at *higher rate* and vice-versa.

4. Cost-Volume Profit Analysis / (P/V) Graph

It is used along with break-even graph. In this profit/loss is plotted on y-axis and quantity on x-axis.

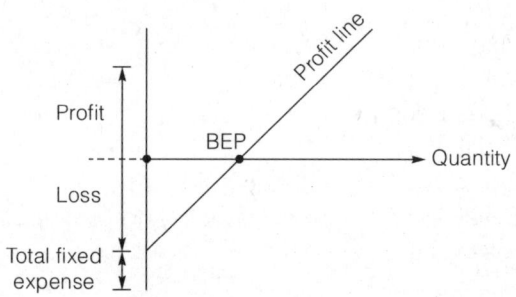

$$\left(\frac{P}{V}\right)_{ratio} = \frac{S-V}{S} = \frac{F+P}{S_{(total)}}$$

S – V is contribution margin/gross margin

Here,   S = Total sales

V = Total variable cost

■ ■ ■ ■

Inventory / Inventory Control

The item which are either stocked for sale or they are in process of manufacturing or they are in the form of material which are yet to be utilized.

1. Reasons to Keep Inventories

- To stabilize production
- To take advantage of price discount
- To meet demand during replenishment
- To prevent loss of order

2. Important Costs Associated With Inventories

- Capital cost / opportunity cost

  It is an amount of capital *not available* for other purchase. If this money is invested somewhere else, a *return on the investment* is expected.

- Ordering Cost / Replenishment Cost

  It is the cost of ordering to get an item into inventory. It is directly proportional with *number of order*.

- Inventory Carrying / Holding Cost

  This cost occur due to holding a given level of inventory and cost increase in direct proportion to the *amount of holding* and *period of holding* in stock.

- **Percentage holding** is sum of (holding) percentage handling, percentage locked-up capital interest percentage insurance, percentage pilferage etc.

- Shortage Cost

  When there are demand but item is not available in stock, the cost occur due to stock out is called shortage cost.

3. Economic Order Quantity (E.O.Q)

There are mainly two costs associated with inventory. Ordering cost and inventory carrying cost. Above two cost is opposite in nature so order quantity must be balanced quantity, to obtain *minimum total cost*. This quantity is known as economic order quantity.

- At EOQ, total cost is minimum.

- At EOQ, ordering cost = Holding cost

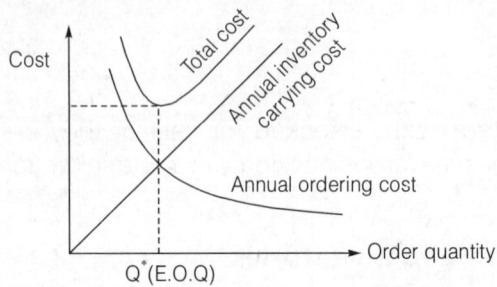

Two type of inventory review system
- Q-system also called 2-Bin system or fixed order quantity system.
- P-system also fixed period system.

Remember

4. Expression for Economic Order Quantity When Stock Replenishment is Instantaneous

- Assumption
  - (i) Demand is *known, deterministic* and *constant.*
  - (ii) Lead time is zero
  - (III) Usage rate is constant
  - (iv) Price of material is fixed
  - (v) Holding cost per unit item per unit time is *fixed*

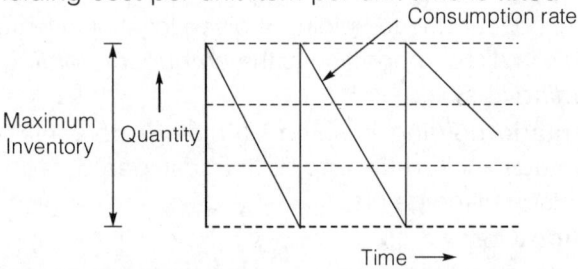

Let,

$$D = \text{Annual Demand}$$
$$O_c = \text{Ordering cost per order}$$
$$h_c = \text{Holding cost (Rs/Unit/Unit time)}$$
$$P = \text{Price per unit}$$
$$Q = \text{Order Quality}$$
$$Q^* = \text{E.O.Q}$$
$$N = \text{Number of order per annum}$$
$$T_c = \text{Total cost per annum}$$

- Annual ordering cost = $\dfrac{D.O_c}{Q}$

- Annual inventory carrying cost = $\dfrac{h_c \cdot Q}{2}$

- Annual total cost (T_c) = $\dfrac{D.O_c}{Q} + \dfrac{h_c.Q}{2}$

- E.O.Q = $Q^* = \sqrt{\dfrac{2D.O_c}{h_c}}$

- Optimum number of order = $N^* = \dfrac{\text{Demand}}{\text{E.O.Q}}$

- Optimum time interval between two order (T^*)

$$= \dfrac{\text{Number of working day in year}}{N^*}$$

- Minimum total yearly inventory cost

$$T_c^* = \dfrac{DO_c}{Q^*} + \dfrac{h_c.Q^*}{2} = \sqrt{2Dh_c.O_c}$$

- Model Sensitivity = $\dfrac{T_c}{T_c^*} = \dfrac{1}{2}\left[\dfrac{Q}{Q^*} + \dfrac{Q^*}{Q} \right]$

5. E.O.Q When Stock Replenishment is Non-Instantaneous

- Assumption
  - Item is consumed at constant rate and it is known.
  - Set-up cost is fixed and does not change with lot size.
  - Increase in inventory is *gradual*

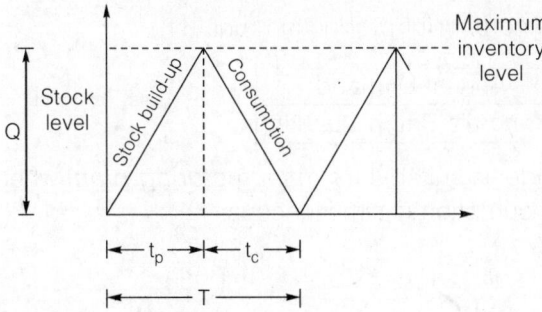

Let,   P = Production rate

S = Consumption rate

S_c = Set-up cost/set-up

Other symbol has usual meaning.

- Annual inventory carrying cost

$$= \frac{Q}{2}\left[1 - \frac{S}{P}\right]h_c$$

Q = Maximum inventory level

Remember: ....

- In production build-up model average inventory get reduced by **factor** $\left(1 - \frac{S}{P}\right)$. Thus reduces annual Inventory cost.

- Annual set-up cost = $\frac{D}{Q} \cdot S_c$

- Total annual cost (T_c) = $\frac{DS_c}{Q} + \frac{Q}{2}\left(1 - \frac{S}{P}\right)h_c$

- Economic production quantity (EPQ)

$$Q* = \sqrt{\frac{2DS_c}{\left(1 - \frac{S}{P}\right)h_c}}$$

- Optimum total cost = $T_c^* = \sqrt{2D.S_c \cdot h_c\left(1 - \frac{S}{P}\right)}$

- In production build-up model optimum total cost get reduced by $\left(\sqrt{1 - \frac{S}{P}}\right)$. Thus saves cost.

- Optimal number of production runs (N*)

$$N^* = \frac{\text{Annual Demand}}{\text{Economic Batch Quantity}}$$

- This model is suitable for *manufacturing plant* where production and consumption is simultaneous.

5. E.O.Q When Shortage are Permitted

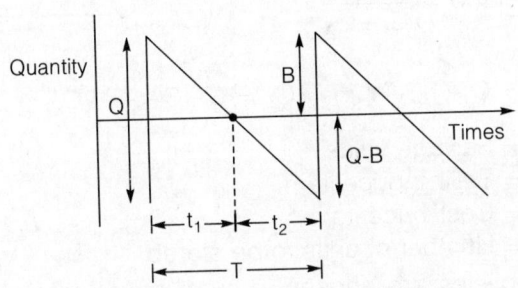

- **Assumption**
    - There is *no loss* of sales due to shortage.

    Here, S_c = Shortage cost per unit/unit time

    $Q-B$ = Number of shortage per order

    B = Balance unit after back order are satisfied

    t_1 = Time period during which inventory is positive

    t_2 = Time period during which shortage exist

    T = Time period between receipt of order

- E.O.Q (Q^*) = $\sqrt{\dfrac{2DO_c}{h_c} \cdot \dfrac{h_c + S_c}{S_c}}$

- Optimal remaining unit after back-order

$$B^* = \sqrt{\dfrac{2DO_c}{h_c} \cdot \dfrac{S_c}{h_c + S_c}}$$

- Optimum amount back ordered

$$Q^* - B^* = Q^* \left[1 - \dfrac{S_c}{h_c + S_c} \right]$$

- Total optimum inventory cost (T^*)

$$T^* = \sqrt{2DO_c \cdot h_c \cdot \left[\dfrac{S_c}{h_c + S_c} \right]}$$

Remember: ....

- Due to shortage, ordering cost *decreases* along with inventory carrying cost.

7. Probabilistic Model

Service level $f(x) = \dfrac{C_2}{C_1 + C_2}$

Here, $C_1 = C + h_c - V$; $C_2 = S - C - \dfrac{h_c}{2} + C_s$

Here,   V = Salvage value
S = Selling price/item
C = Cost price/item
x = Number of units to be stored

- In this policies are chosen in such a way that *expected* cost minimizes rather that actual cost.

Remember
- Some selective inventory control technique: ABC, VED, HML, SOS etc.
- ABC analysis is based on "**Pareto Law**".

■■■■

Forecasting

3

Forecasting is the art and science of predicting event. It includes forward projection of past data. Demand forecast is basically concerned with the estimation of demand.

1. Time Horizon in Forecasting

- Short term forecasting   – 1-3 months.
- Intermediate term forecasting   – 3-12 Months
- Long term forecasting   – More than 1 year

2. Classification of Forecasting Methods

- **Judgment Technique**
  This technique relies on the art of human judgement.
- **Time Series Method**
  It identifies the historical pattern of demand for the product and project.
- **Causal Methods**
  In this, we try to establish a cause and effect relationship between changes in the sales level of the product and set up relevant explanatory variable.

3. Judgemental Technique

- Opinion survey method
- Executive opinion method
- Customer and distributor surveys
- Marketing trials
- Marketing research
- Delphic technique

Remember: ....

- Judgemental technique are *qualitative/subjective* forecasting. It is generally used for *long term* forecasting.

4. Time Series Method

- **Past Average**
  In this method forecast is equal to *average sales* for previous period.

- **Moving Average Method**

    This method uses a past data and calculate a *rolling* average for a *constant period*. Fresh average is calculated at the end of each period by *adding* the demand of the *most recent period* and *deleting* the data of the *old period*.

    $$\text{Moving average} = \frac{\text{Sum of demands for given period}}{\text{Chosen number of period}}$$

- **Weighted Moving Average (WMA)**

    In Simple moving average equal weightage is given to demand of each period but in weighted moving average, *more weightage* is given to *recent demand* and vice-versa.

    Method to give weightage is done,

    $$\therefore \quad \text{Sum} = \frac{N(N+1)}{2}$$

    Weightage given to recent demand and so on.

    $$\frac{N}{\text{sum}}, \frac{N-1}{\text{sum}}, \frac{N-2}{\text{sum}} \dots \frac{1}{\text{sum}}$$

- **Exponential Smoothing Method**

    When the value of 'N' become *very large*, it become difficult to manage large data. So exponential smoothing method is used. It assigns weight to all the previous data and the pattern of weight assigned are of *exponential form*.

    It estimates average forecast for the next period by using the *actual* and the **forecasted demand** for *previous* period.

Remember

- In general, if smoothing constant (α) is NOT given we take

    $$\alpha = \frac{2}{n+1}$$ and equal to weightage assigned to most recent

    data $w_n = \frac{2n}{N(N+1)} = \frac{2}{N+1}$ by **WMA**.

    $$\therefore \quad \boxed{F_T = F_{T-1} + \alpha\left[D_{T-1} - F_{T-1}\right]}, \quad \text{where, } \alpha = \frac{2}{N+1}$$

    Here, F_T = Forecast for the period 'T'

    F_{T-1} = Forecasted demand for the last period

    α = Smoothing constant

    D_{T-1} = Demand for the last period

    N = Number of time period

Remember

- **General form** of 1st order simple exponential smoothing method:

$$F_t = \alpha D_{t-1} + \alpha(1 - \alpha)D_{t-2} + \alpha(1 - \alpha)^2 D_{t-3} + ...$$

- A simple exponential smoothing method is **1st order smoothing** method while trend adjusted is **2nd order exponential smoothing.**

- **Responsiveness and Stability**

  Responsiveness indicates that forecast as calculated have a *fluctuating and swinging* pattern where as *stability* means that the forecast show a labelled or *flat character.*

  As value of N increase the forecast become stable, and lower value of 'N' result in forecasts being more responsive.

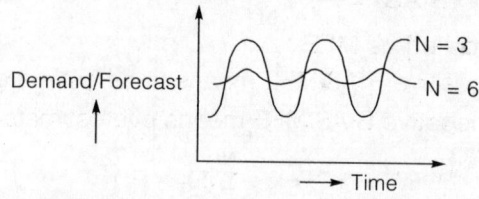

5. Causal Forecasting

It tries to identify the factors which causes the variation of demand and try to establish a relationship between the demand and there factors.

- **Linear Trend Regression**

  For a given demand value for 10 month.

Month (t)	Demand (Y)
1	124
2	132
3	128
4	121
5	170
6	132
7	140
8	137
9	172
10	180

$$Y = a + bt \qquad ...(i)$$
$$Yt = at + bt^2 \qquad ...(ii)$$

$$\Sigma Y = na + b\Sigma t \qquad \text{...(iii)}$$
$$\Sigma Yt = a\Sigma t + b\Sigma t^2 \qquad \text{...(iv)}$$

Solving equation (iii) and (iv), we have

$$a = \frac{\Sigma Y}{n} - \frac{b\Sigma t}{n}$$

$$b = \frac{n\Sigma tY - \Sigma t \cdot \Sigma Y}{n\Sigma t^2 - (\Sigma t)^2}$$

Now by putting value of 'a' and 'b' in equation (i) and (ii).
We can find demand for next month.

6. Error Analysis

- Mean error **(BIAS)** $= \dfrac{\displaystyle\sum_{t=1}^{N}(D_T - F_T)}{N}$

- Bias also called as **MFE**
    - A positive BIAS/MFE means underestimated forecast.
    - A negative BIAS/MFE means overestimated forecast.

- Mean square error **(MSE)** $= \dfrac{\displaystyle\sum_{t=1}^{N}(D_T - F_T)^2}{N}$

- Mean absolute deviation **(MAD)** $= \dfrac{\displaystyle\sum_{t=1}^{N}|D_T - F_T|}{N}$

- Mean absolute percentage error **(MAPE)**

$$MAPE = \frac{\left[\dfrac{\displaystyle\sum_{T=1}^{N}|D_T - F_T|}{D_T}\right]}{N}$$

- Tracking signal **(TS)**

$$TS = \frac{\displaystyle\sum_{T=1}^{N}(D_T - F_T)}{MAD}$$

- If value of TS goes beyond $3\sqrt{MSE}$ then it indicates that model needs to be revised.

Remember

■■■■

Queueing Theory

4

Queueing theory is used in service oriented organization, machine repair shop, in case of semifinished products waiting for finishing operation. The main objective is to maintain the capability for service at levels sufficient to keep the waiting time to certain acceptable average levels.

1. Characteristic of Queueing Model

- **Arrival Pattern of Clients**
  In most of the cases random arrival is observed which is best described by Poisson's process. If the arrivals are governed by Poisson's process then the time between arrivals is **exponentially** distributed.

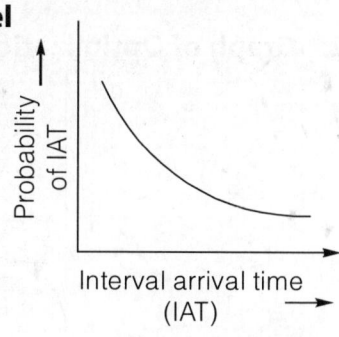

- **Service Discipline**
  It tells about the ways in which the customers are being served in the system.
  - First come first serve
  - Last in, first out
  - Service in random order
- **Number of Servers and Service Channels**
  - Single channel single server
  - Single channel multiple server
  - Multiple channel, single server
  - Multiple channel, multiple server
- **Service Pattern**
  There is no standard probability distribution for service process. Infect in many cases actual data is used for describing the service time. However if the service times are exponentially distributed then the model becomes simple.
- **Maximum number of people allowed in the system**. In most of the cases the capacity of the system is limited.

  Calling population: The entire sample of customer from which a few visit the service facility is known as calling population.

Remember: ....

- The size of calling population can be *finite* or *infinite*.
- Customer Behavior
  - **Balking:** If queue is very long, customer decides not to enter the queue.
  - **Reneging:** Customer leave the *queue without* getting the service.
  - **Jockeying:** When customer switches the queue, called jockeying.

2. Graph of Optimization of Cost

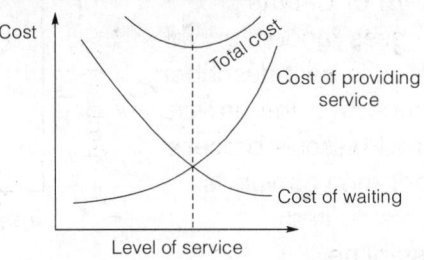

3. Representation of Queueing Model

Kendall Lee Notation:

The general form of notation is

$$(a/b/c) : (d/e/f)$$

Here,   a = Probability distribution of arrival time

b = Probability distribution of service time.

c = Number of servers.

d = Service rule or queue discipline.

e = Maximum number of customers allowed in the system

f = Size of calling population.

4. Queueing Model 1 (M/M/1) : (FIFO/∞/∞)

M = Arrival pattern with Poisson distribution.

M = Service rate is exponentially distributed.

1 = Number of server.

FIFO = Service rule

∞ = Customer number allowed ∞ → size of callying population.

In this, **inter arrival times** are **Poisons distributed**. Service times are **exponentially distributed**.

- **Assumption (λ < μ)**

  When mean arrival rate (λ) is greater than mean service rate, there will be never ending queue and it leads to queue explosion.

- Inter arrival time $= \dfrac{1}{\lambda}$

- Traffic intensity factor or utilization factor or channel efficiency (ρ)

$$\rho = \frac{\lambda}{\mu}$$

- Average number of customer (length) of queue (Lq)

$$Lq = \frac{\lambda}{\mu - \lambda} - \rho$$

- Average number of customer in system (L_s)

$$L_s = \frac{\lambda}{\mu - \lambda}$$

- Waiting time in system (W_s)

$$W_s = \frac{1}{\mu - \lambda}$$

- Average or expected waiting time in queue (W_q)

$$W_s = W_q + \frac{1}{\mu}$$

$$W_q = \frac{1}{\mu - \lambda} - \frac{1}{\mu}$$

- Probability that the service facility is idle $P_0(t)$

$$P_0(t) = 1 - \rho$$

- Probability of 'n' customer in the system at time 't' $P_n(t)$

$$P_n(t) = \rho^n \times P_0(t)$$

- Average length of non-empty queue (L_n)

$$L_n = \frac{1}{1 - \rho}$$

- Probability of 'n' arrival in time 't' P(n, t)

$$P(n, t) = \frac{e^{-\lambda t}(\lambda t)^n}{n!}$$

- Probability that the waiting time in the **queue** is greater than or equal to 't'

$$P(W_q \geq t) = \frac{\lambda}{\mu} e^{-(\mu - \lambda)t}$$

- Probability that waiting time in *system* is greater than or equal to 't'

$$P(W_s \geq t) = e^{-(\mu - \lambda)t}$$

- Probability that waiting time in the system is less than or equal to 't'

$$P(W_s \leq t) = 1 - \rho^{-(\mu - \lambda)t}$$

- In queuing model the assumption is that, if **random variable** is independently and identically distributed random variable then only we can assume it as **Poisson distribution**.

5. Quencing model 2 (M/G/1) : (∞/FIFO)

Here,   G = General service time

       σ = Standard deviation for the service time

       σ^2 = Variance

- $L_q = \dfrac{\lambda^2 \sigma^2 + \rho^2}{2(1 - \rho)}$

- $L_s = L_q + \rho$

- $W_q = \dfrac{L_q}{\lambda}$

- $W_s = W_q + \dfrac{1}{\mu}$

All symbols has usual meaning.

6. Queueing Model 3 (M/D/1) : (∞/FIFO)

Here,

D - Deterministic or constant service times i.e. standard deviation for service time is zero ($\sigma = 0$).

- $L_q = \dfrac{\rho^2}{2(1 - \rho)}$

- $L_s = L_q + \rho$

- $W_q = \dfrac{L_q}{\lambda}$

- $W_s = W_q + \dfrac{1}{\mu}$

■■■■

Line Balancing

5

Line balancing is associated with a product layout in which product are processed as they pass through a line of work sentence.

1. Objective of Line Balancing

The objective of line balancing is to combine various operations on different work stations in such a way that the processing time at each work station is *almost same* and *equal to cycle time*.

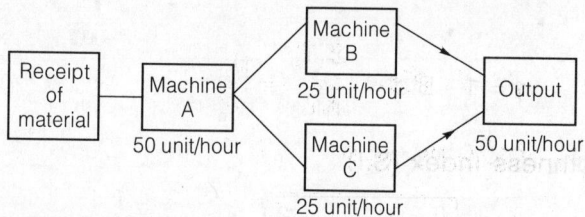

2. Advantage of Line Balancing

- Uniform rate of production
- Less work-in-process
- Effective use of facility
- Easy production control

3. Some Important Term of Line Balancing

- **Work Station**
  It is a location where given amount of work is performed.
- **Cycle Time (T_C)**
  It is the time between two successive products coming out of the production line.
- **Total Work Content (T_{wc})**
  This is equal to sum of processing time of each operation.
- **Station Time (T_{si})**
  It is equal to the sum of processing time of operation on a particular work station.

$$\Sigma T_{si} = T_{wc}$$

- **Balance delay (B.d.)**

$$B.d. = \left(\frac{nT_c - T_{wc}}{nT_c} \right)$$

- **Balance Delay (d)**
  This is the measure of inefficiency of the line

$$d = \left(\frac{nT_c - T_{wc}}{nT_c} \right) \times 100$$

  Here,

  n = Number of work stations

- **Line Efficiency (η)**

$$\boxed{\eta = 1 - B.d = \frac{TWC}{nT_c}}$$

$$\eta = 1 - B.d = \frac{\sum\limits_{i=1}^{n} T_{si}}{nT_c} = \frac{T_{wc}}{nT_c}$$

- **Smoothness Index (S.I)**

$$S.I. = \sqrt{\sum\limits_{i=1}^{n} \left(T_{si_{max}} - T_{si} \right)^2}$$

Remember: ....

- If smoothness index is low, it means line balancing is good.
- **Theoretically Minimum Number of Work Stations (n_{min})**

$$n_{min} = \frac{T_{wc}}{T_c}$$

- **Actual Production Rate**

$$R_{p(actual)} = \frac{1}{T_{c(actual)}}$$

4. Largest Candidate Rule For Line Balancing

- **Step-I**
  List all the elements in decreasing order of elemental time values (T_e).
- **Step-II**
  To assign the element to the first work station, start at the top of the list selecting the first feasible element placement at the station. A feasible element is one that satisfies precedence requirement and does not cause the sum of elemental time (T_e) values at the station to exceed cycle time (T_c).

- **Step-III**
  Continue the process of assigning the work elements to the station as in step-II until further elements can be added without exceeding cycle time T_c.
- **Step-IV**
  Repeat step-II and step-III for other station, until all the elements are assigned.

■ ■ ■ ■

Pert and CPM

6

PERT and CPM are used for planning and scheduling large projects in the fields of construction, maintenance, fabrication and any other area. Pert and CPM is used to minimize the bottlenecks, delays and interruptions by determining the critical factors and co-ordanating various activities.

1. Basic Concept in Network

- **Network**
  It consist of series of activities arranged in a logical sequence and shows the inter relationship between the activities.
- **Activity**
  It is a physically identifiable part of the project, which consume time and resources. It is represented by an arrow ($\rightarrow$).
- **Event**
  It represents start or completion of an activity.

Merge Event   Burst Event

- ## Predecessor and Successor Activities
  All activities, which must be completed before starting the activity under consideration are called its predecessor activities. All activity which have to follow the activity under consideration are called its successor activities.

- ## Dummy Activity
  This activity **does not** consume time or resources but depicts dependency or relationship with other activity. It is used to maintain the **logical sequence**. It is indicated by **dotted** line.

CPM

2. Critical Path Method

In this **activity times** are known with **certainty**. For each activity earliest start time (EST) and latest start times (LST) are computed.
The length of critical path determines the **minimum time** in which the entire project can be completed. The activities on the critical path are called **"critical activities"**.

3. Objective of Time Analysis in Network

- Determine the completion time for the project.
- *Earliest time* when each activity can start.
- *Latest time* when each activity can start *without* delaying the total project.
- To find *float* for each activity.
- Identification of *critical activities* and *critical path*.

4. Forward Pass Computation/Earliest Start Time

This computation gives the *earliest expected* start and finish time for each activity. In this, computation starts from initial node to move to end node.

- **Earliest Start Time (EST)**
  It is the earliest event time of *tail end event*.
- **Earliest Finish Time (EFT)**
  It is the sum of earliest start time and activity time.

  EFT = EST + Activity time
- **Earliest Event Time (EVT)**
  It is the *maximum* of the earliest finish times of all activities ending into that event.

5. Backward Pass Computation/Latest Allowable Time

The latest event times, specifies the time by which all the activities entering into that event must be completed without delaying the total project.

- **Latest Finish Time (LFT)**
  It is that time for an activity equals the latest event time of event.
- **Latest Start Time (LST)**
  Latest completion time-activity time.
- **Latest Event Time (LET)**
  It is the minimum of the latest start time of all activities originating from that event.
- **Slack**
  Latest event time - Earliest event times.
- **Total Float**
  It is the time by which completion of an activity *can be delayed* beyond *earliest expected completion time* without affecting *overall* project duration time.
  Latest start time - Earliest start time.

$$\boxed{\text{Total float} = L_j - E_i - T_{ij}}$$

- **Free Float**

  It is the time by which the completion of an activity can be delayed beyond the earliest finish time without affecting the earliest start of *a subsequent activity*.

$$\boxed{\text{Free float(F.F)} = \text{Total float} - (\text{Head event slack})}$$

- **Independent Float**

  It is the time by which the start of an activity can be delayed without affecting earliest start time of any immediately following activity assuming that the preceding activity has finished at its latest finish time.

$$\boxed{\text{Independent float} = \text{Free float} - (\text{Tail event slack})}$$

Remember: ....

- The events with *zero* slack time are known as critical event.

PERT (PROGRAM EVALUATION AND REVIEW TECHNIQUE)

It is used when the activity times are *not known* with certainty for e.g. In research and development.

6. Time Estimates Associated With Pert

- PERT makes use of three estimates of time
  - **Optimistic time (t_o)**

    It is the shortest possible time, if everything goes perfectly without any complications. It is an estimate of *minimum* possible time to complete the activity under *ideal condition*.
  - **Most likely time (t_m)**

    It is the best estimate of the activity time. This lies between the optimistic and pessimistic time estimates.
  - **Pessimistic time (t_p)**

    It is the *longest* time taking into consideration all odds. This is the time estimate if everything goes wrong.

    The three time estimates t_o, t_p and t_m are combined to develop expected time (t_e) for an activity.

Remember: ....

- The fundamental assumption in PERT is that the three estimates of time follow b (Beta) distribution.

7. Expected Time (t_e) / Standard Deviation (σ)

The expected time (t_e) is given by

$$t_e = \frac{t_o + 4t_m + t_p}{6}$$

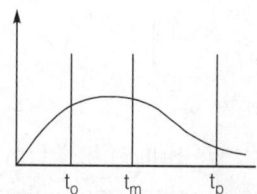

The standard deviation of the time required to complete each activity

Standard deviation (σ) = $\dfrac{t_p - t_o}{6}$

Standard deviation of the time to complete the project

$$\sigma_{C.P.}^2 = \left(\frac{t_{p1} - t_{o1}}{6}\right)^2 + \left(\frac{t_{p2} - t_{o2}}{6}\right)^2 + \cdots + \left(\frac{t_{pn} - t_{on}}{6}\right)^2$$

$$\sigma_{C.P.} = \sqrt{\sigma_1^2 + \sigma_2^2 + \sigma_3^2 + \cdots + \sigma_n^2}$$

$$\sigma_{C.P.}^2 = \left(\frac{t_{p1} - t_{01}}{6}\right)^2 + \left(\frac{t_{p2} - t_{02}}{6}\right)^2 + \cdots + \left(\frac{t_{pn} - t_{0n}}{6}\right)^2$$

Remember

- Expected time of activity (t_e) is assumed to be β-distributed.
- Expected time of project (T_e) is assumed to be normally distributed.

8. Probability of Completion of The Project Within a Scheduled Time

The probability of completion of the project within scheduled time is computed as:

- Calculate the mean of the event time (t_e) by adding the times of the activities along the critical path leading to the event.
- Calculate the variance of the event time by adding up the variances of the activities on the critical path. Take the square root of this variance to get σ (Standard deviation).
- Compute standard normal variate,

$$z = \frac{T_s - T_e}{\sigma}$$

Probability that the project completion time (T) is less than any given value 't' is given by

$$\text{Prob}(T < t) = \text{Prob}\left(z < \frac{t - \overline{T}}{\sigma_{cp}}\right)$$

Here,

z = Normal distribution function

$\overline{T}$ = Expected project completion time

σ_{cp} = Standard deviation of the critical path

9. Crashing

It is assumed that for each activity, there is an activity duration for which the ***direct cost is minimum***. If activity results in more than this time, more resources and, hence more funds are required. For instance a point will be reached beyond which no further reduction in time will be possible irrespective of resources spent.

$$\text{Cost slope (CS)} = \frac{\text{Crash cost} - \text{Normal cost}}{\text{Normal time} - \text{Crash time}}$$

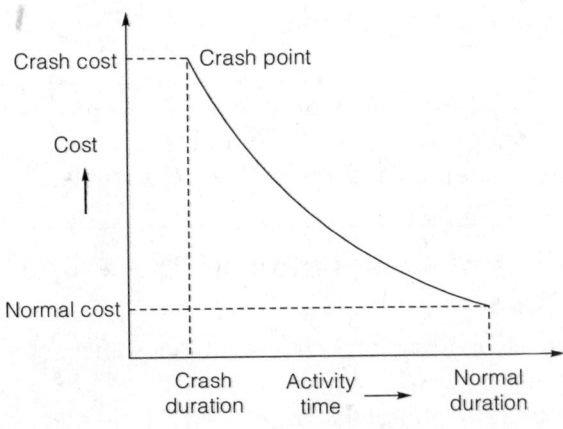

Remember

- It implies reduction of the duration of the entire project.
- Only critical activities are selected for the purpose of crashing.
- Among critical activities, that critical activity is selected for crashing which has ***least cost slope***.

10. Comparison Between CPM and Pert

CPM	PERT
1. It is activity oriented.	It is event oriented.
2. It is used when the activity times are deterministic.	It uses a probabilistic times.
3. One time estimate.	Three time estimates (a) Optimistic (b) Most likely (c) Pessimistic.
4. It directly introduces cost concept analysis.	PERT indirectly currents for costs.
5. It is a planning device.	It is a control device.

■ ■ ■ ■

Quality Control and Analysis

Quality is defined as customers perception about the degree to which a product or a service meets his expectations.

1. Types of Quality

- **Quality of Design**
  It is concerned with the tightness of specification for manufacturing any product.
- **Quality of Performance**
  It is concerned with how well a product gives its performance. It depends upon quality of *design* and quality of *conformance*.

2. Parameters Governing Quality

- Performance
- Range and type of feature
- Reliability and durability
- Maintainability and serviceability
- Sensory characteristics.

3. Quality Cost / Total Quality Management

- **Quality Cost**
  - Failure cost (internal, external)
  - Appraisal cost
  - Prevention cost
- **Effect of Total Quality Management**
  - Total quality cost is reduced
  - Proportion of different cost will change

4. Statistical Quality Control (SQC)

It is defined as the quality control system where statistical techniques are used to control, improve and maintain quality.

Quantitative aspects of quality management

Statistical quality control (Acceptance sampling)   Statistical process (SPC) Control (Process Control Charts)

Remember: ....

- SPC minimizes the defective items produced thus it is a preventive measure.

5. Process Control Chart

It is graphical representation of quality characteristics and indicate whether the process is under control or not. These charts are based on the fact that variability exists in all the process.

- **Variation Due to Chance Cause**

  It conform to *normal distribution* curve and they are not of very high magnitude. They are also known as random variations and can not be eliminated completely.

- **Variation Due to Assignable Cause**

  It do not conform to any standard distribution. It can be found out and can be completely eliminated.

  If in a process *assignable causes* are absent then we say that the process is under control.

6. Control Charts and Their Types

It gives a warning whether there is any variation from the desired level and shows the presence or absence of assignable cause of quality variation.

- **Control Chart for Variable**

  - $\bar{X}$ **Chart:** It shows the centering of the process and shows the variation in the *averages* of individual samples.

  - **R Chart:** It show the variation in the range of the *sample*.

  - **Control Unit:** For plotting control charts *generally* ±3σ limits are selected. Therefore such control charts are known as 3σ control charts.

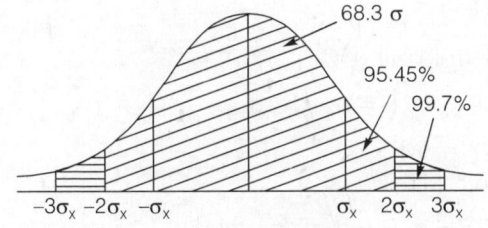

Standard Deviation

- **Calculation Procedure For $\bar{x}$ and R Charts**

  - Calculate the $\bar{x}$ and Range for Each Sample

  - Calculate the grand average $(\bar{\bar{x}})$ and average range $(\bar{R})$.

  Let sample size (n) = 5

S. No.	1 2 3 4 5	$\overline{x}$	R(Maximum-Minimum)
1		$\overline{x}_1$	R_1
2		$\overline{x}_2$	R_2
⋮		⋮	⋮
N		$\overline{x}_N$	R_N

- **For x-Chart**

  Centre line (control limit) $\overline{\overline{x}} = \dfrac{\sum\limits_{i=1}^{N} \overline{x}_i}{N}$

  Calculation of 3σ limits for control charts.
  Upper control limit (UCL)

  $$UCL = \overline{\overline{x}} + 3\sigma_x$$

  Lower control limit (LCL)

  $$LCL = \overline{\overline{x}} - 3\sigma_x$$

  Here,　$\sigma_x = \dfrac{\sigma}{\sqrt{n}}$

  σ_x = Standard mean of sample mean
  σ = Process standard deviation
  n = Sample size

- **For R-Chart**

  $$\overline{R} = \dfrac{\sum\limits_{i=1}^{N} R_i}{N}$$

  Upper control limit (UCL)

  $$UCL = d_4 \overline{R} = d_4 \cdot d_2 \cdot \sigma$$

  $$\overline{R} = \sigma d_2$$

  Lower control limit (LCL)

  $$LCL = d_3 \overline{R} = d_3\, d_2\, \sigma$$

  Here,　$\sigma = \dfrac{\overline{R}}{d_2}$

  If $n < 7$ then $d_3 = 0$.

Remember: ....

- The value of d3 and d4 depends on the sample size.

9. Control Charts for Attribute

These are made for those quality characteristics which can not be measured and can only be classified as good or bad.

- **P-Chart**

  It is also known as fraction defective chart. This is made for the situation, where the sample size is varying.

Sample No.	Sample Size (n)	No. of Defectives	Fraction Defective $p = \dfrac{d}{n}$
1	n_1	d_1	$p_1 = \dfrac{d_1}{n_1}$
2	n_2	d_2	$p_2 = \dfrac{d_2}{n_2}$
3	n_3	d_3	$p_3 = \dfrac{d_3}{n_3}$
$\vdots$	$\vdots$	$\vdots$	$\vdots$
N	n_N	d_N	$p_N = \dfrac{d_N}{n_N}$

Centre line (CL) $= \bar{p} = \sum\limits_{i=1}^{N} p_i$

Average sample size $(\bar{n}) = \dfrac{\sum\limits_{i=1}^{N} n_i}{N}$

- **Upper control limit**

$$UCL = \bar{P} + 3\sqrt{\dfrac{\bar{P}(1-\bar{P})}{\bar{n}}}$$

- **Lower control limit**

$$LCL = \bar{P} - 3\sqrt{\dfrac{\bar{p}(1-\bar{p})}{\bar{n}}}$$

$$\sigma_p = \sqrt{\dfrac{\bar{P}(1-\bar{P})}{\bar{n}}}$$

- **Chart 2-np chart**

  This is known as number of defective chart and is made for the cases where the sample size 'n' is constant.

Sample number	Sample size (n)	No. of defective (d)	$p = \dfrac{d}{n}$
1	n	d_1	$p_1 = \dfrac{d_1}{n}$
2	n	d_2	$p_2 = \dfrac{d_2}{n}$
3	n	d_3	$p_3 = \dfrac{d_3}{n}$
$\vdots$	$\vdots$	$\vdots$	$\vdots$
N	n	d_N	$p_N = \dfrac{d_N}{n}$

$$\bar{p} = \frac{\sum\limits_{i=1}^{N} pi}{N}, \ CL = n\bar{p}$$

$$UCL = n\bar{p} + 3\sqrt{n\bar{p}(1-\bar{p})}$$

$$LCL = n\bar{p} - 3\sqrt{n\bar{p}(1-\bar{p})}$$

- **Chart 3 : C chart (count of defect chart)**

  C chart is made for number of defects which are present in a sample and is made for the situation where the sample size 'n' is constant, 'n' can be equal to 1 or more than one.

Sample number	Sample size (n)	No. of defective (d)	$C = \dfrac{d}{n}$
1	n_1	d_1	$C_1 = \dfrac{d_1}{n}$
2	n_2	d_2	$C_2 = \dfrac{d_2}{n}$
3	n_3	d_3	$C_3 = \dfrac{d_3}{n}$
$\vdots$	$\vdots$	$\vdots$	$\vdots$
N	n_N	d_N	$C_N = \dfrac{d_N}{n}$

$$C_L = \frac{\sum\limits_{i=1}^{N} C_i}{N}$$

$$UCL = \bar{C} + 3\sqrt{\bar{C}}$$

$$LCL = \bar{C} - 3\sqrt{\bar{C}}$$

Number of defects when exceed a particular limit, they make an item defective. For the former case we make c-chart and for the later case we make np chart.

- **Chart 4 : U-chart**

  This chart is used for the situations where sample size is varying.

Sample number	Sample size (n)	No. of defective (d)	$U = \dfrac{d}{n}$
1	n_1	d_1	$U_1 = \dfrac{d_1}{n}$
2	n_2	d_2	$U_2 = \dfrac{d_2}{n}$
3	n_3	d_3	$U_3 = \dfrac{d_3}{n}$
$\vdots$	$\vdots$	$\vdots$	$\vdots$
N	n_N	d_N	$U_N = \dfrac{d_N}{n}$

- $$\bar{U} = \frac{\displaystyle\sum_{i=1}^{N} U_i}{N}$$

- Average sample size

$$\bar{n} = \frac{\displaystyle\sum_{i=1}^{N} n_i}{N}$$

- $$UCL = \bar{U} + 3\sqrt{\frac{\bar{U}}{\bar{n}}}$$

- $$LCL = \bar{U} - 3\sqrt{\frac{\bar{U}}{\bar{n}}}$$

9. Acceptance Sampling

It is the process of evaluating the quality of a large number of items in a batch or a lot based upon the quality of a small sample of items.

- **Types of sampling plan**
  - Single sampling plan

(N,   n,   C)

Batch   Size of   Acceptance
size   Sample   number

  - Double sampling plan

$$\begin{bmatrix} N & & ; & n_2 \\ n_1 & C_1 & & \\ & C_2 & & C_3 \end{bmatrix}$$

Here, N = Batch size

n_1 = Size of sample

c_1, c_2 = Rejection number

  - Sequential sample plan

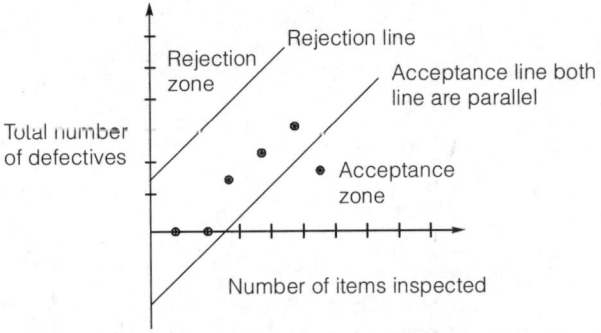

10. Operating Characteristic Curve

It is a tool that is used for the purpose of understanding how well a given sampling plan is effective in differentiating between a good and a bad lot. With every OC curve following four terms are associated

- **Acceptance Quality Level (AQL)**

It represents a good quality level of the lot submitted by the producer for inspection and should be accepted most of the times by the sampling plan.

- **Lot Tolerance Percent Defective (LTPD)**

It indicates a bad quality lot submitted by the producer and such a lot should be rejected most of the times by the sampling plan.

- **Producers Risk (α)**

  It represents a probability of rejecting a good lot having percent defective equal to AQL.

- **Consumers Risk (β)**

  It indicates the probability of accepting a bad lot having percent defective equal to LTPD.

  Ideal OC curve for single sampling

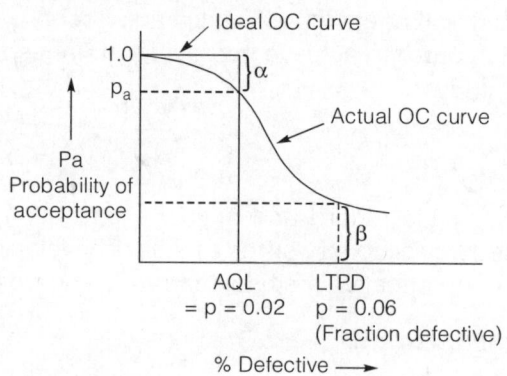

$$p_a = \frac{e^{-np}(np)^x}{x!}$$

P = Fraction defective

x = number of defective item

n = Sample size

11. Average Outgoing Quality

For a lot with p fraction defective if the probability of acceptance under a given sampling plan is P_a then the expected fraction defective entering the customer production system is $p \times p_a$. This quantity is known as AOQ. Taking AOQ over a wide range of fraction defective value we get a curve, the maxima of which is known as AOQ limit or AOQL.

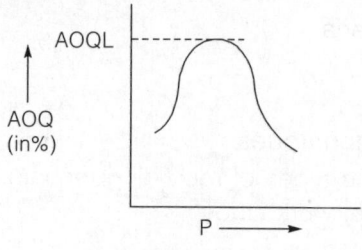

■ ■ ■ ■

Work Study

8

Work study is a generic term for those techniques particularly method study and work measurement which are used in the examinat' . of human work in all its context and which lead systematically to the .nvestigation of all the factors which affect the efficiency and economy of the situation being reviewed in order to achieve improvement. It has two types

- Method study
- Work measurement

1. Method Study

It is the systematic record and critical examination of the existing and proposed way of doing work as a mean of developing and applying easier and more effective methods and reducing cost.

2. Work Measurement

It is the application of techniques designed to establish a standard time for a qualified worker to carry out a specified job at a defined level of performance. Work measurement is thus concerned with calculation of standard time.

3. Steps for Performing Method Study

- S → Select
- R → Record
- E → Examine
- D → Develop
- I → Install

4. Criteria for Selecting a Method for Carrying out Method Study

(i) Economic criteria

(ii) Technical

(iii) Human

- **Recording Techniques**

  There are three types of recording techniques that are available for carrying out work study.
  - Charts
  - Diagrams
  - Photographic films

5. Symbols

Operation	◯
Transportation	⇨
Inspection	▭
Delay	⬭
Storage	▽

6. Charts

- **Operation Process Chart**

  This is also known as outline process chart. This chart gives an overall view of the process and records only principle operation and inspection.

- **Flow Process Chart**

  It is a type of process chart which helps us to go into a greater detail once a general picture of a process has been established. Flow process charts are of three types:
  - Man type chart
  - Material type
  - Equipment type

- **Multiple Activity Chart**

  This chart records the activity of man and/or machine on a common time scale. It is used where it is necessary to steady on one chart the activities of one subject in relation to other.

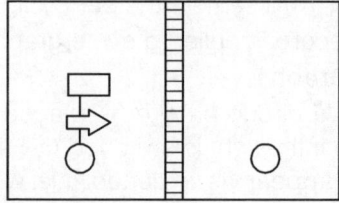

- **Left Hand Right Hand Chart**

  It is also known as two handed process chart and is a form of operation process chart which records the work of operator's two hands simultaneously.

- **SIMO Chart (Simultaneous Motion Chart)**

  It is used to record therbligs (18 nos.) against the time scale which are performed by the operator or operators at work. SIMO chart is the micro motion form of operator process chart. Because SIMO charts are used primarily for operation of short duration often performed with extreme quickness, it becomes necessary to combine then with films. In this chart movements are recorded against time measured in wink and 1 *wink* = $\dfrac{1}{2000}$ the minute.

- **Travel Chart**

  This chart records the movement of material and/or man between different departments.

7. Diagrams

- **Flow Diagram**

  It show the path of man, material and/or components on a scale model of the factory. This is used to supplement the flow process chart.

- **String Diagram**

  It is a scale diagram on which movement in a given area and over a given period of time is plotted by means of a continuous thread. This is done with the objective of sharing the frequency of the movement between various points and for determing the distance covered.

8. Photographic Films

- **Cycle Graph**

  These graphs have been developed for studying the motion path of an operator.   It   is   possible to record the path of motion of an operator by attaching a small electric bulb to the finger or hand of the operator and photograph with a still camera the path of light as it moves through space such a record is called a cycle graph.

- **Chronocycle Graph**

  If an interrupter/a circuit breaker is placed in the electric circuit with the bulb an if the light is switched ON and OFF then the path o the bulb will appear as a dotted line with pear shaped dots indicating the direction of motion. The spots of the light will be spaced according to the speed of movement. They will be closed if operator moves slowly and will be wide apart when the operator moves fast.

9. Principle of Motion Economy

Motion economy is best achieved by work space design. There are three principles governing motion economy.
- Use of human body
- Arrangement of work place
- Design of tools and equipments

10. Steps for Performing Work Measurement

- S → Select
- R → Record
- E → Examine
- M → Measure
- D → Define

11. Techniques for Carrying out Work Measurement

There are three techniques for carrying out work measurement.
(i) Stop watch time study method
(ii) Predetermined motion time standard (PMTS)
(iii) Work sampling or ratio delay

(i) and (ii) techniques are used for short cycle or repetitive operations.
(iii) for long cycles.

- **Stop Watch Time Study Method**
  - Repeat time or snap back
  - Continuous

  Number of cycles or observations to be taken while conducting stop watch time study is given by

  $$N = \left[\frac{K\sqrt{N'(\Sigma x^2) - (\Sigma x)^2}}{\Sigma x} \right]^2$$

Where

N = Total number of cycles

N' = Number of observations already taken

X = Individual observation

K = Constant

     = 40, 95% confidence level, ±5% accuracy

     = 20, 95% confidence level, ± 10% accuracy

Normal time = Elemental time × Rating factor

Standard time = Normal time + Allowances

- Types of Allowances
  - Personal allowances
  - Delay allowances
  - Policy allowances
  - Contingency allowances
  - Fatigues allowances
- Predetermined Motion Time Standard

  Standard time for an operation can also be found out by different system of motion time data. This motion time data has already been developed for various activities on the basis of type of motion involved in those activities some of the most popular of motion time data are
  - Motion time analysis (MTA)
  - Motion time data for assembly work
  - Work factor system
  - Basic motion time study
  - Pre determined human work time
  - Methods time measurement (MTM) → Most popular values in MTM are given in time measurement unit (TMU).

    $$1 \text{ TMU} = 10^{-5} \text{ hr}$$

- Work Sampling or Ratio Delay

  Work sampling is used for long cycle jobs one it has following uses:
  - **Ratio delay:** This delays with determining the percentage of the day the man is working and the percentage time he is idle.
  - **Performance Sampling:** It is concerned with establishing a performance index for the person.
  - To determine standard time

    $$N = \frac{z^2 p(1-p)}{e^2}$$

    N = Number of observation

    p = Proportion of activity

    e = Absolute error

    z = 1.96 for 95% confidence

       = 2.33 for 99% confidence

■■■■

Material Requirement Planning (MRP)

9

Materials requirement planning (MRP) is a technique for determining the quantity and timing for the acquisition of dependent demand items needed to satisfy master production schedule requirement.

MRP is one of the powerful tools that, when applied properly, helps the managers in achieving effective manufacturing control.

1. MRP Objective

- Inventory reduction
- Reduction in manufacturing time
- Reduction in delivery lead time
- Realistic delivery commitments
- Increased efficiency

2. MRP Terminology

- **MRP** : A technique for determining the quantity and timing of dependent demand items.
- **Lot size** : The quantity of items required for an order.
- **Time phasing** : Scheduling to produce or receive an appropriate amount (Lot) of material so that it will be available in the time periods when required.
- **Time bucket** : The time period used for planning purposes in MRP.
- **Gross requirements** : The overall quantity of an item needed at the end of the period to meet the planned output levels.
- **Net requirements** : The net quantity of an item that must be acquired to meet the schedule output for the period.

  Net requirement = Gross requirement – Inventory on hand – Scheduled receipt

- **Requirements explosion** : The breaking down of parent items into component parts that can be individually planned and scheduled.
- **Scheduled receipts** : The quantity of an item that will be received from suppliers as a result of orders that have been placed.
- **Planned order receipts** : The quantity of an item that is planned to be ordered so that it will received at the beginning of the period to meet net requirements for the period. The order has **not yet** been placed.
- **Planned order release** : The quantity of an item that is planned to be ordered or it is plan (quantity and date) to initiate the purchase

or manufacture of materials so that they will be received on schedule after and lead time offset.

- **Lead time offset** : The supply time or number of time buckets between releasing an order and receiving the materials.

3. Master Production Schedule (MPS)

MPS is a series of time phased quantities for each item that a company produces, indicating how many are to be produced and when. The MRP system accepts whatever the master schedule demands and translates MPS end items into specific component requirements.

- **Inventory Status File**
  Every inventory item being planned must have an inventory status file which gives complete and up to date information on the on hand quantities, gross requirement, scheduled receipts and planned order releases for the item. It also includes planning information such as lot sizes, lead times, safety stock levels and scrap allowances.

Remember: ....

- MPS act like nucleus between MRP and CRP. Thus difficult to achieve MPS.

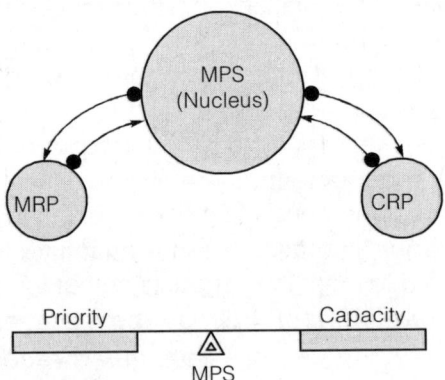

4. Bill of Materials (BOM)

To schedule the production of an end product, the MRP system must plan for all the materials, parts and subassemblies that go into the end product. The Bill of Material file in the computer provide this information.

5. MRP Outputs

A variety of reports can be generated from the information made available by an MRP Program.

- To MPS Planner's
  - Simulation of proposed MPS
  - Researching information for open orders (due to cancellation, delays, shortages)
- To Purchasing and Production
  - Changes to keep priorities valid
  - Order releases (Purchase and shop orders)
  - Planned order releases
- To Capacity Requirements Planning
  - Order release information for load profiles, delays, shortages
- To Management
  - Performance measurement of (Vendors, cost, forecast accuracy)
  - Exception reports (on due dates BOM file, etc.)

6. Capacity Requirement Planning (CRP)

CRP is a technique for determining what personnel and equipment capacities are needed to meet the production objectives embodied in the master schedule and the material requirements plan.

CRP is an effort to develop a match between the MRP schedule and the production capacity of the company.

■■■■

Production, Planning and Control

10

It is the function of management which plans, directs and controls the material supply and processing activities of an enterprise, so that specified product are produced by specified method.

1. Loading, Sequencing and Scheduling

- **Routing**

  It consists of deciding the path the item would take in its transformation from raw material to final product. It involves the creation of the route sheet which contains the following information:

  - Item to be manufactured.
  - Machine on which processing will take place.
  - Details of operation.

- **Loading**

  The work of transformation of raw material has to be converted into work loads on individual machine. This process is known as loading.

- **Balancing**

  It is the process of ensuring that the individual work loads on each of the machine is more or less the same. This would result in the cycle time being almost equal to each of the station time.

- **Scheduling**

  Scheduling answers the question when the work will be done and within what time it will be completed. GANTT Chart, PERT and CPM are tools of scheduling.

- **Dispatch**

  This function of PPC actually starts the production activities. It deals with the smooth introduction of work on to the shop floor. It includes the release of work orders, release of tools, drawings etc. on the shop floor in their required sequence.

- **Follower**

  This function of PPC deals with the control or the feedback part to ensure smooth running of all the previous operations.

2. Parameters Associated with Scheduling of 'N' Job on '1' Machine

- **Job Flow Time**

  The flow time for a job is the time from some starting point until that job is completed.

- **Make Span Time**

  It is the time from when processing begins on first job in the set until the last is completed.

- **Average Number of Jobs in The System**

  For a fixed set of job this will be total flow time for all the jobs divided by the make span time.

$$\text{Average number of job} = \frac{\Sigma \text{ flow time for all jobs}}{\text{Make span time (MST)}}$$

- **Average Tardiness**

  The tardiness of the job is the amount of time after its due date that the job is completed. If the job is completed *before* due date then the *tardiness is zero.*

3. Sequencing Rule of N Jobs on One Machine

- **Stortest Processing Time (SPT)**

  This rule sequences the job in the increasing order of their processing time.

- **Earliest Due Date (EDD)**

  It sequences the job in increasing order of their due dates.

- **Critical Ratio (CR)**

  This rule sequences the job in increasing order of their critical ratios.

$$\text{C.R} = \frac{\text{Due date}}{\text{Processing time}}$$

  Critical ratio for a job is the due date divided by processing time.

- **Slack Time Remaining (STR)**

  This rule sequences the job in increasing order of their slack time remaining where

  STR = Due date – Processing time

Remember
- The SPT *always produces* a sequence that has the *smallest average flow time and average number of job* i.e. why SPT is most popular.
- EDD rule minimizes tardiness better than SPT.
- SPT rule minimizes average number of job and average flow time.

4. Sequencing of 'N' Jobs on Two Machine

In this problem two machines are A and B and the assumption is that the processing order is fixed. Each of the jobs numbered i = 1, 2, 3, $\cdots$,n has processing times A_i and B_i respectively. The problem is to find a processing order for different jobs and to ensure that the ideal time for both the machines is minimized. In other words the aim is to maximize the machine utilization. This is achieved by using Johnson's Algorithm.

- **Step (i):** Find out the minimum of A_i and B_i. If minimum is A_k then process the kth job in the beginning. If minimum is B_r then process the rth job in the end.
- **Step (ii):** When a particular job is placed in a sequence strike it off.
- **Step (iii):** If $A_k = B_r$, then in this case process the kth job first and the rth job last.

  If both equal values occur on machine A then select the one which has lowest value on B first. If both equal values occur on machine B then select the one with the lowest value on A first.

5. Sequencing of 'N' Jobs on Three Machine

A, B, C (A $\rightarrow$ B $\rightarrow$ C)

$$\text{Minimum } A_i \geq \text{Maximum } B_i$$

$$\text{Minimum } C_i \geq \text{Maximum } B_i$$

For applying Johnson's algorithm for J jobs three machine the given conditions have to be checked.

Atleast one of these two conditions has to be satisfied.

6. Sequencing of 'N' Jobs on Four Machine

A, B, C, D (A $\rightarrow$ B $\rightarrow$ C $\rightarrow$ D)

$$\text{Minimum } A_i \geq \text{Maximum } B_i \text{ and Maximum } C_i$$

$$\text{Minimum } D_i \geq \text{Maximum } B_i \text{ and Maximum } C_i$$

7. Sequencing of '2' Jobs on 'N' Machine

If machines are more than '2' then it is converted into 2 machine problem. This done by adding all processing times for a job from machine 1 to (n – 1) machine and let this time be the time for machine 1 and then by adding all processing times for a job from machine 2 to 'n' and let this be the time for machine 2. Then apply Johnson's algorithm.

Remember: ....

• Jhonson's rule is applicable for sequencing of N-job on 2 machine.

■■■■

Linear Programming

11

It is a mathematical technique for determining optimal allocation of resources.
- **Objective Function**
  Clearly identifiable and measurable in quantity e.g. maximization of profit.
- **Constraints**
  These are limited resource with in which we have to obtain the objective function maximum
  $$z = c_1 x_1 + c_2 x_2 + ...$$
  $$a_{11} x_1 + a_{12} x_2 + ... \geq b_1$$
  C_j = profit coefficient
  b_i = resource value
  a_{ij} = technological coefficient
  x_j = variable
- **Special Cases:** Linear programming problems can be solved by
  - Graphical method
- **Infinite Solution**
  Objective function slope equals to one of the constrain which forms the boundary.
- **No Solution**
  It is not possible to find feasible solution which satisfy all the constrain.
- **Unbounded Solution**
  The greatest value of objective function occurs at infinite and it simply means the common feasible region is not bounded by limit on constrain.

SIMPLEX METHOD

1. Procedure
- Right hand side of each constrain that is resource value (bi) should be non-negative.
- Each decision variable of the problem should be non-negative.
- Inequalities of the constrain are converted into equalities.
- Set m = no of equality constrain and n = number of variable.
- Put (n – m) variable equal to zero.
- (n – m) = non basic variable

- m = basic variable
- Initial feasible solution is obtained by putting **(n − m)** non basic variable (NBV) **equal to zero** i.e. $x_1 = 0$, $x_2 = 0$
  Thus slack variable becomes basic variable **(BV = m)**.

2. Special Case

- **Infinite Solution**
  When a non basic variable in an optimal solution has a zero value for Δj row then the solution is not unique.

- **Unbounded Solution**
  When all replacement ratio are either infinite (or) negative then the solution terminates. This indicates the problem has un-bounded solution.

- **Infeasible Solution**
  When in the final solution an artificial variable is in the basis then there is no feasible solution to the problem.

DUALITY IN LINEAR PROGRAMMING

For every LP problem there exist a related unique LP problem involving the same data which also describe and solve the original problem.

Primal	Dual
Maximum	Minimum
No. of variables	No. of constrain
No. of constrain	No. of variables
≤ Type of constrain	Non negative variables
= Type constrain	Unrestricted variable
Unrestricted variable	= Type constraints

1. Big-M Method

In those situation when an identity matrix is not obtain initially another form of simplex method called "Big-M" method is applied. In this method, artificial variables put into the model to obtain an initial solution.

2. Transportation Problem

These problems are used for meeting the supply and demand requirements under given conditions in the best optimal effective manner.

C_{ij} = Cost of transportation of one unit from the ith source to the jth destination

x_{ij} = Quantity to be transported to i^{th} source to the ith destination

Total transportation cost = $\sum\limits_{i=1}^{n}\sum\limits_{j=1}^{m} C_{ij}\, x_{ij}$

- **Feasible Solution**
  A set of non negative individual allocation which also satisfy the given constrain.
- **Basic Feasible Solution (B.f.s)**
  A feasible solution of **m × n [Row × columns]** transportation problem is basic feasible if the total number of allocations is exactly equal to **(m + n − 1)**.
- **Optimal Solution**
  A feasible solution is said to be optimal if it minimizes the total transportation cost.
- **Non Degenerate Basic Feasible Solution**
  A feasible solution of m × n transportation problem is non degenerate if
  - Total number of allocations is exactly equal to (m + n − 1).
  - These allocations are in independent positions.

3. Unbalance Transportation Problem

- If total supply from all the source is equals to the total demand in all destinations it is called balance otherwise unbalanced.
- If given problem is unbalance, balance it with dummy source or destination.

4. Degeneracy

When the number of allocations is less than **(m + n − 1)** then optimality test cannot be performed and such a solution is called degenerate solution.

Transportation is special case of L.P.P. and **(m + n − 1)** allocation at independent position gives non-degenerate B.f.s.

5. Assignment

- These problems are special case of transportation problem where
  - Matrix must be a square matrix.
  - The optimal solution to the problem would always be such that there would be only one assignment in the given row or column.
- Hungarian method is used to solve assignment problem.

- **Steps**
  - Subtract the smallest element of each row to the every element of corresponding row.
  - Subtracting the smallest element of each column to the every element of corresponding columns.
  - Now all zeros are covered with minimum number of lines.
  - If number of lines is equal to the number of rows or columns then optimal solution is obtained.

Remember

- Assignment is special case of transportation and always gives degenerate solution as it has n number of allocation.
- Assignment involves allocation for square matrix with allocation either $x_{ij} = 0$ or $x_{ij} = 1$.

■ ■ ■ ■